ADVANCES IN MATERIALS RESEARCH

Series Editor-in-Chief: Y. Kawazoe

Series Editors: M. Hasegawa   A. Inoue   N. Kobayashi   T. Sakurai   L. Wille

The series Advances in Materials Research reports in a systematic and comprehensive way on the latest progress in basic materials sciences. It contains both theoretically and experimentally oriented texts written by leading experts in the field. Advances in Materials Research is a continuation of the series Research Institute of Tohoku University (RITU).

Yasunori Fujikawa
Kazuo Nakajima
Toshio Sakurai
(Eds.)

Frontiers
in Materials Research

With 199 Figures

 Springer

Professor Yasunori Fujikawa
Professor Kazuo Nakajima
Professor Toshio Sakurai

Institute for Materials Research, Tohoku University
2-1-1 Katahira, Aoba-ku, Sendai 980-8577, Japan
E-mail: fujika-o@imr.tohoku.ac.jp, nakakazu@imr.tohoku.ac.jp, sakurai@imr.tohoku.ac.jp

Series Editor-in-Chief:

Professor Yoshiyuki Kawazoe

Institute for Materials Research, Tohoku University
2-1-1 Katahira, Aoba-ku, Sendai 980-8577, Japan

Series Editors:

Professor Masayuki Hasegawa
Professor Akihisa Inoue
Professor Norio Kobayashi
Professor Toshio Sakurai

Institute for Materials Research, Tohoku University
2-1-1 Katahira, Aoba-ku, Sendai 980-8577, Japan

Professor Luc Wille

Department of Physics, Florida Atlantic University
777 Glades Road, Boca Raton, FL 33431, USA

Advances in Materials Research   ISSN 1435-1889

ISBN 978-3-540-77967-4   e-ISBN 978-3-540-77968-1

Library of Congress Control Number: 2008923541

Typesetting: Data prepared by SPi using a Springer LATEX macro package
Cover: eStudio Calmar Steinen

SPIN: 12184200   57/3180/SPI
Printed on acid-free paper

9 8 7 6 5 4 3 2 1

springer.com

Preface

New advanced materials are being rapidly developed, thanks to the progress of science. These are making our daily life more convenient. The Institute for Materials Research (IMR) at Tohoku University has greatly contributed for to the creation and development of various advanced materials and the progress in the field of material science for almost a century. For example, our early research achievements on the physical metallurgy of iron carbon alloys led to the innovation of technology for making high-quality steels, which has greatly contributed to the advancement of the steel and related industry in Japan and rest of the world. IMR has focused on basic research that can be translated into applications in the future, for the benefit of mankind.

With this tradition, we have established the first high-magnetic field as well as low-temperature technologies in Japan, which were essential to the advancement of magnetism and superconductivity. Recently, IMR has expanded its research in the field of advanced materials including metallic glasses, ceramics, nano-structural metals, semiconductors, solar cell crystals, new opto- and spin-electronics materials, organic materials, hydrogen storage alloys, and shaped crystals.

In the face of the crisis of the destruction of the global environment, the depletion of world-wide natural resources, and the exhaustion of energy sources in the twenty-first century, we all have an acute/serious desire for a better/safer world in the future. IMR has been and will continue the pursuit of research aimed at solving global problems and furthering eco-friendly development.

In these research areas, we strongly believe that it is of utmost importance for us to rapidly develop a new environment-friendly clean energy system using solar energy as the ultimate natural energy source. For instance, most of our natural resources and energy sources will be exhausted within the next 100 years at current consumption rate. Oil, natural gas, and uranium, in particular, constitute a serious problem.

Solar energy is the only renewable ultimate energy source. About 30% of the total solar energy is reflected from the surface, so 70% of the total solar energy is available for us to capture and utilize. The world's energy consumption is about 9,000 Mtoe (M ton oil equivalent), only 0.01% of the available solar energy. If we use solar energy to cover 10% of the world's energy consumption, we need to manufacture 40 GW solar cells every year for the next 40 years. The required Si feedstock is about 400,000 ton per year, which is an attainable target by increasing the world production of Si feedstock by only 27 times the present production. To accomplish this target of using solar energy for 10% of the world's energy consumption, we must develop several key materials to establish a clean energy cycle, taking into consideration the lifespan of various materials. At IMR, we are working on a number of projects aimed at the development of a new environment-friendly clean energy conversion system, including high-quality Si multicrystals to be used in high-efficiency solar cells, hydrogen storage materials using the synthesis of complex hydrides, and advanced fuel cell materials and materials for saving energy such as white lights using ZnO LEDs.

Our ultimate goal is to become the world-premier center of excellence in materials science. We will work hard to excel in fulfilling the needs of our future society including a clean-energy based "green world" and interdisciplinary technologies. Therefore, one of the main purposes of this workshop is to stimulate in-depth discussion of advanced materials that will play a key role in our future society and will offer a brighter future to the next generation.

This is a book based on an international workshop on advanced materials organized by the IMR on March 1, 2007. We thank all the participants attending this workshop, and especially the invited speakers and the special lecturer, Dr. Heinrich Rohrer, the 1986 Physics Nobel Laureate, known as the founding father of nano-science and technology.

Sendai   *Kazuo Nakajima*
March 2008

Contents

Part III Precise Control of Microscopic and Complex Systems

15 Atom Probe Tomography at The University of Sydney
*B. Gault, M.P. Moody, D.W. Saxey, J.M. Cairney, Z. Liu, R. Zheng,
R.K.W. Marceau, P.V. Liddicoat, L.T. Stephenson, and S.P. Ringer*

16 A Study on Age Hardening in Cu-Ag Alloys by Transmission Electron Microscopy
E. Shizuya and T.J. Konno

17 Rubber-Like Entropy Elasticity of a Glassy Alloy
M. Fukuhara, A. Inoue, and N. Nishiyama

List of Contributors

Chihaya Adachi
Center for Future Chemistry
Kyushu University
Nishi, Fukuoka 819-0395
Japan

Toshikazu Akahori
Institute for Materials Research
Tohoku University, 2-1-1 Katahira
Aoba-ku, Sendai 980-8579
Japan

K. Akiyama
Institute for Materials Research
Tohoku University, 2-1-1 Katahira
Aoba-ku, Sendai 980-8577
Japan

A. Al-Mahboob
Institute for Materials Research
Tohoku University, 2-1-1 Katahira
Aoba-Ku, Sendai 980-8577
Japan

T. An
The Institute for Solid State Physics
The University of Tokyo, 5-1-5
Kashiwa-no-ha, Kashiwa 277-8581
Japan

Satria Zulkarnaen Bisri
Institute for Material Research
Tohoku University, 2-1-1 Katahira
Aoba-ku, Sendai 980-8577
Japan

J.M. Cairney
Australian Key Centre
for Microscopy & Microanalysis
The University of Sydney
NSW, 2006
Australia

Arrelaine A. Dameron
Department of Chemistry
Department of Physics
The Pennsylvania State University
104 Davey Laboratory
University Park, PA 16802-6300
USA
Dynamic Organic Light, Inc.
2410 Trade Center Avenue
Longmont CO 80503
USA

Zachary J. Donhauser
Department of Chemistry
Department of Physics
The Pennsylvania State University

104 Davey Laboratory
University Park, PA 16802-6300
USA
Vassar College
Poughkeepsie, NY 12604
USA

T. Eguchi
The Institute for Solid State Physics
The University of Tokyo, 5-1-5
Kashiwa-no-ha, Kashiwa 277-8581
Japan

Dirk Ehrentraut
Institute of Multidisciplinary
Research for Advanced Materials
Tohoku University, 2-1-1 Katahira
Aoba-ku, Sendai 980-8577
Japan

Daniel J. Fuchs
Department of Chemistry
Department of Physics
The Pennsylvania State University
104 Davey Laboratory
University Park, PA 16802-6300
USA
Edwards Vacuum, Inc.
Chanhassen, MN 55317
USA

Y. Fujikawa
Institute for Materials Research
Tohoku University, 2-1-1 Katahira
Aoba-ku, Sendai 980-8577
Japan

Kozo Fujiwara
Institute for Materials Research
Tohoku University, 2-1-1 Katahira
Aoba-ku, Sendai 980-8577
Japan

Tsuguo Fukuda
Institute of Multidisciplinary
Research for Advanced Materials
Tohoku University, 2-1-1 Katahira
Aoba-ku, Sendai 980-8577
Japan
Fukuda Crystal Laboratory
c/o ICR 6-6-3, Minami-Yoshinari
Aoba-ku, Sendai 989-3204
Japan

M. Fukuhara
Institute for Materials Research
Tohoku University, 2-1-1 Katahira
Aoba-ku, Sendai 980-8577
Japan

T. Fukumura
Institute for Materials Research
Tohoku University, 2-1-1 Katahira
Aoba-ku, Sendai 980-8577
Japan

H. Fukuyama
Tokyo University of Science
Kagurazaka, Shinjuku-ku
Tokyo 162-8601

B. Gault
Australian Key Centre
for Microscopy & Microanalysis
The University of Sydney
NSW, 2006
Australia

M. Hamada
The Institute for Solid State Physics
The University of Tokyo
5-1-5 Kashiwa-no-ha
Kashiwa 277-8581
Japan

Y. Hasegawa
The Institute for Solid State Physics
The University of Tokyo
5-1-5 Kashiwa-no-ha
Kashiwa 277-8581
Japan

M. Hisakabe
Institute for Materials Research
Tohoku University, 2-1-1 Katahira
Aoba-ku, Sendai 980-8577
Japan

Akihisa Inoue
Institute for Materials Research
Tohoku University, 2-1-1 Katahira
Aoba-Ku, Sendai 980-8577
Japan

Yoshihiro Iwasa
Institute for Material Research
Tohoku University, 2-1-1 Katahira
Aoba-ku, Sendai 980-8577
Japan
CREST, Japan Science and
Technology Corporation (JST)
Kawaguchi 332-0012
Japan

T. Kawabata
Institute for Materials Research
Tohoku University, 2-1-1 Katahira
Aoba-ku, Sendai 980-8577
Japan

Masashi Kawasaki
WPI Advanced Institute
for Materials Research
Institute for Materials Research
Tohoku University, 2-1-1 Katahira
Aoba-ku, Sendai 980-8577
Japan
CREST, Japan Science Technology
Agency (JST), Kawaguchi 332-0012
Japan

Hisamichi Kimura
Institute for Materials Research
Tohoku University, 2-1-1 Katahira
Aoba-Ku, Sendai 980-8577
Japan

A. Kitora
Institute for Materials Research
Tohoku University, 2-1-1 Katahira
Aoba-ku, Sendai 980-8577
Japan

Toyohiko J. Konno
Institute for Materials Research
Tohoku University, 2-1-1 Katahira
Aoba-ku, Sendai 980-8577
Japan

Penelope A. Lewis
Department of Chemistry
Department of Physics
The Pennsylvania State University
104 Davey Laboratory
University Park, PA 16802-6300
USA
American Chemical Society
1155 16th St. NW
Washington, DC 20036
USA

P.V. Liddicoat
Australian Key Centre
for Microscopy & Microanalysis
The University of Sydney
NSW, 2006
Australia

Z. Liu
Australian Key Centre
for Microscopy & Microanalysis
The University of Sydney
NSW, 2006
Australia

Dmitri V. Louzguine-Luzgin
Institute for Materials Research
Tohoku University 2-1-1 Katahira
Aoba-Ku, Sendai 980-8577
Japan

Brent A. Mantooth
Department of Chemistry
Department of Physics
The Pennsylvania State University
104 Davey Laboratory
University Park, PA 16802-6300
USA
Edgewood Chemical Biological
Center
AMSRD-ECB-RT-PD
5183 Blackhawk Road
Aberdeen Proving Ground, MD
21010-5424
USA

R.K.W. Marceau
Australian Key Centre
for Microscopy & Microanalysis
The University of Sydney
NSW, 2006
Australia

M. Matsuo
Institute for Materials Research
Tohoku University, 2-1-1 Katahira
Aoba-ku, Sendai 980-8577
Japan

M.P. Moody
Australian Key Centre
for Microscopy & Microanalysis
The University of Sydney
NSW, 2006
Australia

Amanda M. Moore
Department of Chemistry
Department of Physics
The Pennsylvania State University
104 Davey Laboratory
University Park, PA 16802-6300
USA

Masaaki Nakai
Institute for Materials Research
Tohoku University, 2-1-1 Katahira
Aoba-ku, Sendai 980-8579
Japan

Kazuo Nakajima
Institute for Materials Research
Tohoku University, 2-1-1 Katahira
Aoba-ku, Sendai 980-8577
Japan

Y. Nakamori
Institute for Materials Research
Tohoku University, 2-1-1 Katahira
Aoba-ku, Sendai 980-8577
Japan

M. Nakano
Institute for Materials Research
Tohoku University, 2-1-1 Katahira
Aoba-ku, Sendai 980-8577
Japan

Mitsuo Niinomi
Institute for Materials Research
Tohoku University, 2-1-1 Katahira
Aoba-ku, Sendai 980-8579
Japan

T. Nishihara
Institute for Materials Research
Tohoku University, 2-1-1 Katahira
Aoba-ku, Sendai 980-8577
Japan

N. Nishiyama
RIMCOF, Tohoku University
Laboratory, R&D Institute
of Metals and Composites
for Future Industries
Sendai 987-8577
Japan

Akiko Nomura
Institute for Materials Research
Tohoku University
2-1-1 Katahira, Aoba-ku
Sendai 980-8577
Japan

Yoshitaro Nose
Institute for Materials Research
Tohoku University
2-1-1 Katahira, Aoba-ku
Sendai 980-8577
Japan

A. Ohtomo
Institute for Materials Research
Tohoku University, 2-1-1 Katahira
Aoba-ku, Sendai 980-8577
Japan

S. Orimo
Institute for Materials Research
Tohoku University, 2-1-1 Katahira
Aoba-ku, Sendai 980-8577
Japan

Wugen Pan
Institute for Materials Research
Tohoku University
2-1-1 Katahira, Aoba-ku
Sendai 980-8577
Japan

J.B. Qiang
Institute for Materials Research
Tohoku University
2-1-1 Katahira, Aoba-Ku
Sendai 980-8577
Japan

S.P. Ringer
Australian Key Centre
for Microscopy & Microanalysis
The University of Sydney
NSW, 2006
Australia

H.H. Rohrer
1986 Nobel Laureate in Physics
Advisor to IFCAM
Institute for Materials Research
Tohoku University, 2-1-1 Katahira
Aoba-ku, Sendai 980-8577
Japan

J.T. Sadowski
Institute for Materials Research
Tohoku University, 2-1-1 Katahira
Aoba-ku, Sendai 980-8577
Japan

T. Sakurai
Institute for Materials Research
Tohoku University, 2-1-1 Katahira
Aoba-ku, Sendai 980-8577
Japan

D.W. Saxey
Australian Key Centre
for Microscopy & Microanalysis
The University of Sydney
NSW, 2006
Australia
Department of Materials
Oxford University
UK

Toetsu Shishido
Institute for Materials Research
Tohoku University
2-1-1 Katahira, Aoba-ku
Sendai 980-8577
Japan

Eiji Shizuya
Department of Materials Science
Tohoku University, Sendai 980-8579
Japan

Rachel K. Smith
Department of Chemistry
Department of Physics
The Pennsylvania State University
104 Davey Laboratory
University Park, PA 16802-6300
USA
Materials Science Division
Lawrence Berkeley National
Laboratory
Berkeley, CA 94720
USA

L.T. Stephenson
Australian Key Centre
for Microscopy & Microanalysis
The University of Sydney
NSW, 2006
Australia

Y. Taguchi
Institute for Materials Research
Tohoku University, 2-1-1 Katahira
Aoba-ku, Sendai 980-8577
Japan

Isao Takahashi
Institute for Materials Research
Tohoku University
2-1-1 Katahira, Aoba-ku
Sendai 980-8577
Japan

Tetsuo Takahashi
Institute for Material Research
Tohoku University, 2-1-1 Katahira
Aoba-ku, Sendai 980-8577
Japan

T. Takano
Institute for Materials Research
Tohoku University, 2-1-1 Katahira
Aoba-ku, Sendai 980-8577
Japan

Taishi Takenobu
Institute for Material Research
Tohoku University, 2-1-1 Katahira
Aoba-ku, Sendai 980-8577
Japan
CREST, Japan Science
and Technology Corporation (JST)
Kawaguchi 332-0012
Japan

Masatoshi Tokairin
Institute for Materials Research
Tohoku University
2-1-1 Katahira, Aoba-ku
Sendai 980-8577
Japan

H. Toyosaki
Institute for Materials Research
Tohoku University, 2-1-1 Katahira
Aoba-ku, Sendai 980-8577
Japan

A. Tsukazaki
Institute for Materials Research
Tohoku University, 2-1-1 Katahira
Aoba-ku, Sendai 980-8577
Japan

T. Tsutaoka
Graduate School of Education
Hiroshima University
Higashi-hiroshima, 739-8524
Japan

K. Ueno
Institute for Materials Research
Tohoku University, 2-1-1 Katahira
Aoba-ku, Sendai 980-8577
Japan

Noritaka Usami
Institute for Materials Research
Tohoku University, 2-1-1 Katahira
Aoba-ku, Sendai 980-8577
Japan

Z.T. Wang
Institute for Materials Research
Tohoku University, 2-1-1 Katahira
Aoba-ku, Sendai 980-8577
Japan

Paul S. Weiss
Department of Chemistry
Department of Physics
The Pennsylvania State University
104 Davey Laboratory
University Park, PA 16802-6300
USA

Guoqiang Xie
Institute for Materials Research
Tohoku University
2-1-1 Katahira, Aoba-Ku
Sendai 980-8577
Japan

Masayuki Yahiro
Center for Future Chemistry
Kyushu University
Nishi, Fukuoka 819-0395
Japan

K. Yamada
Institute for Materials Research
Tohoku University, 2-1-1 Katahira
Aoba-ku, Sendai 980-8577
Japan

Y. Yamada-Takamura
Institute for Materials Research
Tohoku University, 2-1-1 Katahira
Aoba-ku, Sendai 980-8577
Japan

T. Yamasaki
Institute for Materials Research
Tohoku University, 2-1-1 Katahira
Aoba-ku, Sendai 980-8577
Japan

G. Yoshikawa
Institute for Materials Research
Tohoku University, 2-1-1 Katahira
Aoba-ku, Sendai 980-8577
Japan

Wei Zhang
Institute for Materials Research
Tohoku University
2-1-1 Katahira, Aoba-Ku
Sendai 980-8577
Japan

R. Zheng
Australian Key Centre
for Microscopy & Microanalysis
The University of Sydney
NSW, 2006
Australia

1

Science, for the Benefit of Mankind

H.H. Rohrer

1.1 Introduction

The crucial role of a comprehensive education for science at all levels for the general progress of mankind, for economic development, for political maturity, and for worldwide stability is widely recognized – no education, no progress, no development, no maturity, no stability. Yet, reality is lagging far behind and we are still far away from a world wide education for all, even what an education on the most rudimentary level is concerned. Looking at the development aide efforts, it appears to me more like "Development, for the benefit of the developed ones" or "Education, for the benefit of the educated ones." And here we get readily lost in the thicket of conflicting economic and political interests. Finding acceptable and practical ways out lies, I believe, beyond this workshop.

Education comes to bear on mankind in many ways. On the top level, education provides the foundation of science; thus the benefit of education is mediated to mankind by science: Science, for the benefit of mankind. In the following, Science refers mainly to University research and education and to research at dedicated and mostly publicly funded research institutions.

Alfred Nobel was one of the first convinced believers in "Science, for the Benefit of Mankind" by bequeathing his tremendous fortune to honor scientific achievements for their benefit on mankind. In his legacy, we read "...Shall be annually distributed in the form of prizes to those who shall have conferred the greatest benefit on mankind." And he continues: "...one part to the person who shall have made the most important discovery or invention within the field of physics...."

1.2 Which Benefit? Which Mankind?

The first part of the last century until well after World-War II experienced a tremendous rise in the popularity of science in spite of some critical issues during the two world wars. The technical progress, including medicine, was simply too fantastic, and society, economy, industry, and politics enjoyed an incredible rise in living quality and an abundance of new opportunities, respectively. Nobel's conviction was happily shared by all those who were in reach of scientific progress and all its "goodies." That was also the time when science foundations and National research organizations were called into being where science for the benefit of mankind was an implicit objective. The confidence in science was limitless; science was thought to be the remedy for all and everything and became the hope on an ever better future. However, also the expectations and demands on science grew accordingly. Science was increasingly burdened with responsibilities for nearly everything and became the scapegoat for lack of making decisions. Science can play an important role in a decision making process, but it does not make decisions obsolete. Even in science itself decisions have to be taken constantly, whether opportune or not will be confirmed as one proceeds.

Society, economy, and politics started to loose a clear line of what science can do and what science should do. Science became indispensable for society at large and the financial support is constantly growing. But growing were also confusions about the roles to be played by the various players – the benefactors and beneficiaries – in the science concert, particular interests, fear of undesired and uncontrollable developments, misunderstanding of the mission of and mechanisms in science, hostile attitudes and distrust towards science, and in particular a selfishness down to the personal level leading to contradictory expectations and demands. It appears that everybody – be it nation, region, company, interest groups, or individual – has a very strong and selfish notion of what science should do and should not do. As a consequence, these diverse interests let the noble Nobel statement shrink to a multiple question mark: Which Science? For which benefit? Of which mankind?

1.3 Mankind in Science

Science was the first global enterprise, long before economy realized the advantages of globalization. To scout the vast unknown, to understand what is not yet understood, to achieve what could not be done before, to think about the hitherto unthinkable, to go beyond limits, and to open new frontiers, science has to rely on each person who does science, wants to do science, and can offer science. Thus anyone who has science of quality to offer is a most welcome member of the worldwide scientific community. The "science market" is open and accessible to anybody without any restrictions: both what jobs and products are concerned. A particular strength of a global science was the worldwide exposure, quality control, free exchange of ideas, and the global dissemination of scientific achievements and results for free use by everybody who understood the scientific, industrial, economic, and social significance of novel scientific findings. To arrive at any level in science is by no means easier than in the business or political world. The standard is a unique global standard set by the first rate scientists worldwide, undiluted by particular interests of local, national, or other nature. Of course, also in science there exist insider groups and local matadors which, however, hardly affect the high standard of worldwide science and are irrelevant for the scientific progress. That is very different from politics where the local matadors are the norm and world politics, and therefore, is what it is. It is also different from economy, where locally biased leaders are not the exception and where globalization of markets lacks a good deal of credibility.

It is quite remarkable that science has long been living global practices out of necessity and even more remarkable is the consistency in all respects with which science has been acting in accordance with this insight. It is likewise but unfortunately remarkable that economic globalization is changing the once globally open science, which was serving to the benefit of a global mankind into bits and pieces of nationally dictated benefits with a chauvinistic division of mankind. Besides what was said above, there are very practical reasons to let science act again in a less restricted, more global way. If a country's scientific research becomes increasingly focused on domestic research objectives, it starts to loose its breadth and generally its innovation vitality. Second, giving back more to world's science than one draws from it is a top-scientist's unequaled and challenging stimulus for top scientific performance.

1.4 Mankind and Benefits in Society

It is understandable that in a world of increasing competition and performance pressure all the beneficiaries of science would like to have a preference status and have science working for their own narrow benefits. Correspondingly, society at large, including economy, politics, and public, developed quite a restricted notion of mankind. Everybody tries to satisfy best his own desires and mankind decays in innumerable "benefit" groups. Unfortunately, science policy yields too easily to the various pressures and measures of the diverse interest groups. A sound balance between the long term scientific research and the short term, customer- and development-oriented research at Universities and in Institutional research is severely offset. The primary mission, beside higher education, of universities and research Institutions I consider is to foster the fertile environment for novel ideas and insight, procedures, technologies, and processes, which will be central to solving also the unknown issues and problems of tomorrow. Likewise, today's activities of science in universities and industry for solutions of today's problems rest to a great extent on the scientific efforts, insights, findings, and results of yesterday, be they 1 year or 100 years back. The rapid rise of all the particular selected benefits and needs of selected populations down to the personal level is not simply a sign of lacking solidarity with present humankind, but even more so with future generations. We enjoy the benefits of yesterday's science; let future mankind also enjoy the benefits of today's science.

1.5 Science, for the Benefit of Mankind

Science does not serve to the benefit of mankind sheer by excellent scientific achievements. Science provides the scientific insight, knowledge, know-how, processes, and materials from which the products needed and desired by mankind can be made. Which products and whether to the benefit of mankind or not does not lie in the hands of science but is the responsibility of the beneficiaries that include all the customers along the line, from industry as customer of science to the individual buyer of the end products. The beneficiaries decide what to make, to sell, and to buy. Nothing is made and marketed which is not finally bought. This is the great power of society. Of course, this requires consent among the buyers and the support by those who were elected exactly for such support. Such a message will be heard even in that part of science which is closer to a possible product chain.

For most of today's major problems, scientific solutions could well exist if there would be a political will and corresponding actions to have the problems solved. Experience shows that scientific solutions towards sustainability of nearly whatever result by no means automatically in a satisfactory overall reduction of the consumption of resources without additional accompanying measures. The question is always whether improved sustainability is intended for reduction of overall consumption or for a wider market. In the latter case, overall consumption can even increase dramatically.

Once all this is understood, the present futile interference of economy, society, and politics with the objectives of science should come back to an acceptable level, such that science can again enjoy the freedom necessary to perform to the benefit of mankind, of today's mankind, and of that of tomorrow.

1.6 Freedom in Science

Freedom is the key to the creative vitality of science; to question accepted insights, knowledge, thinking, beliefs, and the way things are done; to solve what could not be solved before; to do what could not be done before; simply to go beyond horizons. Nobody knows when, where, and which quantum leaps will come up; not even the scientists, and even much less everybody else. Scientific progress is not the result of long-term planning in some bureaucrat's office or of some wishful thinking of some political or economic bodies; it is the scientists, driven by curiosity, novelty, and eagerness for change, who determine both direction and speed of the scientific advancements.

Discovery means finding the unanticipated. Discovering, however, does not simply come easy; it is usually the unexpected result of a serious effort. It is also not just playing roulette. As Pasteur said, "Chance favors only the prepared mind." A prepared mind is a most important asset of a very good scientist and it requires the scientist's utmost freedom in many respects. It requires the freedom to venture into whatever a scientist considers worth the risk, the freedom to interact with any scientist he deems appropriate, the freedom and the obligation to make available his results to the scientific community,;nd in particular the freedom to make mistakes.

Scientific freedom I consider the most endangered species in the world of science. It is not research or travel budgets, machines or instrumentation, or infrastructure and appropriate buildings. Neither is it the young brilliant minds, the Achilles tendon of science, although they are admittedly increasingly harder to get motivated for science.

Nowhere does freedom come for free. Freedom it is not something you have, you have to fight for it, defend it, and guard it. In the case of scientific freedom, all the ones involved do not a great job, neither the insiders nor the science promoters.

As to the insiders, I miss the determination of the scientific community to fight for scientific freedom, to regain what got lost over the past 50 years and in particular to expand the freedom necessary to meet the future challenges. The ever higher demands on learning, thinking, knowledge, insight, and creativity require more freedom, not less. Instead, the scientists even deprive themselves of some of their freedom by introducing unscientific practices into science. Just to name three of them. First, introducing competition for replacing standards. If anywhere then it is in science where standards can be set, for example, novelty, originality, intellectual challenge, complexity and simplicity, and elegance. Competition is a cheap measure of whatever performance. Better does not even mean good. Freedom gets lost on focusing on others: by betting on priority claims instead of concentrating on the sustainable value of results to be published. Competition creates also biased minds, shaping possibilities to one's mind instead of vice versa. And it is the most efficient way to miss the surprises and discoveries which are the source of novelty. Novelty itself is rarely competitive with decades of refinement, but novelty always carries the future. It took, for instance, the transistor more than 16 years before it replaced only half of the radio tubes. Second, scientists started to operate too much with promises, be it for money, be it for recognition, be it for attention, instead of operating with achievements. Living up to promises deprives you of choosing adventurous approaches which carry really the surprises. Finally, scientists' minds fall too often to the temptation of programs, or of rich but narrowly focused agencies. They adapt their interests to the money sources instead of getting resources for what they consider worthwhile doing. There are scientists who do that for what they are paid for and there are scientists who are paid for what they are doing. You certainly can guess my preference.

The external threat of scientific freedom comes mainly from the science promotion institutions and from the science political bodies.

The science promotion institutions are staffed predominantly with active scientists or with ones of a scientific past, hopefully not with bureaucrats. Since the thinking does hardly change when becoming a member of a promoting institution, unfortunately the same mistakes are made, with harder consequences though. All in all, promoting institutions have to become much more adventurous. Even they should have the freedom to make mistakes. The best measures to advance science are a hard hand in discontinuing old fashioned research and generous support of proposals of high and promising standard without topical constraints. Programs or consideration of former instrumental and infrastructure investments are very often severe handicaps for the selection of first class science. Promoting and supporting changes of scientists' interest areas should encourage the scientists to become more problem- and less instrument-oriented and create unbiased minds in the respective problem areas. Such minds are very often the key to disruptive advances in science. Careful evaluation of project results with correspondingly tough measures requires as much efforts by experts as the screening project proposals.

The major three tasks of the science political bodies I consider is to make strategic decisions that aim at promoting the best science possible, to provide the best possible conditions for science to stay well ahead of economic and social developments and to conquer the future for a later day worthy present of society, and to establish the channels that the scientific findings get to serve to the benefit of mankind. For this purpose, they should understand that the mechanisms in science are distinctly different from that in economy and from those in politics. Pushing the frontiers of science is not a democratic affair, nor a result of management decisions. It is the power of individual brains that determine the direction of science and the new promising areas. They should also understand that the major task of universities and university-level research organizations is to push the frontier of science and to educate scientists for their positions in industry and the economy. The universities are not the playground of industrial research and development, nor responsible for start-ups and industrial parks. Third, science policy should exercise much restraint in directly steering science, be it by defining topical research objectives, for instance through programs, or be it by having their scientists engage in domestic, across-the-border and intercontinental cooperation for the cooperation's sake.

1.7 Reform of Science

The ever-increasing complexity of issues and problems require serious reforms of scientific approaches. Universities and other scientific organizations have little changed over the years of their existence. The considerable inertia of scientific institutions is very astonishing in view that all of them have novelty and change on their flag, and also on their tongue. The basic challenges for a reform of science I consider are the issues in context with interdisciplinarity and holistic approaches, the control of the information flood, the disentanglement of the old-fashioned belief in a unity of teaching and research, and improvement of evaluation of scientific performance.

Interdisciplinarity comes in many shades, but so far we have done poorly in most cases. Interdisciplinarity requires an extra substantial effort. Interdisciplinarity is not just "instead of" but is in "addition to" disciplines. It appears to me that so far every effort towards interdisciplinarity degraded in a very short time into separate bits and pieces, into disciplinary expertise, instead of a discipline embracing creativity. Even fields like material science, which should by itself be exemplary for interdisciplinary thinking and performing, got very rapidly divided up into specialized material science branches with little connections even between them; actually not astonishing in view of their separation into specialized institutes in separate buildings. The "inter" or the "in between" seems to be an omnipresent problem, but it carries the promise and potential in every area of life, not just in science. From interdisciplinarity to interfaces, to interaction, and to international, we do not do that well. Maybe common research buildings could help, as it is apparently the case in research establishments of big companies. But new separate buildings seem to have their special status significance in science. Whatever, the most important guarantee for interdisciplinarity are, however, open minds.

1.8 The Benefits of the Future?

Society at large has to decide constantly which achievements of science they want to exploit for the benefit of mankind. In the future, the decisions will weigh considerably heavier. Scientific and technical feasibility are no longer the central question; they appear in many cases limitless. As Richard Feynman said, "Everything is feasible, except you can prove the contrary". And that's obviously very difficult. Therefore, many important decisions have to be made under the aspects of affordability, not feasibility: "Can we afford it, do we want to afford it, and do we have to afford it?"
Some questions are the following

Can we afford to push the fantastic progress in life science and medicine all the way to immortality, to a stagnant world?
Do we want to lose the pleasure of thinking to computers, and become merely technical actors?
Do we want to leave the enjoyment of creating to smart machines?
Do we want to reduce cultural diversity to a Peta bit mobile phone, and live in a culturally flat desert?
Do we want to replace reading and writing by hearing and speaking?
And finally, what happens with all those people, societies, countries who will not be able to follow the tremendous pace of scientific progress, and thus cannot adequately participate as beneficiaries of science?
From time to time, we should also think about these aspects of "Science, for the Benefit of Mankind."

Part I

Novel Materials for Electronics

2

Trends of Condensed Matter Science: A Personal View

H. Fukuyama

Summary: A personal view on the development of the research activities in condensed matter sciences is presented, with special emphasis on the carrier doping into insulators, which have been classified into four categories: band insulators, Mott insulators, charge ordering, and Anderson localization. Depending on these parent insulators, doped carriers behave essentially differently as typically seen in semiconductors (doped band insulators) in contrast to high T_c cuprates (doped Mott insulators) as bulk materials. Thanks to the establishment of experimental capabilities to probe local properties initiated by the success of scanning tunneling microscope (STM), more attentions are naturally being paid to local structures and associated electronic properties, spectroscopy in particular, which eventually govern material properties in macroscopic scales. Typical research targets from this viewpoint may include (1) strongly correlated electron systems, (2) surfaces, interfaces, and contacts, and (3) molecular assemblies. With more detailed explanation of the recent remarkable progress of the understanding of molecular solids belonging to above three, a hope is expressed that time is ripe to develop studies on bio-related materials, such as proteins and DNA, based on the well established technique in condensed matter science just as natural extensions of those on molecular solids. A tentative list of concrete research targets along this line of bio-material science has been proposed.

2.1 Introduction

Progress in microscopic understanding of material properties is very impressive. Materials consist of atoms whose electronic states are simple and have been fully understood by now. However, properties of materials, which are condensed forms of atoms in one way or another, are "emergent" [1] in the sense that they are completely different from those of individual constituents and truly diverse and have infinite possibilities. This has been very properly phrased as "More is different" by Anderson [2]. An amazing fact is that the properties of crystals, which are typical forms of condensed matter with periodic array of atoms, are understood quantum mechanically based on the knowledge of each constituent and the spatial arrangements in unit cells (lattice structures). Here the existence of lattice periodicity and Bloch bands is crucial. Actually one sheet of paper showing a band structure has information of 10^{23} atoms. This procedure of band theory is to know 10^{23} atoms from one. (In Japanese, there is a saying for a very sharp person that he (she) understands 10 once he (she) hears one!) Thanks to the healthy relationship between experiments and theories in material science, which is not always the case in some of the disciplines of science now unfortunately, this powerful step bridging between microscopic and macroscopic scales differing more than 20 orders of magnitude in the number of constituents are exercised very commonly without noticing, i.e., theoretical predictions being checked by experiments and new experimental findings leading to new concepts. By this band theory, the most remarkable difference of material properties, metals (or conducting states) and insulators, is naturally understood. The parameters characterizing conducting states, the effective masses, and g-factors around symmetry points in the Brillouin zone are derived by k-p perturbation theory by Luttinger and Kohn [3] to a high accuracy to be compared with experiments. These efforts have resulted in detailed understanding of electronic states in crystalline solids, especially those of semiconductors and semi-metals, resulting in reliable devising of applications of present days. These studies on band structures have indicated at the same time that there exist "neutrino" in graphite [4] (more recently in graphene [5] and also in one type of molecular solid, α-ET$_2$I$_3$, [6]) and "Dirac electrons" in Bi [7] as effective ones described by the similar types of Hamiltonian, which clearly indicates that physics is the science of concept.

The difference of metals and insulators in band theory is, in actuality, more simply understood if looked locally in space, i.e., in terms of quantum tunneling of electrons (finite transfer integrals) between atomic orbitals and Pauli principle for up and down spins. (Such a local view called tight-binding approximation (TBA) has a potential to be extended to various forms of materials, as will be discussed in the following in the context of molecular solids.) The insulating state in the band theory corresponds to the case where two electrons, up and down spins, occupy the same orbital. In contrast, the stability of metallic states is due to odd number of electrons in a unit cell or the overlapping bands in the case of even number of electrons in a unit cell. Besides this type of insulators called band insulator, there exist other types of insulators as listed below:

1. Band insulators
2. Mott insulators
3. Charge ordering
4. Anderson localization

(1), (2), and (3) are intrinsic, while (4) is due to extrinsic disorder, causing interference effects of scattered waves of electrons [8, 9]. Among these, (1) and (4) are possible for noninteracting electrons, but (2) and (3) are due to mutual

Coulomb interactions between electrons onsite (U) and between sites (nearest neighbor V for a quarter-filled case), respectively. The latter insulating states, (2) and (3), are the results of strong Coulomb interaction compared with the kinetic energies due to transfer integrals, i.e., "strong correlations." (Insulating states due to nesting frequently observed in quasi-one-dimensional systems can be classified as (1).) The onset of these insulating states due to strong correlations is easy to understand physically in local space but very hard to describe theoretically in momentum space.

In the Mott insulating state, which is realized for half-filled band and in which electronic spins exist on each lattice site, the effective Hamiltonian describing the low energy excitations is the Heisenberg spin Hamiltonian, familiar in the studies of statistical mechanics, where spins are coupled together by exchange interaction. It is to be noted clearly that in (2) and (3) there exist sometimes antiferromagnetism (AF), but that AF is not the cause of insulating states as deduced from the fact that insulating states survive even if AF disappears (by finite temperature or frustration).

In the case of charge ordered states, there are various possible spatial patterns of electronic charge distribution and associated magnetic properties as will be explicitly discussed later.

These facts concerning metals and insulators are very basic and well understood. Historically, studies on electronic properties of materials located near the boundary of metals and insulators have attracted much interest since they can be useful in practical applications. The typical examples are semiconductors where small amount of carriers, electrons or holes, are doped into band insulators [10]. Actually, there have been strong motive forces in research studies on materials near such metal–insulator boundary. Depending on the types of insulators as mentioned before, the metallic (or conducting) states near the metal–insulator boundary have distinct features. In this short note, trends of research activities in material science will be briefly introduced from this point of view of realizing conducting states out of insulating ones.

2.2 Carrier Doping into Insulators

2.2.1 Doped Band Insulators

Insulators are "dead" electrically, and a dramatic change happens if a small number of carriers are doped. This fact leads to controllability of electric properties by external means, which is the basic principle behind the semiconductor technology. Originally the control of carrier density was by impurity doping but later by external electric field, i.e., gate voltage, which is much easier and has resulted in field effect transistors (FET). This development is important not only for application but has played key roles in the development of basic science of two-dimensional electron gases (2DEG) [11]. The surprising findings

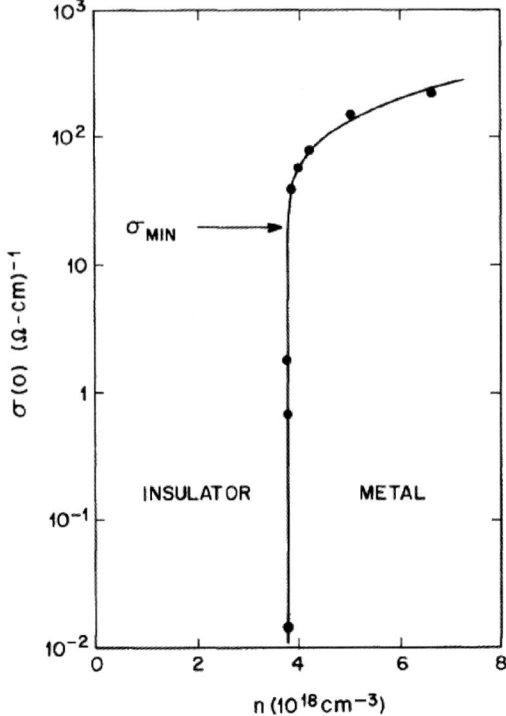

Fig. 2.1. Concentration (n) dependence of conductivity exterpolated to absolute zero, $\sigma(0)$, in P doped Si [15]

of quantum and fractional quantum Hall effects have been achieved in these systems [12–14]. Here basic and applied sciences have been hand in hand.

In these doped band insulators, the ground state can be insulating even if a finite macroscopic density of carriers are introduced as typically seen in Fig. 2.1 for Si doped by P [15]. Here the insulating state in the region of low carrier density is due to Anderson localization [8]. In the case of doped Si, the insulating states are transformed into conducting state as the carrier density is increased. In the case of B-doped diamonds, however, the superconductivity appears next to the insulating state [16, 17]. In this case of doped diamonds, the onset critical temperature, T_c, to superconducting state is very high in view of the presence of strong disorder intrinsic to impurity doping [18] and T_c will be raised appreciably if degree of disorder is reduced by some means because of the expected strong coupling of valence electrons with the stretching modes of carbon atoms [19], which has very high Debye temperature. Actually, targeting at high T_c by doping carriers into covalent bonds is an interesting possibility [20].

Likewise, in the case of polyacetylene, (CH)x, carriers have been doped resulting in conducting films [21], which has led to new possibilities in applications. In this case, conducting paths are one-dimensional together with strong electron–phonon interactions leading to Peierls transition (bond dimerization), and transport properties are proposed to be supported by soliton bands (spin and charge solitons) [22].

2.2.2 Doped Mott Insulators

Copper oxides (cuprates), which is the stage of high temperature superconductivity (High T_c) [23], show the familiar phase diagram on the plane of hole density (δ) and temperature (T) as shown in Fig. 2.2 for $La_{2-x}Sr_xCuO_4$ (LSCO) [24, 25]. The parent materials, $\delta = 0$, are Mott insulators, and antiferromagnetism (AF) is realized below the Neel temperature, $T_N = 325\,K$. (To be precise, they are the charge transfer type insulators, since not only copper (Cu) but also oxygen (O) orbitals play important roles in doping processes [26].) In the region of small doping, the ground states are insulating and superconductivity sets in directly from the insulating states once the hole density exceeds some critical value as in the case of B-doped diamonds. However, electronic properties of doped Mott insulators are completely different from those of doped band insulators, since the nature of parent insulating states are essentially different: nonmagnetic in band insulators whereas Heisenberg spin system in the case of Mott insulators. In fact metallic states before the onset of superconductivity in low doping region (underdoped region) are *anomalous* and unprecedented. Typical features of this anomalous metallic state include the existence of pseudo gap [27] (originally found as spin gap [28]), a large signal of Nernst effects resulting from the charge degree of freedom [29], and strong interplay between superconductivity (SC) and antiferromagnetism (AF) as evidenced by the formation of staggered magnetic

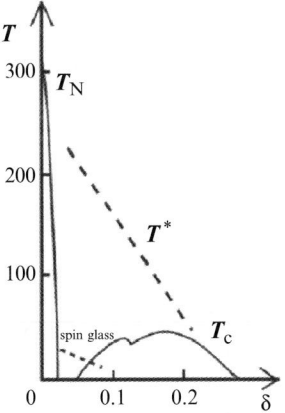

Fig. 2.2. Phase diagram on the plane of doping rate ($\delta = x$) and temperature (T) of $La_{2-x}Sr_xCuO_4$[25]; T_c = superconducting critical temperature, T^* = spin (pseudo) gap temperature, T_N = Neel temperature

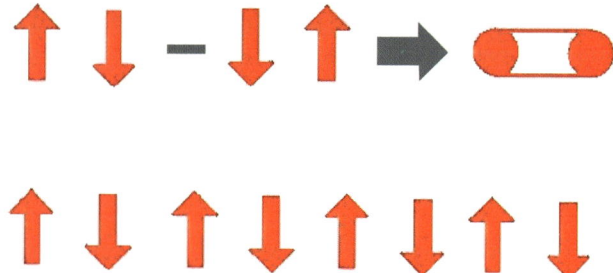

Fig. 2.3. Roles of superexchange interaction, J, leading to the singlet state for 2 spins, and the antiferromagnetic state in the bulk system

moments in the superconducting states once spatial inhomogeneity (e.g., impurities, vortices) is introduced. These will be related with each other [30] as a natural consequence of intrinsic feature of antiferromagnetic superexchange interaction, J, as schematically shown in Fig. 2.3. Hence, in the studies on disorder effects in the doped Mott insulators, which are very challenging, the existence of magnetism has to be taken into account from the beginning. Such may be called "Mott physics."

Not only high temperature superconductivity but also the colossal magneto-resistance observed typically in Mn oxides is another particular phenomenon seen in doped Mott insulators [31].

In strongly correlated electron systems, there are many cases with *ferroic* responses [32] to external perturbations, such as ferromagnetic, ferroelectric, ferroelastic, etc. Recent interests are in the large cross responses, called *multiferroic* [33], for example, magnetic field leading to large dielectric responses [34, 35], which are interesting as basic science and can be useful in future applications as well [36].

2.3 From Bulk to Local

So far electronic properties of bulk crystals are mainly focused upon. However, developments of experimental capability as typically demonstrated by scanning tunneling microscopy (STM) [37] and scanning tunneling spectroscopy (STS) have opened ways to probe local properties. It has turned out that *"local structure (morphology) and associated electronic properties, especially spectroscopy"* are important keywords to look at material properties, since the bulk material properties are quite often governed by such local properties in an essential way, which will be core targets of *nanoscience*. From this view point, one may list the following interesting examples:

(a) Strongly correlated electrons
(b) Interface and surfaces
(c) Molecular assemblies

The characteristic features of above (a) and (b) may be briefly summarized as follows, while those of (c) will be discussed in the following section separately.

(a) In strongly correlated electron systems such as transition metal oxides consisting of perovskite unit introduced in the preceding section as examples of doped Mott insulators, the distortion of each unit together with the number of d-electrons are the determining factors of bulk properties [38]; tetragonally distorted perovskites, resulting in nondegenerate half-filled band, derived from $d_{x^2-y^2}$ orbital in La_2CuO_4 with d^9 configuration are the parent compounds of high temperature superconductivity, while $LaMnO_3$ with cubic perovskites having d^4 electrons of Mn^{+3} in the degenerate e_g bands, which are Hund coupled to t_{2g} spins, is the source of colossal magneteoresistance under doping. These are schematically shown in Fig. 2.4.

(b) Surfaces, interfaces, and contacts, which are boundaries between different constituents and through which electrons move, are crucial in any device applications. Examples include (1) surfaces and interfaces of oxides in the context of possible types of electronics, e.g., resistive random access memory [39–41]; (2) surfaces and interfaces between ferromagnetic metals or those with insulating layers, in view of spintronics [42–45]; (3) interfaces and contacts of molecular materials in the context of possible organic transistors and single molecule transistors [46]; (4) bulk glasses,

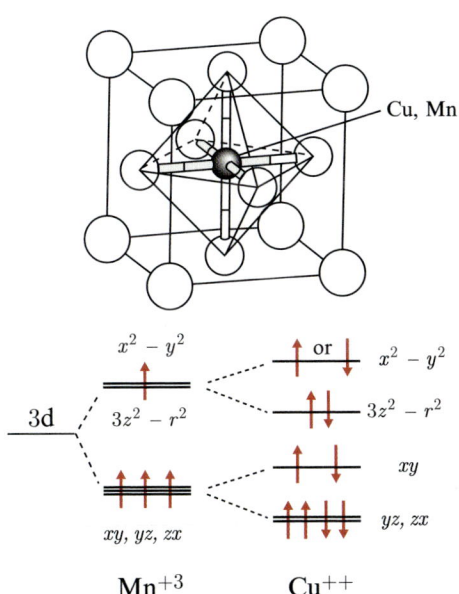

Fig. 2.4. The structure of perovskite unit and crystal field splitting together with occupation of d electrons in parent compounds of cuprates (Cu^{++}) and manganites (Mn^{+3})

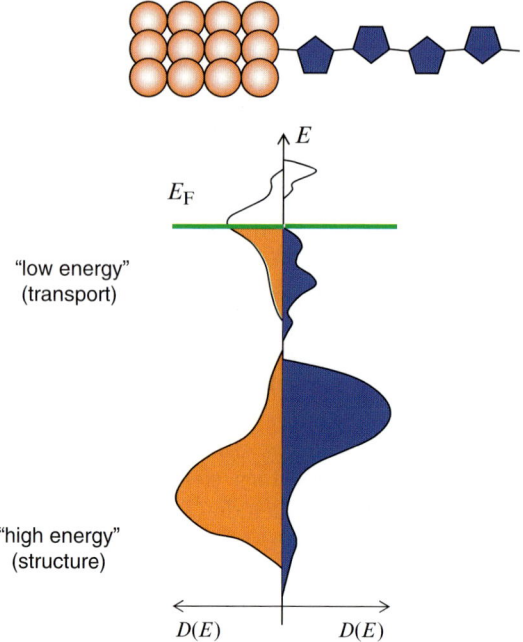

Fig. 2.5. Schematic representation of two energy scales involved in the "Contact Problem"

either metallic [47] or organic [48], where several different kinds of atoms or molecules meet together.

These practically important subjects offer at the same time various interesting and challenging problems in basic sciences as well. These may be phrased as "Contact Problem" as a whole where physics of two different energy scales are involved and have to be matched: "high energy" states (in the sense of away from the Fermi energy) governing the stability of local structures and mechanical properties, and "low energy" ones near the Fermi energy contributing to transport properties, as schematically shown in Fig. 2.5.

2.4 Electronic Properties of Molecular Solids

Similar to doped semiconductors, there have been pioneering activities in molecular (organic) materials in the pursuit for conducting states in dyestuffs [49, 50] and doping Br into perylenes [51]. Such processes of carrier doping has been developed to synthesize charge transfer salts where two different kinds of molecules jointly form crystals working as donors (cations) and acceptors (anions). Since around 1970, there have been many examples of realizing metallic states in such charge transfer salts, whose typical examples

NMP-TCNQ	Mott insulator	1d AF
TTF-TCNQ	1d semimetal	Peierls transition
(CH)x	Peierls insulator (Bond alternation)	Doped carriers in soliton bands
TMTCF$_2$X	Metal	SDW, SC, AF, Spin-Peierls, CO, Ferroelectricity
DCNQI$_2$X	Metal	CO
BEDTTTF$_2$X (ET$_2$X)	Metal	Mott Insulator, CO, SC, Spin liquid
BETS$_2$X	Metal	Mott insulator, AF, SC, Field-Induced SC
M(tmdt)$_2$ M=Ni, Au	Semi-metal	First single component molecular metal

Fig. 2.6. A short list of typical conducting molecular solids so far realized. The first and second columns indicate constituting molecules and nature of conducting states, respectively, while third indicates characteristic electronic states (AF = antiferromagnetism; SDW = spin density wave; SC = superconductivity; CO = charge ordering)

are shown in Fig. 2.6. Not only conductors but also many superconductors now exist in this category [52]. A remarkable achievement in this context is the realization of metallic states in molecular solids consisting of a single kind of molecule [53], single component molecular metals, [M(tmdt)2] with M = Ni, Au. The existence of the Fermi surface has been indeed identified by de Haas van Alphen oscillations [54] whose result agree well with LDA band structure calculations [55].

2.4.1 Strongly Correlations in Molecular Solids

Among these, molecular solids consisting of BEDT-TTF molecules (abbreviated as ET, shown in Fig. 2.7a) are of particular interest, which have usually a particular composition ET$_2$X with another molecules, X, working as anions with valence −1. In this family of ET$_2$X, which show two-dimensional conductions, there exist many possible spatial arrangements of ET molecules in the unit cell called polytypes as shown in Fig. 2.7b and corresponding diversity of ground states have been revealed experimentally. Basis for theoretical understanding of these facts should be the description of energy bands. The unambiguous identification of Bloch bands, however, in molecular solids is not easy, since there exist many atoms in a unit cell in molecular solids, and there were confusions in early stage. This difficulty has been overcome by the group of Kobayashi, who have realized the validity of TBA based on the molecular orbitals [56]. A most convincing example will be DCNQI$_2$Cu where TBA [57], LDA [58], and de Haas van Alfven experiments [59] have all agreed. It turns out that the transfer integrals in such TBA are very anisotropic, representing those of each molecular orbital. Based on very solid understanding

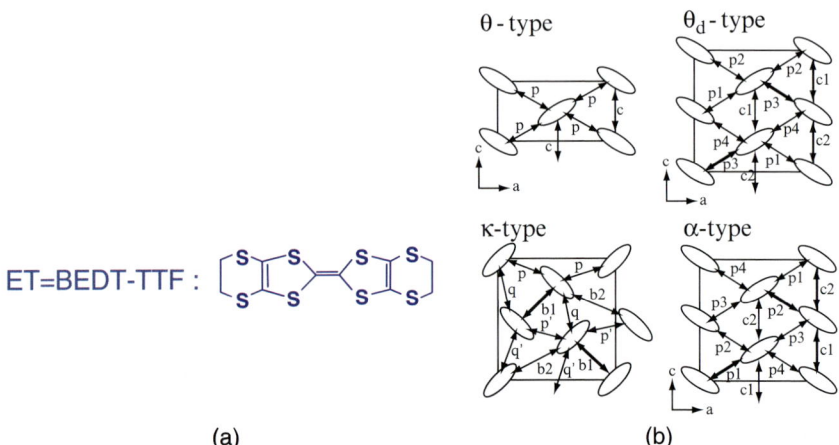

Fig. 2.7. (a) Structure of an ET molecule, (b) typical polytypes of ET_2X solids

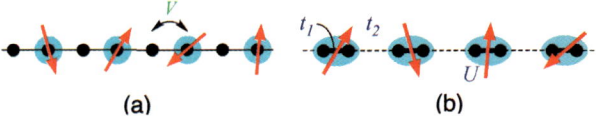

Fig. 2.8. Two characteristic insulating states in the quarter-filled bands depending on the degree of dimerization; charge ordering or dimer Mott for weak or strong dimerization, respectively

of one-particle states, effects of strong correlation have been studied by a Hamiltonian similar to the Hubbard model but with more spatially extended Coulomb interactions (extended Hubbard model), leading to very systematic studies on the variety of ground states [60–64]. In the present case of ET_2X with a specific 2:1 composition where the energy bands are quarter-filled if ET molecules are equivalent, one can expect interesting competition between different types of insulating states resulting from strong correlation, charge ordering, and Mott insulators, depending on the degree of dimerization as schematically shown in Fig. 2.8. Moreover, it turned out that there is another key parameter [62], degree of band splitting reflecting the existence of variety of transfer integrals within the unit cell in favor of band insulators, needed for classifying the whole families. The result is shown in Fig. 2.9. In the cases where the band splitting parameter is small (i.e., left side in Fig. 2.9), important transfer integrals are those shown in Fig. 2.10, which characterizes the degree of frustrations: square lattice for $t'/t \ll 1$ and triangular one for $t'/t \sim 1$. In the latter case of triangular lattice together with strong dimerization, that is, in frustrated dimer Mott insulators, the spin liquid state, which has been predicted by Anderson many years ago [65] and has been long wanted, has first been observed experimentally in κ-ET_2X [66, 67]. The identification of this fascinating new state of matter is a stimulating issue at

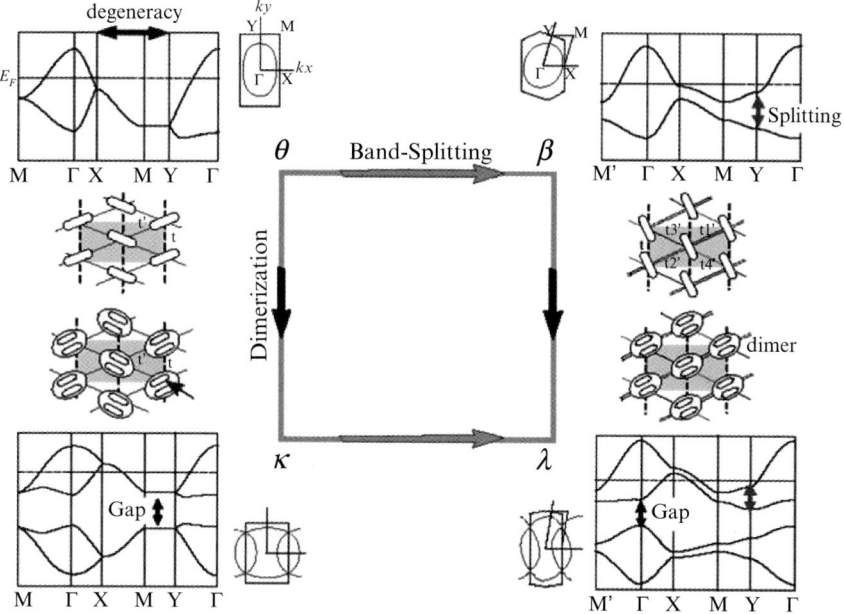

Fig. 2.9. A global classification of ground states of ET$_2$X crystals

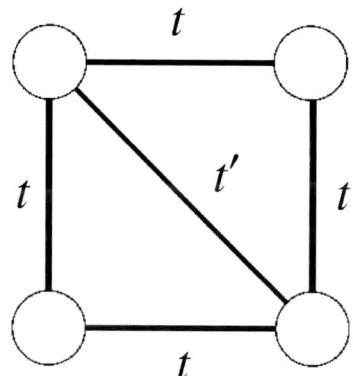

Fig. 2.10. Basic transfer integrals, t and t', bridging between square ($t'/t \ll 1$) and triangular ($t'/t \sim 1$) lattices. Each circle stands for either an ET molecule or its dimer

present. On the other hand, in polytypes with weak dimerization, the phase transitions to charge ordered states have been observed with variety as a function of temperature, for example, no visible change or sharp drop in magnetic susceptibility through the charge ordering phase transition temperatures as schematically shown in Fig. 2.11a, b, respectively. These are now understood as a result of the onset of one-dimensional (stripe) type charge ordered states

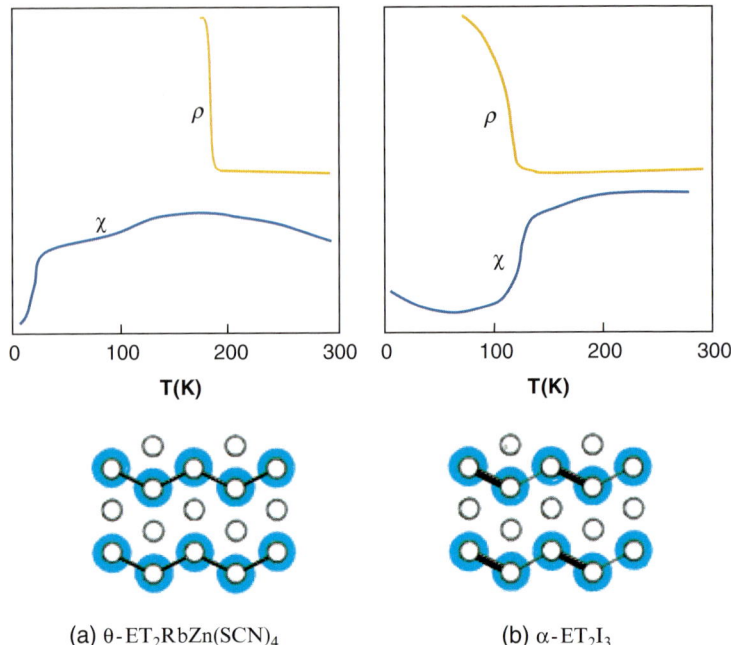

Fig. 2.11. Schematic representation of typical temperature dependences of resistivity (ρ) and magnetic susceptibility (χ) of (**a**) θ-ET$_2$RbZn(SCN)$_4$ and (**b**) α-ET$_2$I$_3$

together with different types of expected exchange interactions resulting from crystal structures [68]. In the latter case, α-ET$_2$I$_3$, the superconductivity has been observed under uniaxial pressure [69] even in the presence of finite (but reduced degree of) charge ordering, which has been theoretically analyzed [70] as due to the interaction processes between spins (physically in terms of super-exchange interaction J similar to cuprates) along the stripes. Another interesting possibility has theoretically been indicated in this polytype, α-ET$_2$I$_3$, under pressure, that is, massless fermions described by the "tilted Weyl equation" [6] (the 2×2 Weyl equation for neutrinos together with nonvanishing momentum dependent diagonal term), which are expected to resolve the long standing experimental mysteries observed in transport properties [71].

2.4.2 p-d Systems

In the cases of ET$_2$X introduced above, π electrons are strongly correlated resulting in charge ordering and (dimer) Mott insulators by themselves. In some molecular solids, conducting π electrons are coupled to transition metal ions, which are called p-d systems, where the effects of Coulomb interactions can be even more enhanced depending on the nature of the wave functions of the d-orbitals. The d electrons act like localized spins if the wave functions are localized, while valences will fluctuate if the wave function of d orbitals are spatially extended and mix strongly with conducting π electrons.

As typical examples of the former, one may note the following examples:

A-1. Two-dimensional λ-BETS$_2$(FeCl$_4$) with Fe^{+3} in high spin state, $S = 5/2$, which is antiferromagnetic insulator, shows superconductivity under high magnetic field applied precisely parallel to conducting layers [72], considered to be due to Jaccarino–Peter mechanism.

A-2. One-dimensional phthalocyanine containing Fe, TTP[FePc(CN)$_2$]$_2$, with similar Fe^{+3} but in the low spin state, $S = 1/2$, where localized spins exist on every lattice site (periodic Kondo systems) but density of carriers (holes) are half, could lead to interesting interplay between charge localization and magnetic ordering [73, 74].

On the other hand, examples of the latter with fluctuating valence may include the following:

B-1. DCNQI$_2$Cu, which is introduced in Sect. 2.3(1), shows very strong metal–insulator transition as a function of temperature [57]. In metallic states, the average valence of Cu is known to be 4/3 from photoemission experiment [75]. The insulating state sets in by the charge ordering of Cu ions with spatial pattern of valence $(+1 + 1 + 2)$, which works together with the nesting transition of π electrons with the same periodicity. This concomitant ordering will be the origin of strong first order metal–insulator transition [76].

B-2. The single component molecular metals introduced in Sect. 2.3, [M(tmdt)$_2$] belong to this category. The transition metals, M, are located at the center of this molecule as shown in Fig. 2.12 and LDA calculations have disclosed the fact that d states are strongly mixed with molecular orbitals of tmdt's located on both sides of M. In the case of M = Au a magnetic phase transition has been identified by NMR experiment [77], whose origin is the target of intensive studies at present.

It is to be noted that the spatial relationship between metallic ions and surrounding π orbitals in B-2 is similar to those of phthalocyanine [78], MPc, shown in Fig. 2.13, whose crystal structure varies depending on M, affecting the stacking pattern of this planar molecule.

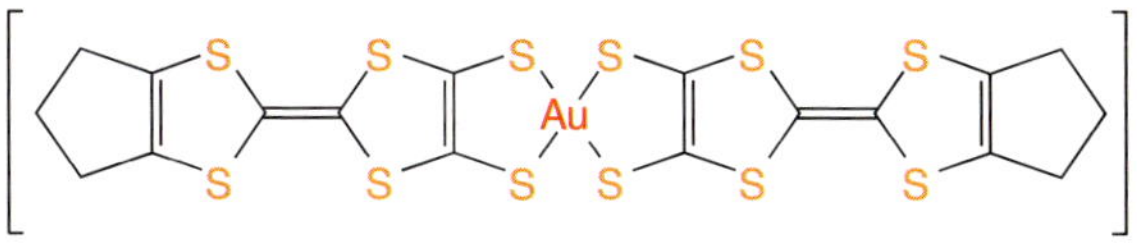

Fig. 2.12. The molecular structure of [M(tmdt)$_2$] with M = Au

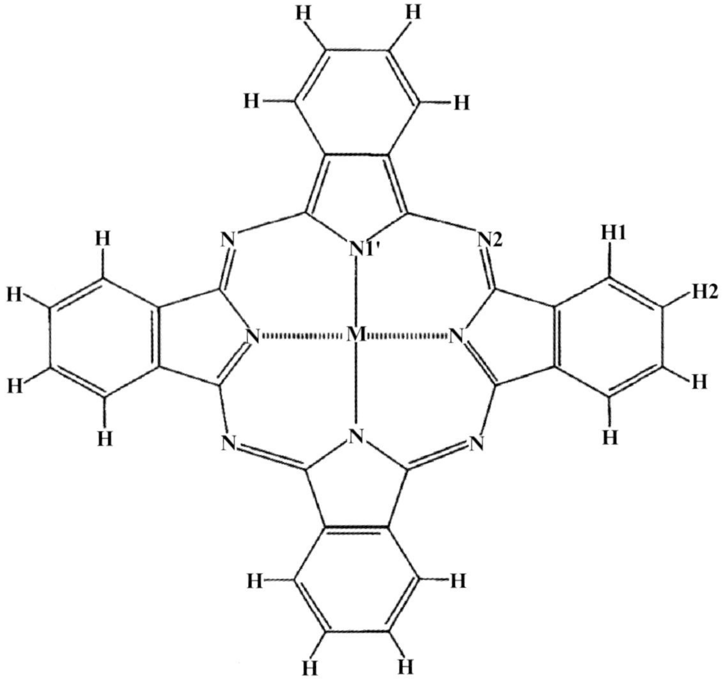

Fig. 2.13. A phthalocyanine molecule

2.5 Electronic Properties of Molecular Assemblies

It is interesting to note that the characteristic features of the relative spatial location of metallic ions and π electrons in Fig. 2.13 are similar to those of myoglobin as shown in Fig. 2.14, a typical and important protein containing metallic ion (metalloproteins) at the action center. Hence, understanding electronic properties of p-d molecular solids will give important clues for the understanding of functionalities of some kinds of proteins. Here knowledge of *local structure and associated electronic spectroscopy* are crucial. Now firm experimental results of the energy scheme of d-states in myoglobins have been reported for the first time by Shin and collaborators [80], who have observed d-d transitions in soft X-ray emission spectroscopy (SXES) and succeeded in fitting the data by well-established ligand field theory of Tanabe and Sugano [81] as a first trial, with properly chosen Racah parameters. This is very encouraging and expected to open a new direction in condensed matter science, which may be called "Bio-material science" [81].

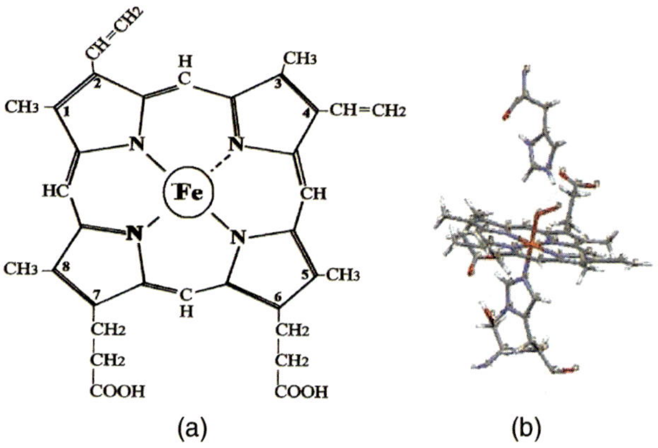

Fig. 2.14. A schematic representation of myoglobin in the vicinity of a Fe ion (a) in the plane of a Fe ion and (b) in three dimension for the case of O_2 binding [79]

(a)   (b)

2.6 Some Concrete Targets of *Bio-Material Science*

It is true that bio-materials, whose typical elements are proteins and DNA, are in general condensed matters in the sense that they consist of molecular segments that are coupled together rather strongly. Their structures, however, do not have spatial periodicity. Even in such cases without lattice periodicity one may hope, encouraged by the success in molecular solids, that the electronic states of bio-related materials can be studied once a proper version of TBA is introduced. To proceed in this direction it is crucial, though not easy, to identify important and at the same time feasible research targets to be attacked based on the viewpoint of condensed matter science. Preliminary results of efforts in this direction are as follows [64, 82]:

Local structures and electronic states of

1. metallic ions in proteins as explained above
2. metallic clusters, like Fe_4S_4, in proteins
3. functional molecules, like retinals, in proteins
4. DNA

Studies will now be possible on electronic states of these molecular assemblies in ways similar to conventional condensed matter sciences, leading to Bio-material science. For example, a possible way to dope carriers (holes) into DNA has been proposed recently [83]. However, to substantiate efforts, collaborations over various disciplines, physics, chemistry, biology, and electro-engineering, etc., are necessary.

Acknowledgements. Thanks are due to many colleagues for stimulating and illuminating discussions over the years. Discussions in the Forum "Electronic Properties of Molecular Assemblies" are acknowledged especially.

References

1. P.W. Anderson, in *Twentieth Century Physics*, ed. by L.M. Brown, A. Pais, B. Pippard (IOP, New York, 1995), p. 2017
2. P.W. Anderson, Science **177**, 393 (1992)
3. J.M. Luttinger, W. Kohn, Phys. Rev. **97**, 869 (1955)
4. J.W. McClure, Phys. Rev. **119**, 606 (1960)
5. K.S. Novoselov et al., Science **306**, 666 (2004)
6. S. Katayama, A. Kobayashi, Y. Suzumura, J. Phys. Soc. Jpn. **75**, 054705 (2006); A. Kobayashi, S. Katayama, Y. Suzumura, H. Fukuyama, J. Phys. Soc. Jpn. **76**, 034711 (2007)
7. P.A. Wolff, J. Phys. Chem. Solids **25**, 1057 (1964)
8. P.W. Anderson, Phys. Rev. **102**, 1008 (1958)
9. E. Abrahams, P.W. Anderson, D.C. Licciardello, T.V. Ramakrishnan, Phys. Rev. Lett. **42**, 673 (1979)
10. N.F. Mott, *Metal – Insulator Transition* (Taylor and Francis, London, 1974)
11. T. Ando, A.B. Fowler, F. Stern, Rev. Mod. Phys. **54**, 437 (1982)
12. K.V. Klitzing, G. Dorda, M. Pepper, Phys. Rev. Lett. **45**, 494 (1980)
13. H.L. Stormer, A. Chang, D.C. Tsui, J.C.M. Hwang, A.C. Gossard, W. Wiegmann, Phys. Rev. Lett. **50**, 1953 (1983)
14. R.B. Laughlin, Phys. Rev. Lett. **52**, 1583 (1984)
15. T.F. Rosenbaum, K. Andres, G.A. Thomas, R.N. Bhatt, Phys. Rev. Lett. **45**, 1723 (1980)
16. E.A. Ekimov et al., Nature **428**, 542 (2004)
17. Y. Takano et al., Appl. Phys. Lett. **85**, 2851 (2004)
18. T. Shirakawa et al., J. Phys. Soc. Jpn. **76**, 014710 (2007)
19. M. Hoesch et al., Phys. Rev. B **75**, 140508 (2007)
20. H. Fukuyama, J. Supercond. Novel Magn. **19**, 201 (2006)
21. C.K. Chiang, C.R. Fincher Jr., Y.W. Park, A.J. Heeger, H. Shirakawa, E.J. Louis, S.C. Gau, A.G. MacDiarmid, Phys. Rev. Lett. **24**, 1098 (1977)
22. W.P. Su, J.R. Schrieffer, A.J. Heeger, Phys. Rev. Lett. **42**, 1698 (1979)
23. J.G. Bednorz, K.A. Muller, Z. Phys. B **64**, 189 (1986)
24. P.W. Anderson, *The Theory of Superconductivity in the High-Tc Cuprates* (Princeton University Press, Princeton, NJ, 1997)
25. B. Keimer et al., Phys. Rev. B **46**, 14034 (1992)
26. F.C. Zhang, T.M. Rice, Phys. Rev. B **37**, 3759 (1988)
27. A. Damascelli, Z. Hussain, Z.X. Shen, Rev. Mod. Phys. **75**, 473 (2003)
28. H. Yasuoka, T. Imai, T. Shimizu, in *Strong Correlation and Superconductivity* ed. by H. Fukuyama, S. Maekawa, A.P. Malozemoff, (Springer, Berlin Heidelberg, New York, 1989), p. 254
29. Y. Wang, L. Li, N.P. Ong, Phys. Rev. B **73**, 024510 (2006)
30. M. Ogata, H. Fukuyama, Rep. Prog. Phys. **71**, 036501 (2008)
31. Y. Tokura (ed.), *Colossal Magnetoresistive Oxides* (Gordon and Breach Science, New York, 2000)
32. K. Aizu, J. Phys. Soc. Jpn. **27**, 387 (1969)
33. H. Schmid, Ferroelectrics **162**, 317 (1994)
34. T. Kimura et al., Nature **426**, 55 (2003)
35. H. Katsura, N. Nagaosa, A.V. Balatsky, Phys. Rev. Lett. **95**, 057205 (2005)
36. Y. Tokura, Science **312**, 1481 (2006); J. Magn. Magn. Mater **310**, 1145 (2007)

37. G. Binnig, H. Rohrer, Ch. Gerber, E. Weibel, Phys. Rev. Lett. **50**, 120 (1983)
38. Y. Tokura, N. Nagaosa, Science **288**, 462 (2000)
39. A. Ohtomo, H.Y. Huang, Nature **427**, 423 (2004)
40. A. Tsukazaki et al., Science **315**, 1388 (2007)
41. S. Thiel et al., Science **313**, 1942 (2006)
42. G. Binasch et al., Phys. Rev. B **39**, 4828 (1989)
43. M.N. Baibich et al., Phys. Rev. Lett. **61**, 2472 (1988)
44. T. Miyazaki, N. Tezuka, J. Magn. Magn. Mater. **139**, 231 (1995)
45. S. Yuasa et al., Jpn. J. Appl. Phys. **43**, 588 (2004)
46. M.A. Read, J.M. Tour, Sci. Am. **282**, 86 (2000)
47. A. Inoue, Acta Mater. **48**, 279 (2000)
48. Y. Shirota, J. Mater. Chem. **15**, 75 (2005)
49. D.D. Eley, Nature **162**, 819 (1948)
50. H. Akamatsu, H. Inokuchi, J. Chem. Phys. **18**, 810 (1950)
51. H. Akamatsu, H. Inokuchi, Y. Matsunaga, Nature **173**, 168 (1954)
52. For a review, H. Mori, J. Phys. Soc. Jpn. **75**, 051003 (2006)
53. H. Tanaka et al., Science **291**, 285 (2001); A. Kobayashi, Y. Okano, H. Kobayashi, J. Phys. Soc. Jpn. **75**, 051002 (2006)
54. H. Tanaka et al., J. Am. Chem. Soc. **126**, 1051 (2004)
55. S. Ishibashi et al., J. Phys. Soc. Jpn. **74**, 843 (2005)
56. T. Mori et al., Bull. Chem. Soc. Jpn. **57**, 627 (1984)
57. R. Kato, H. Kobayashi, A. Kobayashi, J. Am. Chem. Soc. **111**, 5224 (1989); R. Kato, Bull. Chem. Soc. Jpn **73**, 515 (2000)
58. T. Miyazaki, K. Terakura, Y. Morikawa, T. Yamasaki, Phys. Rev. Lett. **74**, 5104 (1995)
59. S. Uji et al., Phys. Rev. B **50**, 15597 (1994)
60. H. Kino, H. Fukuyama, J. Phys. Soc. Jpn. **65**, 2158 (1996)
61. H. Seo, H. Fukuyama, J. Phys. Soc. Jpn. **66**, 1249 (1997)
62. C. Hotta, J. Phys. Soc. Jpn. **72**, 840 (2003)
63. H. Seo, C. Hotta, H. Fukuyama, Chem. Rev. **104**, 5005 (2004)
64. H. Fukuyama, J. Phys. Soc. Jpn. **75**, 051001 (2006)
65. P.W. Anderson, Mater. Res. Bull. **8**, 153 (1973)
66. Y. Shimizu et al., Phys. Rev. Lett. **91**, 107001 (2003)
67. K. Kanoda, Hyperfine Interact. **104**, 235 (1997); J. Phys. Soc. Jpn. **75**, 051007 (2006)
68. H. Seo, J. Phys. Soc. Jpn. **69**, 805 (2000)
69. N. Tajima et al., J. Phys. Soc. Jpn. **71**, 1832 (2002); N. Tajima et al., J. Phys. Soc. Jpn. **75**, 051010 (2006)
70. A. Kobayashi et al., J. Phys. Soc. Jpn. **73**, 3135 (2004)
71. K. Kajita et al., J. Phys. Soc. Jpn. **61**, 23 (1992)
72. S. Uji, J.S. Brooks, J. Phys. Soc. Jpn. **75**, 051014 (2006)
73. T. Inabe, H. Tajima, Chem. Rev. **104**, 5503 (2002)
74. C. Hotta, M. Ogata, H. Fukuyama, Phys. Rev. Lett. **95**, 216402 (2005)
75. I.H. Inoue et al., Phys. Rev. **45**, 5828 (1992)
76. H. Fukuyama, J. Phys. Soc. Jpn. **61**, 3452 (1992)
77. W. Suzuki et al., J. Am. Chem. Soc. **125**, 1486 (2003)
78. C.C. Leznoff, A.B.P. Lever (ed.), *Phthalocyanines, Properties and Applications* (VCH Publishers, New York, 1993)

79. Coordinates around heme extracted from PDB entry 1A6M [private communications, from S. Shin]
80. S. Shin, private communications; Y. Harada et al., (to be submitted)
81. Y. Tanabe, S. Sugano, J. Phys. Soc. Jpn. **9**, 753, 766 (1954); S. Sugano, Y. Tanabe, H. Kamimura, *Pure and Applied Physics*, vol. 33 (Academic Press, New York, 1970)
82. A report of the Forum "Electronic Properties of Molecular Assemblies" (2006); Symposium on "Structure and Electronic Properties of Molecular Assemblies – Toward Bio-Materials Science" (RIKEN), (2007)
83. H. Kino et al., J. Phys. Soc. Jpn. **73**, 2089 (2004)

3

Measurements and Mechanisms
of Single-Molecule Conductance Switching

A.M. Moore, B.A. Mantooth, A.A. Dameron, Z.J. Donhauser, P.A. Lewis,
R.K. Smith, D.J. Fuchs, and P.S. Weiss

Summary: We have engineered and analyzed oligo(phenylene-ethynylene) (OPE)
derivatives to understand and to control the bistable conductance switching exhibited by these molecules when inserted into saturated alkanethiolate and amide-
containing alkanethiolate self-assembled monolayers (SAMs) on Au{111}. By
engineering the structures of the OPE derivatives, we have shown conductance
switching to depend on hybridization changes at the molecule–substrate interface. In
addition, we have demonstrated bias-dependent switching controlled by interactions
between the dipole of the OPEs and the electric field applied between the scanning
tunneling microscope tip and the substrate. These interactions are stabilized via intermolecular hydrogen bonding between the OPEs and host amide-containing SAMs.

3.1 Introduction

The ultimate goal for molecular electronics is to fabricate and to control functional devices using single-molecule components, and to integrate these devices into addressable nanoscale assemblies [1–3]. This goal presents many
challenges, including the ability to fabricate structures and to characterize
assemblies at the nanometer scale, the ability to control the local place-
ment of molecular components into stable configurations, and the ability to
understand and to control conductance through single molecules. Many molecular components containing high degrees of π-conjugation have been investigated as potential future devices due to their electron delocalization, leading
to low barriers for electron transport [4, 5]. Highly conjugated molecules
studied include polythiophenes, polyphenylenes, polyphenylenevinylenes, and
oligo(phenylene-ethynylene)s (OPEs) [1, 4–16]. Engineering conjugated molecules exploit the inherent ability of synthetic chemistry to create structures
with modifiable, controllable, and reproducible properties. This ability, coupled with progress in self- and directed assembly to form controlled local
geometries on a variety of substrates with high structural order, has driven
molecular electronics forward [17].

Molecular device candidates have been studied using a variety of techniques analyzing ensembles of molecules down to even single molecules [1, 17]. Bundles of molecules have been studied using nano-rod junctions [18–21], crossed-wire junctions [22–24], nanopores [25–28], mercury drop junctions [29–33], and tip-end junctions [34–36], while few to single-molecules have been studied using break junctions [37–41], electromigration junctions [42–47], particle bridges [48–50], and scanning probe techniques [8–16, 51–53]. Scanning tunneling microscopy (STM) enables the study of the physical and electronic properties of molecules on an individual basis. Using STM, a circuit is created with the molecule of interest chemically attached to one electrode (the substrate), while the scanning tunneling microscope tip acts as the second electrode (although detached, given the tunneling gap). In addition to structural information, this enables the electronic properties of individual molecules to be measured.

Here, we discuss investigations in which we have engineered and studied OPE derivatives isolated in host self-assembled monolayers (SAMs) to understand the bistable conductance switching exhibited by individual and small bundles of OPE molecules. All OPE derivatives we have studied have exhibited conductance switching. Our experimental evidence determines contributions from both the host SAMs (Fig. 3.1a) and the OPE derivatives (Fig. 3.1b) important to conductance switching [8–11]. Many mechanisms have been proposed to describe conductance switching, including reduction of functional groups [54], rotation of functional groups [55], backbone phenyl ring rotations [56], neighboring molecule interactions [57, 58], bond fluctuations [16], and changes in bond hybridization [59, 60], We tested these hypotheses via molecular design and single-molecule experiments and conclude that bond hybridization change is the only mechanism consistent with all the data [12, 14]. Finally, we discuss our ability to drive the bistable conductance switching through control of the dipole of the OPE derivatives acting in response to the electric field between the scanning tunneling microscope tip and substrate stabilized by hydrogen bonding available in host amide-containing SAM matrices [10, 11].

3.2 Methods

3.2.1 Sample Preparation

Commercially obtained Au{111} on mica substrates (Molecular Imaging, Phoenix, AZ) were annealed using a hydrogen flame prior to exposure to a 1 mM ethanolic/SAM solution (typically 24 h deposition). Host SAM matrices included n-alkanethiolates (**1**), 3-mercapto-N-nonylpropionamide (**2**) [61, 62], and decanoic acid(2-mercaptoethyl)amide (**3**) (Fig. 3.1a). The n-alkanethiolate chains varied in length from 8 to 12 carbons, using only even number of carbon due to more favorable van der Waals interactions and the orientation of the

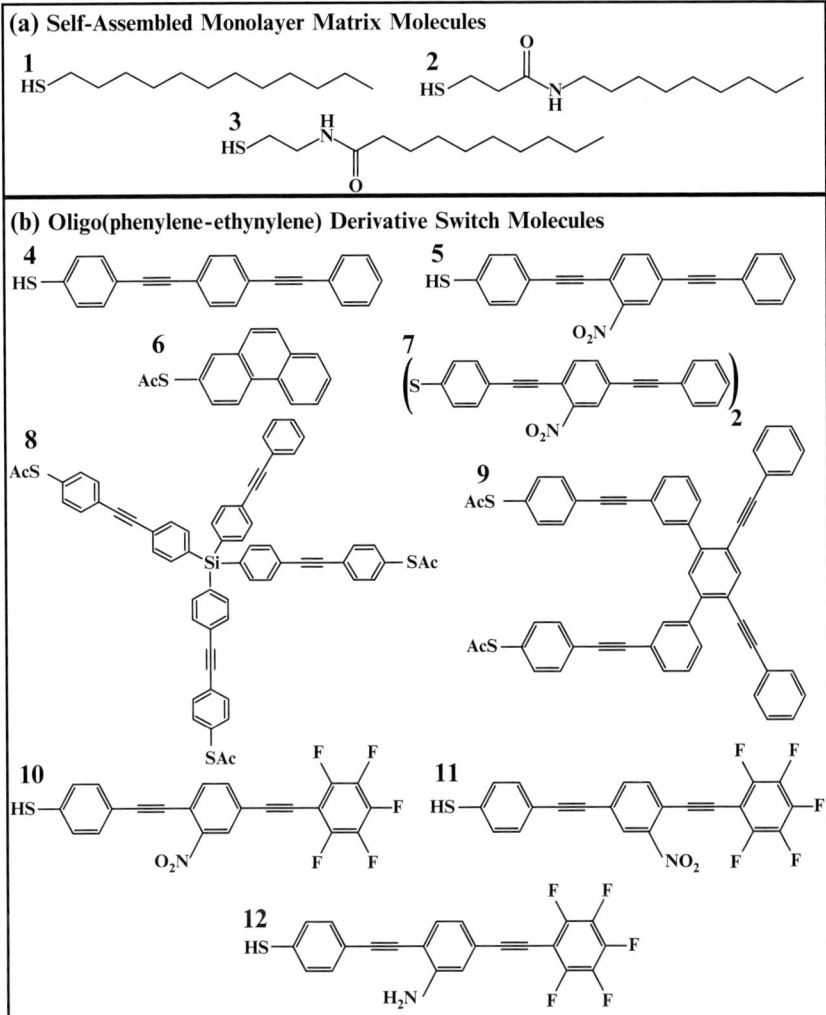

Fig. 3.1. (a) Thiol-terminated molecules used for self-assembled monolayer matrices: **1**, alkanethiol; **2**, amide-functionalized; and **3**, reverse amide-functionalized. (b) Thiol- or thioacetyl-terminated oligo(phenylene-ethynylene) derivatives: **4**, unfunctionalized; **5**, nitro-functionalized; **6**, phenanthrene-based; **7**, nitro-functionalized disulfide; **8**, caltrop; **9**, two-contact; **10**, nitro-functionalized pentafluorinated; **11**, orthogonal nitro-functionalized pentafluorinated; and **12**, amino-functionalized pentafluorinated

terminal methyl group. [63] After immersion, the samples were rinsed using neat ethanol and dried under nitrogen.

OPE derivatives (Fig. 3.1b) were inserted from solution into preformed host SAMs (Fig. 3.2a) by immersing the preassembled SAMs into 0.1 μm

32   A.M. Moore et al.

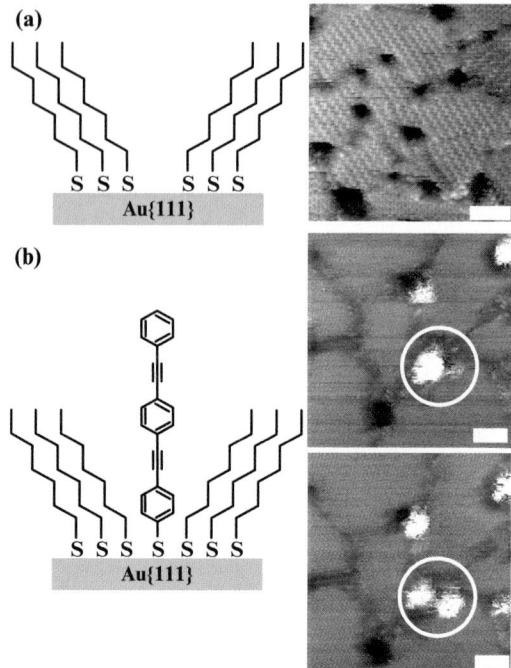

Fig. 3.2. (a) Schematic of an alkanethiolate SAM formed by solution deposition (*left*) and STM image (*right*) of a dodecanethiolate SAM (scale bar: 50 Å; $V_{sample} = -1.0\,V$; $I_{tunnel} = 2\,pA$). (b) Schematic of an unfunctionalized OPE molecule inserted into the host SAM at a domain boundary in the host matrix (*left*). Scanning tunneling microscopy images (*right*) of nitro-functionalized OPE molecules inserted into a host octanethiolate SAM (scale bar: 50 Å; $V_{sample} = -1.0\,V$; $I_{tunnel} = 2\,pA$). A switching event is shown for the circled molecules, *top*: one molecule is in the "on" conductance state; *bottom*: rightmost molecule switches to the "on" conductance state

tetrahydrofuran/OPE solutions under a nitrogen environment. Insertion of the OPE molecules occurs at the defect sites (step edges and domain boundaries) of the host matrices (Fig. 3.2b) [6]. For molecules synthesized with a thioacetyl protecting group (indicated in Fig. 3.1 with the abbreviation SAc), the thiol was generated *in situ* by adding aqueous ammonia to hydrolyze the acetate. SAMs were exposed to inserting solutions for times ranging from a few seconds to 0.5 h, allowing more time for molecules containing thioacetyl end groups and larger, branched molecules than for molecules with thiol end groups. Following insertion, the substrates were rinsed with neat ethanol and dried under nitrogen. The OPE derivatives included 4-(4-phenylethynyl phenylethynyl) benzenethiol (**4**), 4-(2-nitro-4-phenylethynyl phenylethynyl)benzenethiol (**5**), 2-thioacetyl-phenanthrene (**6**), 4-[4-(2-nitro-4-phenylethynyl phenylethynyl) benzenethiol]disulfide (**7**), caltrop (**8**) [64], thioacetic acid S-{4-[3″-(4-acetyl-sulfanyl-phenylethynyl)-4′-6′-bis-phenylthynyl-[1, 1′, 3′, 1″]terphenyl-3-ylethy-nyl]-phenyl}ester (**9**), 4-(2-nitro-4-pentafluorophenylethynyl phenylethynyl) benzenethiol (**10**), 4-(3-nitro-4-pentafluorophenylethynyl phenylethynyl)ben-zenethiol (**11**), and 4-(2-amino-4-pentafluorophenylethynyl phenylethynyl) benzenethiol (**12**).

Vapor annealing was performed on some samples to tighten the host matrix around the inserted molecules, thereby favoring insertion of only single OPE molecules [8, 9, 65]. To do this, a neat solution of n-alkanethiol was placed into a v-vial (Wheaton Science Products, Millville, NJ) with the inserted SAM matrix suspended above the solution. The vial was sealed and held at 80°C for 2 h, exposing the sample to the vapor-phase alkanethiol. Annealing added molecules from the gas phase into defects in the original matrix, while mini-mizing the exchange processes that occur during solution-phase annealing, so that inserted molecules would not be displaced from the sample [65].

3.2.2 Imaging

The OPE-inserted SAMs were imaged using ambient-condition, custom-built, beetle-style scanning tunneling microscopes with high mechanical stability and high current sensitivity, described previously [6]. OPE molecules were shown to be isolated in the host matrix using STM images recorded with molecular resolution of the surrounding SAM. The STM topography images recorded for the OPE molecules are convolutions of the geometries and con-ductivities of the molecules; therefore, the molecules appear as protrusions from the SAM matrix, since they are both more conductive and physically longer than the host matrix is thick. Each inserted molecule appears with the same shape when imaged, indicative of the feature being more conductive than the host SAM and smaller than the tip [6, 7]. Larger bundles of inserted OPEs are also imaged and appear broader, demonstrating we have inserted both single molecules and bundles of molecules. The number of molecules in-serted together can be selected to an extent by controlling the quality of the host matrix [7].

Series of time-lapse images were acquired for many hours (8–30 h), with individual image recording time intervals of 2.5–6 min, over single areas (1,000 × 1,000–3,000 × 3,000 Å^2) to follow the switching of OPE molecules. We use these image sizes to characterize 10–50 switches per frame over hun-dreds of image frames so that we can track the activity of many molecules, while maintaining single-molecule resolution. At this scale, the STM tip was scanned over a molecule 2–4 times per image, averaging 5–10 image pixels representing each molecule (pixels are sampled on a millisecond time scale); thus, the switching of a single molecule was not recorded during the majority of each image, but multiple measurements of each molecule were collected in each image. In addition, many high-resolution time-lapse series of images have been recorded. Piezoelectric drift from thermal fluctuations could change the area imaged, and so a tracking algorithm was developed to correct for drift during acquisition and for postacquisition analyses [13, 66].

3.2.3 Apparent Height Determination

We determine the conductance of each individual molecule or bundle over hundreds of frames from the imaging field of view, using a drift track calculated by cross-correlation of fast Fourier transform spectra of sequential images [66]. From this, we determine the apparent heights for each molecule in the field of view as a function of time. Apparent heights of the inserted molecules are defined as the difference between the imaged height of the OPE in each frame and the average of the imaged height of the host matrix. For each molecule in each frame we extract two areas for analysis. First, we extract the area around the molecule and take the median of the top five pixel values giving our raw molecule topographic height, and second, we extract an independent area of the host matrix close to the molecule (free of defect sites, including substrate vacancy sites and step-edges) for our matrix background. The mean value of the extracted background is subtracted from the median raw topographic height of the molecule to calculate the apparent height of the inserted molecule [13]. By tabulating apparent height distributions for all of the inserted molecules over all of the frames, we find bimodal distributions of the apparent heights, indicating that the molecules switch between bistable states [8–12, 14, 66].

Apparent height is indicative of the conductance state of the inserted OPE derivatives; thus, we are able to determine whether the inserted switch was in a high or low conductance state while the image was recorded. Note that apparent height is dependent on the thickness of the host matrix; a switch located in a host matrix of octanethiolate will have a greater apparent height than a switch in a dodecanethiolate host matrix. From our apparent height calculations, we have also differentiated between molecules that exhibit motion up and down substrate step edges and molecules that exhibit conductance switching, and have shown that molecules can exhibit either event independently from the other [13]. It is important to note that our apparent height calculations favor molecules in the high conductance state. This is due to the fact that we record apparent height information for all molecules that ever appear as protrusions (high conductance state) during imaging, but we do not include molecules that remain in the low conductance state (never appearing as a protrusion from the host) throughout imaging.

3.2.4 Dipole Moment Calculations

Dipole moments, listed in Table 3.1 for OPE derivatives used in our bias-dependent switching studies, were calculated semiempirically using CAChe Worksystem Pro (version 6.1.1) with PM5 parameterization. The values presented are for the free-thiol OPEs, and are a qualitative comparison between

Table 3.1. Calculated dipole moment

OPE	Dipole (total)	Dipole component (parallel to OPE axis)
4	4.6	2.4
5	1.6	1.1
10	6	2.6
11	−8.1	−5.2
12	−4.4	−3.9

Values give are for the free-thiol OPEs

the OPE species bound to the Au substrate. We expect the dipole magnitudes would change upon adsorption to a Au{111} substrate; however, for qualitative purposes we expect a comparison of the free thiols are relevant.

3.3 Results and Discussion

3.3.1 Stochastic Conductance Switching

We have determined the OPE derivatives to be more conducting than the host SAM matrices in which they were inserted [7]. However, when inserted into host alkanethiolate SAM matrices and imaged using STM (Fig. 3.2), OPE derivatives exhibit stochastic conductance switching, imaging in both high "on" and low "off" conductance states [8, 9, 12, 14, 16]. A molecule observed in an "on" state appears as a protrusion from the host SAM using STM, while a molecule in the "off" conductance state appears to protrude only slightly from the matrix or not at all, depending on the thickness and location of the molecule in the SAM. Figure 3.2b exhibits a switching event for the circled molecules. In the top STM image, only one OPE molecule switch appears as a protrusion, and is in the "on" conductance state. In the bottom image, two molecules appear as protrusions in the circled area. The switching between these two states for OPE derivatives occurs spontaneously and reversibly throughout imaging when OPEs are inserted into alkanethiolate host matrices.

To test how the film quality of the alkanethiolate matrix affects the observed switching for derivatives **4** and **5**, we compared three different SAM sample preparations methods: decreased SAM deposition time (5 min) followed by OPE insertion, typical SAM deposition time (24 h) followed by OPE insertion, and vapor annealing into a preformed, OPE inserted 24 h SAM [8, 9, 65]. These methods change the defect density and film quality from the typical (24 h) matrix by both increasing defect density and lowering film quality (5 min SAM) and decreasing defect density while improving film quality (vapor annealing). The shorter allowed time (5 min) for SAM deposition allows for initial SAM formation that occurs rapidly, but film restructuring and further reordering takes longer times.

Using our apparent height extraction methods, we found the molecules in the vapor-annealed SAMs more often exhibited apparent heights in the on state than in the off state, with fewer molecules exhibiting switching [8, 9]. This does not indicate that molecules were not present in the off state, rather molecules were stable in only one conductance state and were limited in their ability to switch to the opposite conductance [8, 9]. Conversely, the OPE molecules imaged in the high defect density, low quality SAMs, exhibit a greater number of switching events, and throughout imaging more molecules appear in the "off" conductance state; thus, we have determined that the quality of the host matrix is critical to the switching activity we observe on the time scale of STM images [8, 9].

3.3.2 Conductance Switching Mechanisms

The conductance switching observed for OPE derivatives inserted into host SAM matrices has been compared with the negative differential resistance observed in nanopores and some other testbeds [19, 26, 27, 48, 67, 68]; yet, these two processes are different [15]. Many switching mechanisms have been proposed to describe the bistable conductance switching we observe (Fig. 3.3), including mechanisms suggested for the nanopore junction. Seminario et al. suggested that applying a sufficient potential to nitro- and amino-functionalized molecules can reduce the molecules, thereby extending the lowest unoccupied molecular orbital (LUMO) over the molecule and thus increasing conduction (Fig. 3.3a) [54]. Lang and coworkers predicted that the rotational plane of the ligands with respect to the phenyl ring to which it is bonded would change the highest occupied molecular orbital (HOMO) to LUMO gap, thereby affecting molecular conductance [55]. Similar calculations for molecular rotations by Brédas and coworkers suggested that the rotation of the middle phenyl ring controls conductance and that constriction of the rotation would inhibit changes in conductance (Fig. 3.3b) [56]. Several groups have based predictions on calculations looking at reduced conductance when current flows through adjacent molecules (Fig. 3.3c) [58, 69]. Lindsay and coworkers have performed experiments similar to ours, analyzing conductance switching for inserted α, ω-dithiol OPE molecules with nanoparticles attached to the pendant thiols at the outer film surface. They inferred the molecules detaching from the surface caused the observed conductances changes where the nanoparticle was not present in the image (Fig. 3.3d) [16]. Finally, we have hypothesized and tested that hybridization changes between the molecule and substrate affect conductance switching (Fig. 3.3e) [8–12, 14, 59, 60, 70]. To test each of these mechanisms, we engineered a variety of OPE derivatives, independently assessing the ability of each mechanism to exhibit switching [12].

Molecules **4** and **5** have shown similar stochastic switching characteristics; **4** is an unfunctionalized OPE while **5** contains a nitro group on the middle phenyl ring of the OPE (although **5** can be driven from one state to another [8, 9], **4** cannot be [10]). Both of these derivatives have exhibited stochastic

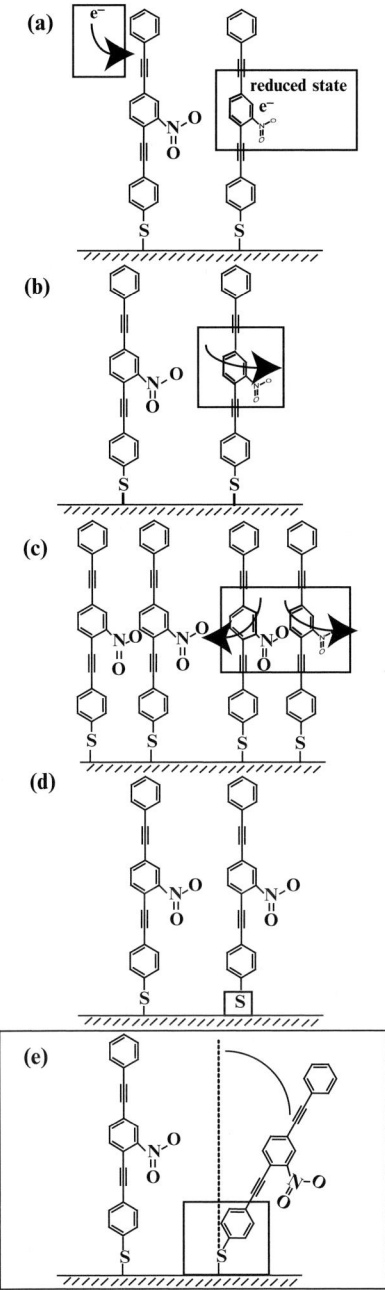

Fig. 3.3. Several mechanisms have been used to describe conductance switching. Suggested mechanisms include (**a**) functional group reduction, (**b**) ring-rotation, (**c**) concerted motions, (**d**) bond-fluctuations, and (**e**) hybridization changes causing conductance switching. Hybridization change is boxed since it is the only mechanism consistent with all data

bistable conductance switching; thereby, eliminating a functional group on the OPE derivatives as being necessary for switching to occur. This eliminates mechanisms requiring the reduction of functional groups (Fig. 3.3a) [54], or the rotation of functional groups [55] from being the primary mechanism for conductance switching. Furthermore, our applied potentials of ±1.0 V sample bias used for imaging are nonreductive potentials.

To test the mechanism suggesting that ring rotation is responsible for switching (Fig. 3.3b) [56], we studied phenanthrenethiol **6**, containing fused phenyl rings held in the same plane, precluding rotation around a central bond. An excess of 6 eV would be necessary for internal rotation to occur; this energy is not available at room temperature or through our tunneling conditions. Since these molecules exhibited bistable stochastic conductance switching similar to that observed for **4** and **5**, it is inconsistent with our data that changes in alignment of the phenyl rings cause the observed conductance switching [14].

Using STM, neighboring molecule interactions are more difficult to detect since the imaged OPE derivatives all have the same shape, indicative of our tip shape; highly conductive molecules in close proximity are difficult to resolve [6, 7]. Therefore, to test this mechanism, we engineered a disulfide OPE derivative **7**. Nuzzo et al. have shown that disulfide molecules dissociatively chemisorb as thiolates on the Au{111} surface [63]. We thus expect that when we insert **7**, after dissociative chemisorption, one or both halves of the molecule adsorb, favoring pairs of molecules to insert in close proximity over insertion of single thiolate molecules. No significant differences were observed for molecules inserted as pairs, spontaneously forming pairs, or those existing as individuals; thus eliminating neighbor molecule interactions as the mechanism for conductance switching (Fig. 3.3c) [12]. The observation of not requiring neighboring molecule interactions is also consistent with other experimental data [22, 51, 71].

We studied in detail the bond-fluctuation mechanism (Fig. 3.3d) proposed by Lindsay and coworkers, since they performed experiments using an OPE derivative similar to our studies [16]. This OPE derivative (2,5-di(phenylethynyl-4′, 4″-thio)benzene) has the same structure as **4**, but also includes a thiol on the pendant end of the molecule, to which they attached a gold nanoparticle. These nanoparticle–molecule–substrate assemblies were imaged via STM and stochastic conductance switching was observed. They reasoned that the nanoparticles were too large to be buried in the surrounding matrix (i.e., not imaged), and therefore a hybridization change through a molecular tilt or substrate reconstruction was not possible. However, this reasoning does not take into account the fact that even if tilting alone was the primary cause for conductance switching, rather than the hybridization changes we have shown, in a (~30°) tilted conformation, most of our OPE derivatives would still protrude from the host matrix [12].

The bond fluctuation mechanism is also unlikely due to the strength of the S–Au bond (~1.6 eV for alkanethiolates) [72, 73]. In our experiments, the samples were in ambient conditions or ultrahigh vacuum and there is not enough energy for this bond breaking to occur, although it is more likely in their system due to the fact that their samples were imaged under toluene and solution displacement of alkanethiols is known [74]. For the bond fluctuation hypothesis, Lindsay and coworkers assumed that if a nanoparticle does not have a covalent tether to the surface, the conductivity will be lowered; thus causing the nanoparticle not to appear from STM images [16]. However, we have shown many STM images of nanoparticles that are not covalently attached to a substrate, but are nevertheless somewhat immobilized on an alkanethiolate SAM matrix via van der Waals interactions between the nanoparticle ligand shell and the SAM [12, 75]. This ability to image noncovalently attached nanoparticles, Lindsay and coworkers' definition of a broken-bond (switched-off) system, renders this assumption false [12, 16]. The conditions used in their imaging experiments were apparently too perturbative and their instruments too insensitive to resolve these nanoparticles.

In addition, we have imaged $Au_{11}(S(CH_2)_9CH_3)_8$ clusters adsorbed for 20 min on a SAM containing inserted decanedithiol tethers (Fig. 3.4). Figure 3.4b depicts the intermittent appearance (imaged for a few lines, disappears and reappears) of nanoparticle (indicated by arrow) that was not present in Fig. 3.4a. This nanoparticle remained stable through Fig. 3.4c, and then "disappears" after being imaged for a few lines in Fig. 3.4d. The streak in Fig. 3.4d (indicated by arrow) is indicative of the nanoparticle being pushed around by the tip or that the nanoparticle hopped from the surface to the tip. Finally, in Fig. 3.4e, the nanoparticle was no longer present. We assert that in the system studied by Lindsay and coworkers, when the nanoparticles disappeared (reappeared), they had detached from (reattached to) the tethering molecule and either transferred to (released from) the tip, been swept across the surface, or released into (attached from) the toluene solution. Lindsay and coworkers believed that nanoparticle detachment/reattachment was not possible because the incubation time for nanoparticle reattachment was too long; however, incubation times on the minute time scale are adequate to observe many nanoparticles bind to dithiol tethers. We have observed that exposing dithiol-functionalized SAMs to thiol-functionalized nanoparticles for deposition times as short as 20 min produces surfaces with substantial coverages of nanoparticles (Fig. 3.4). Finally, Lindsay and coworkers have previously reported that nanoparticles attach to the Au surface through multiple alkanedithiol tethers [51]. For the proposed bond-fluctuation mechanism, multiple tethering molecules should influence the switching rate; if a nanoparticle has to break all tethers before switching off, this would slow switching. In addition, the presence of multiple tethers should influence conductivity; that is, the apparent height of the nanoparticle may scale with the number of tethers. Neither of these observations was reported [16].

A hybridization change between the molecule and the substrate (Fig. 3.3e) is the only proposed mechanism consistent with all the data. The contacts between organic molecules and electrodes can have a large influence on the

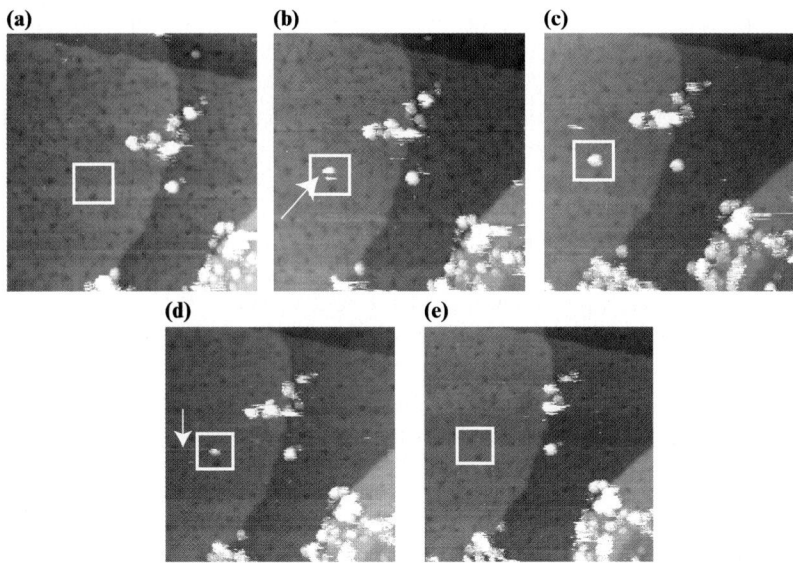

Fig. 3.4. (a–e) Series of STM images of $Au_{11}(S(CH_2)_9CH_3)_8$ clusters adsorbed for 20 min on a SAM of decanethiolate with decanedithiol tethers. These images display the mobility of (initially) dithiolate-tethered clusters. The boxed area shows (**a**) no cluster attached; (**b**) intermittent appearance of a cluster that was imaged for a few lines, "disappeared", and then reappeared for the remainder of the image (indicated by the arrow); (**c**) a stable cluster; (**d**) a streak in the image before the cluster was imaged (indicated by the arrow) indicates that the cluster was being pushed by the STM tip, or that the cluster hopped from the surface to the tip; and (**e**) no cluster attached. Image conditions $1{,}500 \times 1{,}500$ Å^2, $V_{\mathrm{sample}} = -0.5\,\mathrm{V}$, $I_{\mathrm{tunnel}} = 5\,\mathrm{pA}$

conductance observed through the molecule [5, 76–78]. In each of the molecules engineered to test the hypotheses for conductance switching, we have not changed the S–Au bonding scheme. For the hybridization change, the metal–molecule system must have enough degrees of freedom to change the structure of the metal–molecule bond. This agrees with our data from the stochastic switching experiments analyzing the effects of the host monolayer packing [8, 9].

Note that we cannot rule out substrate atom rearrangements that occur simultaneously with bond rehybridization as possible contributors to changes in conductance. Even for simple alkanethiolate SAMs, there is significant controversy over S–Au bonding and binding sites [79, 80]. When thiolates and other electronegative species bind to Au and other coinage metals, the substrate surface atoms are known to relax and to become more mobile [81, 82]. Alkaneselenolates induce substantial substrate rearrangements on Au{111} and adopt a variety of binding sites [76]. Also, atoms in a linear chain moved from their equilibrium positions change conductance [83]. Determining the roles of substrate rearrangements remains central to understanding self-assembly and conductance switching.

We have attempted to restrict the motion of the OPE derivatives by engineering **8** and **9**, containing multiple possible contacts to the surface, thus reducing switching activity if each contact was bound [12]. If multiple thiols attached to the gold, this would inhibit switching unless hybridization changes occurred at each attachment. Multiple attachments would also give insight into how the conductance is affected by multiple conduction pathways, similar to the neighboring molecule interaction predictions. However, from our experiments, we infer that it is most likely that only one attachment was able to form because the spacing between the thiols does not adequately match the binding site spacing between gold substrate atoms. We would expect the registry to be close to that of the host alkanethiolate monolayer preassembled on the Au{111} substrate. It is unlikely (but in principle possible) to have a defect in the host SAM large enough to accommodate molecule **8** with several contacts to the substrate. Design and syntheses of new molecules with two or more contacts matching the gold lattice spacing are in progress.

3.3.3 Driven Conductance Switching

With a tested, working hypothesis of how conductance switching occurs in our system, we next stepped from the stochastic switching described above to the driven switching regime [10, 11]. Here, using molecular design, we have been able to analyze two contributions to drive bias-dependent switching: the dipole of the OPE derivatives (relating these to the electric field they experience within the STM junction) and the hydrogen-bonding interactions possible with appropriately designed amide-containing SAMs (Fig. 3.5). We have studied OPEs **4, 5, 10, 11**, and **12** inserted into host matrices of **2** and **3**.

Matrices composed of **2** or **3** (differing by the amide group orientation, **2** is inverted as compared to **3**) have stronger intermolecular forces between individual molecules composing the SAM than their *n*-alkanethiolate counterparts. This is due to the hydrogen-bonding interactions arising from the amide functionality contained in these molecules [62, 84]. In addition, OPE derivatives containing nitro- or amino-functional groups have the ability to hydrogen bond with the appropriate amide-containing SAMs.

To test the mechanism for driven bias-dependent conductance switching, we have systematically altered the sign and magnitude of the dipole moment of the OPEs by engineering a series of fluorinated OPE derivatives to control the polarity of the bias dependence (Table 3.1). Non-fluorinated derivatives **4** and **5** have intrinsic dipole moments of +1.6 and +4.6 D, respectively. These dipoles are designated as positive for the purpose of this discussion and indicate that the positive end of the dipole is presented at the pendant end of the molecule (film/air or film/vacuum interface) when the OPEs are inserted into the host SAM. Important to our study is the dipole component parallel to the OPE axis that will interact with the electric field between the STM tip and substrate. For **4** and **5**, the parallel components of the dipoles are +1.1 and +2.4 D, respectively. Derivative **10** is the fluorinated analogue of **4** and

(a)

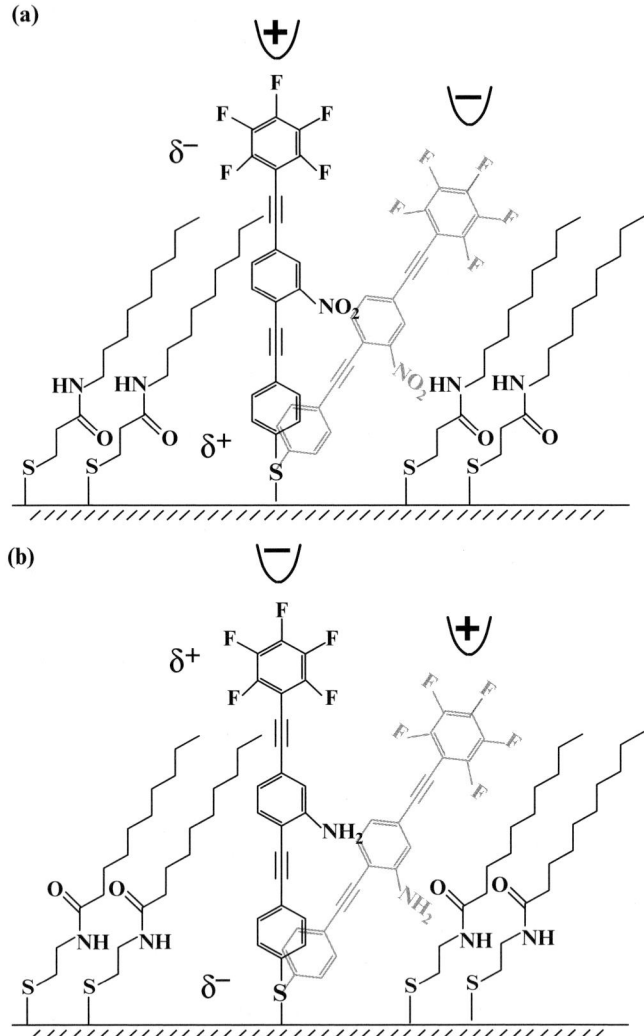

(b)

Fig. 3.5. (a) Molecule **10** inserted into a matrix of **2**, the positive dipole aligns when the tip has a positive bias (negative sample bias) driving the molecule into the "on" state. When the tip has a negative bias, electrostatic repulsion between the tip and positive dipole of the molecule turns the molecule "off." (b) The dipole for molecule **12** is inverted compared to molecule **10**, and the amide group has been reversed from matrix molecule **2** to matrix molecule **3**; therefore, the interaction for this molecule with the tip is opposite that found in (a). A negative tip bias (positive sample bias) drives the molecule to the "on" state while a positive tip bias turns the switch "off"

has a parallel dipole moment of $+2.6$ D. If we change the position of the nitro-functionalization of this molecule on the central ring phenyl ring, as in **11**, we find that the parallel component of the dipole moment is in the opposite direction at -5.2 D. Similarly, a molecule that replaces the nitro group in **4** with an amine group, as in **12**, the dipole moment is also negative by our convention, -3.9 D. The fluorinated derivatives with a negative parallel dipole component allowed us to test whether the polarity of the bias-dependent switching could be reversed from that of the derivatives containing positive parallel dipole moments [10, 11].

When **4** was inserted into a matrix of **2**, no bias-dependent switching was observed; however, when **5** or **10** was inserted into **2**, bias-dependent switching was observed, where the high conductance state was favored at positive sample bias and the low conductance state was favored at negative bias [10, 11]. The directionality of the bias dependence aligns with the electric field of our STM tunneling junction (Fig. 3.5). Derivatives **5** and **10** both have stronger dipole interactions and the possibility of hydrogen bonding as compared to **4**; thus, we had to analyze other OPEs to determine the importance of each of these contributing factors. Using **11**, we were able to test the dipole dependence of polarity. Comparing **11** to **10**, the position of the nitro-functionality on the middle phenyl ring is the only modification. However, this chemical modification changes the direction (positive to negative) of the intrinsic dipole moment. When molecule **11** was inserted into a host matrix of **2**, bias-dependent switching was observed, but with the opposite polarity of **10** inserted into **2**; thus, the direction of the dipole moment contributes to and determines the polarity of the observed bias-dependent switching.

Molecule **12** is another fluorinated OPE designed to have a negative relative dipole moment for bias-dependent switching. The amine group on this molecule serves two purposes: to contribute to the dipole moment and to enable hydrogen bonding. When **12** was inserted into **2**, no bias-dependent switching was observed, presumably because hydrogen bonding could not stabilize this system. To enable hydrogen bonding between the amine group and the matrix, we inserted **12** into a preformed SAM of **3**. In this sample, bias-dependent switching was observed; thus, mediated switching depends both on the dipole moment of the molecule and hydrogen-bonding interactions between the OPE derivative and the host SAM (Fig. 3.5) [10, 11].

3.4 Conclusions

We have engineered and studied the conductance switching of OPE derivatives inserted into host alkanethiolate and amide-containing alkanethiolate SAMs. We have tailored the behavior of molecular switching through the preparation and control of the host alkanethiolate matrix and determined the mechanism through which the molecules switch conductance by engineering the properties of the OPE derivatives. The only mechanism consistent with all data is that of a hybridization change at the molecule–substrate contact. As described, there may be additional contributions due to substrate atom rearrangements. Finally, we can drive conductance switching via interactions between the electric field of the STM junction and the dipole of the OPE derivatives, tailoring the interactions with the host matrix. We conclude that designing the local chemical and structural environment of molecular electronic components is crucial to optimizing their function.

Acknowledgements. The authors thank Drs. J.M. Tour, J.W. Ciszek, F. Maya, and Y. Yao who synthesized the OPE derivatives, J.E. Hutchison, C.E. Inman, and G.H. Woehrle who synthesized the amide molecules and Au nanoclusters, Dr. J. Sofo for helpful discussions on theoretical modeling of the OPE molecules and for the use of the CAChe Software, and Drs. D.L. Allara, J.G. Kushmerick, M.A. Ratner, R. Shashidhar, and R. van Zee for helpful discussions. We gratefully acknowledge the continued support from the Army Research Office, Defense Advanced Research Projects Agency, National Institutes for Standards and Technology, National Science Foundation, Air Force Office of Scientific Research, and the Office of Naval Research.

References

1. B.A. Mantooth, P.S. Weiss, Proc. IEEE **91**, 1785 (2003)
2. M.A. Reed, Proc. IEEE **87**, 652 (1999)
3. J.M. Tour, Accounts Chem. Res. **33**, 791 (2000)
4. M. Magoga, C. Joachim, Phys. Rev. B **56**, 4722 (1997)
5. J.M. Seminario, C.E. de la Cruz, P.A. Derosa, J. Am. Chem. Soc. **123**, 5616 (2001)
6. M.T. Cygan, T.D. Dunbar, J.J. Arnold, L.A. Bumm, N.F. Shedlock, T.P. Burgin, L. Jones II, D.L. Allara, J.M. Tour, P.S. Weiss, J. Am. Chem. Soc. **120**, 2721 (1998)
7. L.A. Bumm, J.J. Arnold, M.T. Cygan, T.D. Dunbar, T.P. Burgin, L. Jones II, D.L. Allara, J.M. Tour, P.S. Weiss, Science **271**, 1705 (1996)
8. Z.J. Donhauser, B.A. Mantooth, K.F. Kelly, L.A. Bumm, J.D. Monnell, J.J. Stapleton, D.W. Price, A.M. Rawlett, D.L. Allara, J.M. Tour, P.S. Weiss, Science **292**, 2303 (2001)
9. Z.J. Donhauser, B.A. Mantooth, T.P. Pearl, K.F. Kelly, S.U. Nanayakkara, P.S. Weiss, Jpn. J. Appl. Phys. **41**, 4871 (2002)
10. P.A. Lewis, C.E. Inman, F. Maya, J.M. Tour, J.E. Hutchison, P.S. Weiss, J. Am. Chem. Soc. **127**, 17421 (2005)
11. P.A. Lewis, C.E. Inman, Y.X. Yao, J.M. Tour, J.E. Hutchison, P.S. Weiss, J. Am. Chem. Soc. **126**, 12214 (2004)
12. A.M. Moore, A.A. Dameron, B.A. Mantooth, R.K. Smith, D.J. Fuchs, Y.X. Yao, J.W. Ciszek, F. Maya, J.M. Tour, P.S. Weiss, J. Am. Chem. Soc. **128**, 1959 (2006)
13. A.M. Moore, B.A. Mantooth, Z.J. Donhauser, F. Maya, D.W. Price, Y. Yao, J.M. Tour, P.S. Weiss, Nano Lett. **5**, 2292 (2005)
14. A.A. Dameron, J.W. Ciszek, J.M. Tour, P.S. Weiss, J. Phys. Chem. B **108**, 16761 (2004)

15. A.S. Blum, J.G. Kushmerick, D.P. Long, C.H. Patterson, J.C. Yang, J.C. Henderson, Y.X. Yao, J.M. Tour, R. Shashidhar, B.R. Ratna, Nat. Mater. **4**, 167 (2005)
16. G.K. Ramachandran, T.J. Hopson, A.M. Rawlett, L.A. Nagahara, A. Primak, S.M. Lindsay, Science **300**, 1413 (2003)
17. A.M. Moore, D.L. Allara, P.S. Weiss, *NNIN Nanotechnology Open Textbook* (2007)
18. L.T. Cai, M.A. Cabassi, H. Yoon, O.M. Cabarcos, C.L. McGuiness, A.K. Flatt, D.L. Allara, J.M. Tour, T.S. Mayer, Nano Lett. **5**, 2365 (2005)
19. I. Kratochvilova, M. Kocirik, A. Zambova, J. Mbindyo, T.E. Mallouk, T.S. Mayer, J. Mater. Chem. **12**, 2927 (2002)
20. J.K.N. Mbindyo, T.E. Mallouk, J.B. Mattzela, I. Kratochvilova, B. Razavi, T.N. Jackson, T.S. Mayer, J. Am. Chem. Soc. **124**, 4020 (2002)
21. Y. Selzer, L.T. Cai, M.A. Cabassi, Y. Yao, J.M. Tour, T.S. Mayer, D.L. Allara, Nano Lett. **5**, 61 (2005)
22. J.G. Kushmerick, J. Naciri, J.C. Yang, R. Shashidhar, Nano Lett. **3**, 897 (2003)
23. J.G. Kushmerick, S.K. Pollack, J.C. Yang, J. Naciri, D.B. Holt, M.A. Ratner, R. Shashidhar, Ann. NY. Acad. Sci. **1006**, 277 (2003)
24. J.G. Kushmerick, C.M. Whitaker, S.K. Pollack, T.L. Schull, R. Shashidhar, Nanotechnology **15**, S489 (2004)
25. J. Chen, L.C. Calvet, M.A. Reed, D.W. Carr, D.S. Grubisha, D.W. Bennett, Chem. Phys. Lett. **313**, 741 (1999)
26. J. Chen, M.A. Reed, A.M. Rawlett, J.M. Tour, Science **286**, 1550 (1999)
27. J. Chen, W. Wang, M.A. Reed, A.M. Rawlett, D.W. Price, J.M. Tour, Appl. Phys. Lett. **77**, 1224 (2000)
28. C. Zhou, M.R. Deshpande, M.A. Reed, L. Jones II, J.M. Tour, Appl. Phys. Lett. **71**, 611 (1997)
29. R.E. Holmlin, R. Haag, M.L. Chabinyc, R.F. Ismagilov, A.E. Cohen, A. Terfort, M.A. Rampi, G.M. Whitesides, J. Am. Chem. Soc. **123**, 5075 (2001)
30. M.A. Rampi, O.J.A. Schueller, G.M. Whitesides, Appl. Phys. Lett. **72**, 1781 (1998)
31. M.A. Rampi, G.M. Whitesides, Chem. Phys. **281**, 373 (2002)
32. K. Slowinski, H.K.Y. Fong, M. Majda, J. Am. Chem. Soc. **121**, 7257 (1999)
33. K. Slowinski, M. Majda, J. Electroanal. Chem. **491**, 139 (2000)
34. T. Graves-Abe, Z. Bao, J.C. Sturm, Nano Lett. **4**, 2489 (2004)
35. T. Graves-Abe, J.C. Sturm, Appl. Phys. Lett. **87**, 133502 (2005)
36. N.B. Zhitenev, H. Meng, Z. Bao, Phys. Rev. Lett. **88**, 226801 (2002)
37. C.J. Muller, J.M. Krans, T.N. Todorov, M.A. Reed, Phys. Rev. B **53**, 1022 (1996)
38. C.J. Muller, B.J. Vleeming, M.A. Reed, J.J.S. Lamba, R. Hara, L. Jones II, J.M. Tour, Nanotechnology **7**, 409 (1996)
39. M.A. Reed, C. Zhou, C.J. Muller, T.P. Burgin, J.M. Tour, Science **278**, 252 (1997)
40. J. Reichert, R. Ochs, D. Beckmann, H.B. Weber, M. Mayor, H. von Löhneysen, Phys. Rev. Lett. **88**, 176804 (2002)
41. J. Reichert, H.B. Weber, M. Mayor, H. von Lohneysen, Appl. Phys. Lett. **82**, 4137 (2003)
42. W. Liang, M.P. Shores, M. Bockrath, J.R. Long, H. Park, Nature **417**, 725 (2002)

46   A.M. Moore et al.

43. H. Park, A.K.L. Lim, A.P. Alivisatos, J. Park, P.L. McEuen, Appl. Phys. Lett. **75**, 301 (1999)
44. H. Park, J. Park, A.K.L. Lim, E.H. Anderson, A.P. Alivisatos, P.L. McEuen, Nature **407**, 57 (2000)
45. J. Park, A.N. Pasupathy, J.I. Goldsmith, C. Chang, Y. Yaish, J.R. Petta, M. Rinkoski, J.P. Sethna, H.D. Abruña, P.L. McEuen, D.C. Ralph, Nature **417**, 722 (2002)
46. Y. Selzer, M.A. Cabassi, T.S. Mayer, D.L. Allara, J. Am. Chem. Soc. **126**, 4052 (2004)
47. Y. Selzer, M.A. Cabassi, T.S. Mayer, D.L. Allara, Nanotechnology **15**, S483 (2004)
48. I. Amlani, A.M. Rawlett, L.A. Nagahara, R.K. Tsui, Appl. Phys. Lett. **80**, 2761 (2002)
49. A. Bezryadin, C. Dekker, G. Schmid, Appl. Phys. Lett. **71**, 1273 (1997)
50. D.P. Long, C.H. Patterson, M.H. Moore, D.S. Seferos, G.C. Bazan, J.G. Kushmerick, Appl. Phys. Lett. **86**, 153105 (2005)
51. X.D. Cui, A. Primak, X. Zarate, J. Tomfohr, O.F. Sankey, A.L. Moore, T.A. Moore, D. Gust, G. Harris, S.M. Lindsay, Science **294**, 571 (2001)
52. X.D. Cui, A. Primak, X. Zarate, J. Tomfohr, O.F. Sankey, A.L. Moore, T.A. Moore, D. Gust, L.A. Nagahara, S.M. Lindsay, J. Phys. Chem. B **106**, 8609 (2002)
53. S.M. Lindsay, Jpn. J. Appl. Phys. **41**, 4867 (2002)
54. J.M. Seminario, A.G. Zacarias, J.M. Tour, J. Am. Chem. Soc. **122**, 3015 (2000)
55. M. Di Ventra, S.G. Kim, S.T. Pantelides, N.D. Lang, Phys. Rev. Lett. **86**, 288 (2001)
56. J. Cornil, Y. Karzazi, J.L. Brédas, J. Am. Chem. Soc. **124**, 3516 (2002)
57. N.D. Lang, P. Avouris, Phys. Rev. B **62**, 7325 (2000)
58. J.M. Seminario, A.G. Zacarias, J.M. Tour, J. Am. Chem. Soc. **120**, 3970 (1998)
59. H. Sellers, A. Ulman, Y. Shnidman, J.E. Eilers, J. Am. Chem. Soc. **115**, 9389 (1993)
60. Y.T. Tao, C.C. Wu, J.Y. Eu, W.L. Lin, Langmuir **13**, 4018 (1997)
61. P.A. Lewis, Z.J. Donhauser, B.A. Mantooth, R.K. Smith, L.A. Bumm, K.F. Kelly, P.S. Weiss, Nanotechnology **12**, 231 (2001)
62. R.K. Smith, S.M. Reed, P.A. Lewis, J.D. Monnell, R.S. Clegg, K.F. Kelly, L.A. Bumm, J.E. Hutchison, P.S. Weiss, J. Phys. Chem. B **105**, 1119 (2001)
63. R.G. Nuzzo, L.H. Dubois, D.L. Allara, J. Am. Chem. Soc. **112**, 558 (1990)
64. Y. Yao, J.M. Tour, Org. Chem. **64**, 1968 (1999)
65. Z.J. Donhauser, P.S. Weiss, D.W. Price, J.M. Tour, J. Am. Chem. Soc. **125**, 11462 (2003)
66. B.A. Mantooth, Z.J. Donhauser, K.F. Kelly, P.S. Weiss, Rev. Sci. Instrum. **73**, 313 (2002)
67. A.M. Rawlett, T.J. Hopson, I. Amlani, R. Zhang, J. Tresek, L.A. Nagahara, R.K. Tsui, H. Goronkin, Nanotechnology **14**, 377 (2003)
68. A.M. Rawlett, T.J. Hopson, L.A. Nagahara, R.K. Tsui, G.K. Ramachandran, S.M. Lindsay, Appl. Phys. Lett. **81**, 3043 (2002)
69. N.D. Lang, P. Avouris, Phys. Rev. Lett. **84**, 358 (2000)
70. C.W.J. Bauschlicher, A. Ricca, N. Mingo, J. Lawson, Chem. Phys. Lett. **372**, 723 (2003)
71. B. Xu, N.J. Tao, Science **301**, 1221 (2003)

72. L.H. Dubois, R.G. Nuzzo, Annu. Rev. Phys. Chem. **43**, 437 (1992)
73. A. Ulman, Chem. Rev. **96**, 1533 (1996)
74. J.B. Schlenoff, M. Li, H. Ly, J. Am. Chem. Soc. **117**, 12528 (1995)
75. R.K. Smith, S.U. Nanayakkara, G.H. Woehrle, T.P. Pearl, M.M. Blake, J.E. Hutchison, P.S. Weiss, J. Am. Chem. Soc. **128**, 9266 (2006)
76. J.D. Monnell, J.J. Stapleton, J.J. Jackiw, T.D. Dunbar, W.A. Reinerth, S.M. Dirk, J.M. Tour, D.L. Allara, P.S. Weiss, J. Phys. Chem. B **108**, 9834 (2004)
77. J.M. Beebe, V.B. Engelkes, L.L. Miller, C.D. Frisbie, J. Am. Chem. Soc. **124**, 11268 (2002)
78. N.D. Lang, C.R. Kagan, Nano Lett. **6**, 2955 (2006)
79. P. Fenter, A. Eberhardt, P. Eisenberger, Science **266**, 1216 (1994)
80. H. Basch, R. Cohen, M.A. Ratner, Nano Lett. **5**, 9 (2005)
81. S.J. Stranick, A.N. Parikh, D.L. Allara, P.S. Weiss, J. Phys. Chem. **98**, 11136 (1994)
82. D.J. Trevor, C.E.D. Chidsey, J. Vac. Sci. Technol. B **9**, 965 (1991)
83. M. Di Ventra, S.T. Pantelides, N.D. Lang, Phys. Rev. Lett. **84**, 979 (2000)
84. P.A. Lewis, R.K. Smith, K.F. Kelly, L.A. Bumm, S.M. Reed, R.S. Clegg, J.D. Gunderson, J.E. Hutchison, P.S. Weiss, J. Phys. Chem. B **105**, 10630 (2001)

4

Exploration of Oxide Semiconductor Electronics Through Parallel Synthesis of Epitaxial Thin Films

M. Kawasaki

Summary: A review is given for advanced thin film technology based on pulsed laser deposition of oxide thin films. Atomically regulated epitaxy has been extended to various kinds of oxides to construct artificially made heterostructures. Computer controlled masking and target switching in pulsed laser deposition have made it possible to integrate number of thin films with different layered structures and/or various compositions. This technique has been applied to search for new materials and functionalities in such oxide semiconductors as ZnO and TiO_2. The highlights of magnetic, photonic, and electronic devices are introduced with examples of TiO_2 ferromagnetic junctions, ZnO light emitting diodes, and (Mg, Zn)O/ZnO quantum Hall devices, respectively.

4.1 Introduction

The discovery of high critical temperature (T_c) superconducting oxides [1] triggered intensive research in solid state physics, chemistry, and electronics. The understanding of microscopic mechanism for superconductivity has been an open question after two decades of intensive research and still needs very difficult challenges. The material physics focusing on these and related oxides have made considerable progress to understand the properties and electronic states of strongly correlated electron systems [2]. Besides these physics, this class of materials gave us a big challenge in terms of solid-state chemistry. The high T_c superconducting oxides generally have layered structures that are constructed naturally, where basic block builders of Perovskite ABO_3 having 0.4 nm in size stack each other as two-dimensional molecular sheets. These components have different rare-earth or alkaline-earth elements in A site and different transition-metal elements in B site. Important degree of freedom here is the tunability of valence state of B both by the mixing ratio of trivalent rare-earth and divalent alkaline-earth elements at A site as well as by the choice of B element; thereby the filling of d orbital in B atom can be changed almost freely from 0 to 10 electrons. In addition to this, the degree of freedom in stacking sequence gave us infinite variation in compound diversity. Therefore, accumulation of Perovskite compounds in an atomic scale by thin film technology has attracted considerable interest for making new state of matter as well as possible exploration of "oxide electronics" [3].

Making oxide thin films had been widely studied even before high T_c superconductors in areas such as transparent conducting thin films of In_2O_3 doped with SnO_2 [4]. However, most of the work had been carried out for polycrystalline thin films on glass or quartz substrates. Some work employed single crystalline substrates for growing epitaxial films, but there had been no demand to make atomically flat surfaces and interfaces implemented in oxide thin films and heterostructures. The discovery of high T_c superconductors has changed this situation drastically. The active device element of superconducting electronics is Josephson junction that is composed of two superconducting electrodes inserted with an insulating tunnel barrier. The success of Nb-based superconducting electronics relied on the presence of oxidized Al barrier for which thickness can be controlled in a self-organized manner without suffering from pinholes. The reliability and reproducibility of the Josephson junctions are so excellent that one can make very large scale integrated circuits with very low power consumption and with very high speed. Therefore, it was natural to try to make Josephson junctions out of high T_c superconductors [5, 6]. However, there has not been discovered any process to passivate the surface of high T_c superconducting films in an atomic scale, allowing epitaxial growth of over layer. Hence, one needed to grow heteroepitaxial tunnel junction structures. That gave us a challenge of atomically controlled epitaxy of oxides. Although this sounded very difficult, many solid-state chemists jumped to this field because one can overlook very exciting and new research field of solid-state chemistry, that is, artificial construction of oxide superlattices that may exhibit new properties and functionalities, which cannot be attained in any other materials subsets such as conventional semiconductors and metals.

After overcoming technological aspects in making oxide superlattices, one confronted another or even tougher problem. In semiconductor superlattice research, there exists well understood guiding principle of band theory. In most of the semiconductor materials and devices, the properties and dynamics of charge carriers can be described very well by using single particle approximation. Therefore, one can predict the properties of materials and performance of devices by computer simulation. In case of oxides, correlated electrons cannot be theoretically treated well even now because of many body interactions; it is very difficult to predict the physical properties even for single component single crystalline materials. When one makes heterostructures composed of correlated electron oxides having different ground states, there is no way to make systematic research except for preparing all the samples with different compositions and combinations. This was very time consuming and one desired much more efficient way. Same problem existed in the research field of synthetic chemistry for bioscience and drug discovery. People used to design molecules as potential candidates for drug and then total-synthesize them for testing. This was also or much more time consuming than oxide interface problem. What people developed in the mid 1990s to solve this problem was the downsizing of synthetic chemistry tools to integrate the equipments aided with computers, enabling the synthesis of thousands of molecules having slightly different components and structures in a parallel fashion. Since all the combinations are realized in a carpet bombing manner, this technique was named as "combinatorial chemistry" [7–10]. The group of the author imported this concept for oxide lattice engineering and realized a neat way of epitaxially synthesizing number of thin film samples on a single substrate. By this technique, one can now make very systematic experiments to understand physical properties or to optimize device structures. Even more important fact is that one can make very challenging search for new materials having rapid screening tools.

In this article, the key technologies for the progress on atomically controlled oxide epitaxy are introduced and then the way of parallel synthesis and characterization of integrated thin film libraries are explained. Examples of material discovery, systematic characterization, and device optimization are given for an important subset of oxides, i.e., oxide semiconductors. This review article is designed for scientists in other fields rather than the experts in oxide thin films.

4.2 Atomically Regulated Oxide Epitaxy

For making high T_c superconducting oxide thin films, various techniques were applied such as sputtering, molecular beam epitaxy (MBE), and pulsed laser deposition (PLD). Three major difficulties were recognized at the early stage. One is stoichiometry control of cationic elements. Usually, high T_c superconducting oxides contain three or more cationic elements. Sputtering has drawbacks of reverse-sputtering of growing surface and preferential sputtering of some of the elements, yielding in the necessity of compositional compensation in the target. MBE requires precise control of atomic flux of individual elements. Later, atomic absorption flux monitoring was developed to feed-back control the flux very precisely [11]. PLD has an advantage of stoichiometric transfer of elements from the target to the film, which is believed to be due to explosive evaporation of surface layer of the target. Second difficulty was oxidation of components used in substrate heaters. Conventional high melting point metals such as W, Mo, Ta are very harmful choices because they are easily oxidized to form low melting point or high vapor pressure oxides, so that the films are usually contaminated with these elements. Halogen lump equipped with gold-plated and water-cooled mirror jacket has been widely used with Ni-based alloy (Inconel) components as shown in Fig. 4.1a. Later, infrared laser was proved to be a neat way to heat only the substrate holder to very high temperature as shown in Fig. 4.2c and Fig. 4.3 [12, 13]. Third difficulty was the compromise in the choice of high oxygen pressure or high

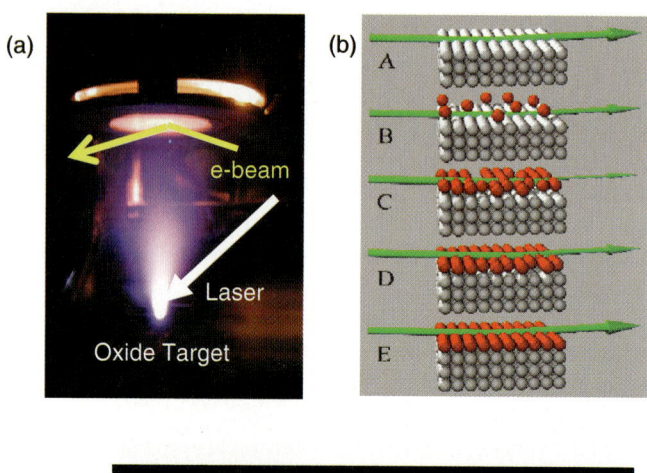

Fig. 4.1. (a) A picture of a pulsed laser deposition process. (b) Schematics for layer-by-layer growth and electron beam reflection. (c) Atomic force microscope image of a $SrTiO_3$ surface treated with an etching solution [17]

energy electron beam diffraction (RHEED). Transition metals such as Cu requires rather high pressure of a few hundreds mTorr of oxygen to get the right valence states of +2, but in this case one cannot readily apply RHEED to in-situ monitor the surface crystallinity and growth rate. Later, two-stage differential pumping system was developed so that one can use RHEED up to 1 Torr environment [14]. Through the developments above, MBE and PLD are believed to be the best choices. MBE is excellent for making large area films with highest quality but has a drawback of time consuming optimization at each time when the choice of elements are changed. PLD has an advantage of easy exchange of target materials but uniformity of film thickness was rather limited. The basic research in which one usually wants to test various kinds of oxide compounds, PLD is the better choice.

Figure 4.1a shows typical laser ablation process of oxide target by the irradiation of focused laser pulses. Typical substrate–target distance is 3–5 cm

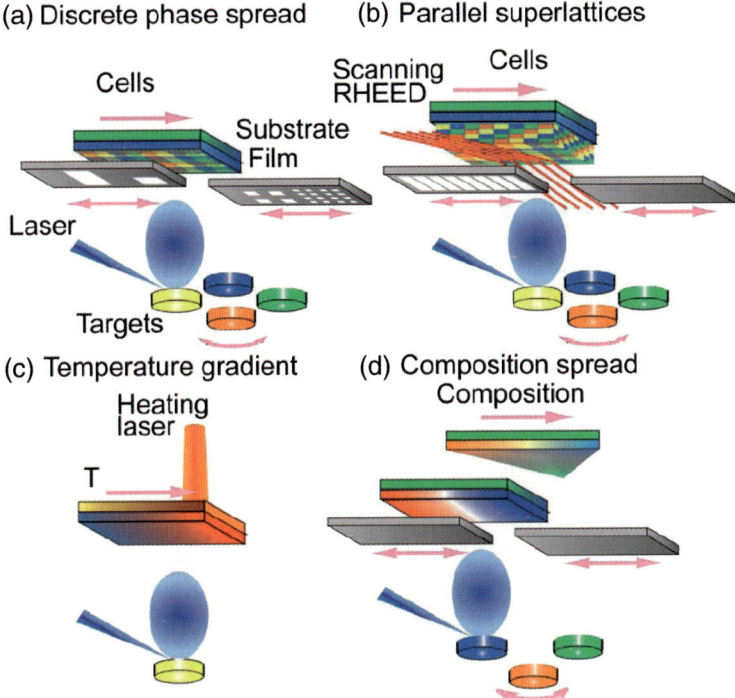

Fig. 4.2. Schematics of four different schemes in epitaxial thin films integration on a substrate [27]

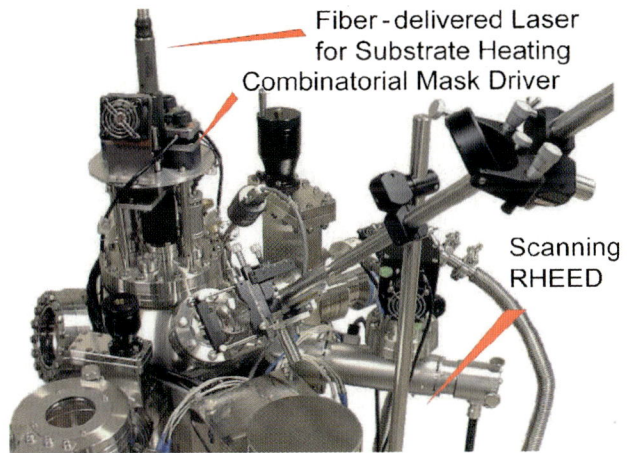

Fig. 4.3. A picture of a compact pulsed laser deposition system equipped with all the capabilities shown in Fig. 4.2 [27]

and laser power is usually tuned so that about several tens to a few hundreds of pulses are required to form one unit cell (0.2–0.4 nm) thick films. While making films, electron beam is impinged with a glazing incidence to observe RHEED pattern. Figure 4.1b is a schematic of intensity oscillation of RHEED due to repeated change of atomic scale roughness on the surface. When the surface of the substrate or film with integer layers thickness is atomically smooth, the electron beam is specularly reflected with high intensity. In between the completed molecular layers deposition with fractional coverage, the electron beam reflection is diffused, yielding in low intensity at the specular spot.

Although RHEED intensity oscillation was observed for high T_c and related oxide thin film growths [15, 16], the oscillation was not regular and the oscillation amplitude decayed quickly in the earlier studies. A breakthrough was made in finishing the substrate surface atomically flat and RHEED intensity oscillation persisted long time for the epitaxy on this surface [17]. Figure 4.1c shows an atomic force microscope image of such a substrate surface. $SrTiO_3$ substrate was wet-etched by a pH controlled buffer HF–HH_4F solution. The trick was the selective etching of SrO atomic layer from $SrTiO_3$ surface. The Perovskite crystal lattice can be regarded as an alternating stuck of AO and BO_2 atomic layers to form ABO_3 molecular layer. Usually, AO is basic oxide and BO_2 is acidic one. Therefore, one can find a window of pH for the etching solution to selectively dissolve one of the two atomic layers. Weak acid (pH 4.5) accomplish SrO etching to leave TiO_2 terminated surface. Once the epitaxy is started from such an atomically smooth surface, the crystal growth can proceed in a layer-by-layer mode to form flat surfaces and interfaces.

4.3 Integration of Oxide Epitaxy

Number of oxide superlattices have been prepared to study, in many cases, the interactions of electronic states between component layers [18–20]. However, there have been very few examples that succeeded in making completely new materials having properties that cannot be realized in similar compounds, which are thermodynamically stable [21]. One of the reasons is the low throughput of experiments. One has to optimize the growth conditions of component layers and has to find a window of conditions to accumulate these layers. Then, one needs to prepare number of superlattices with different layer thicknesses. To solve this problem, a concept of parallel epitaxy has been developed. Figure 4.2 shows schematics of parallel epitaxy schemes accommodated in the PLD system shown in Fig. 4.3. Depending on the mask patterns inserted between the targets and the substrate or the way of heating the substrate, four different schemes are shown for fabricating samples with different diversity.

In Fig. 4.2a, through the masks having openings with different sizes and locations, thin films are deposited in selected areas on the substrate. The exchanges of mask patterns and targets are synchronized so that various sequences and amounts of component films are deposited in the areas defined by the mask opening. If one repeats these processes with sub-monolayer depositions, quasi-alloy films having different compositions are prepared. This technique was first examined for $Mg_xZn_{1-x}O$ solid solution alloys [22] and then extended to the search for ferromagnetic semiconductors [23, 24] described in Sect. 4.1. It should be noted that similar technique was reported earlier by Xian et al. for making combinatorial libraries of phosphor and superconducting phases [25]. However, his method adopted the so-called "precursor method," where the accumulation of component layers was not repeated and the thicknesses of them were a few hundreds of nanometers. Usually, the deposition was carried out at room temperature, resulting in amorphous precursor layers. To obtain uniformly mixed crystalline samples, the accumulated precursors were then annealed in a furnace. In this case, the resulting phases are limited to those that are thermodynamically stable. Therefore, in most cases, the resulting samples are mixed phases of two or more compounds. The diversity tends to result in the variation of mixing ratio of fixed phases. In contract, the epitaxial in situ alloying method allows us to make metastable phases due to epitaxial stabilization. In fact, the $Mg_xZn_{1-x}O$ alloy spread extends x well beyond the thermodynamic solubility limit and this fact gave us confidence to move into the experiments looking for completely new phases in $TM_xZn_{1-x}O$, where TM stands for transition metal elements, to search for ferromagnetic semiconductors. The precursor method would have resulted in boring results of tracing existing phase diagram.

Figure 4.2b shows the scheme of making many superlattices having, for instance, different layer thicknesses. The mask motion and target switching are similar to those of discrete phase spread shown in Fig. 4.2a. However, the present case utilizes a specially designed RHEED system [26, 27]. The electron gun has two sets of deflection coils so that the electron beam can be scanned on the substrate by keeping the same incident and azimuthal angles. Typically, the scanning frequency is about $30\,Hz$ and the diffraction patterns are stored through a charge coupled device (CCD) to a computer so that RHEED oscillations are obtained in a real time fashion for many positions on the substrates. Therefore, mask motion could be synchronized with the RHEED oscillation to control digitally the component film thicknesses. Figure 4.4 shows an example of RHEED intensity oscillations taken for three different positions on a substrate. Here, $SrTiO_3$ and $BaTiO_3$ were selected as targets and 2 unit cells (u.c.)/2 u.c., 4 u.c./4 u.c., and 6 u.c./6 u.c. superlattices were fabricated on a $SrTiO_3$ substrate. When the sliding mask hided the positions b and c, RHEED intensity oscillation appeared only for the position a. After observing two periods, sliding mask was moved to hide only position c. Again counting two periods, all the positions were exposed, and another two periods were counted and the target was switched. By repeating these procedures, aforementioned integrated superlattices could be made.

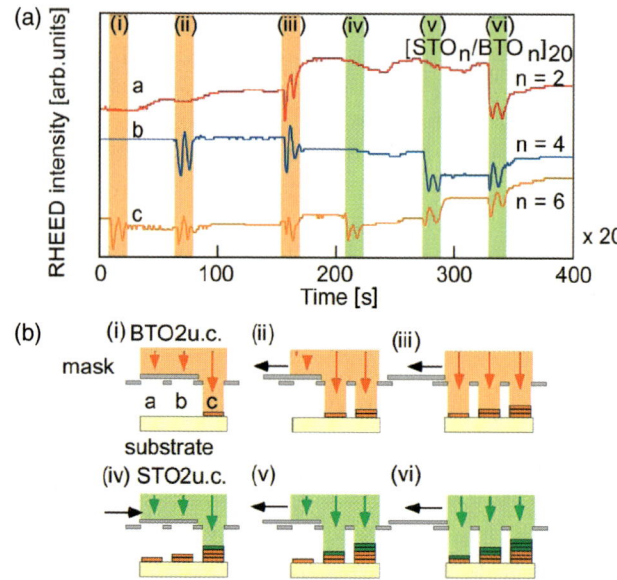

Fig. 4.4. RHEED intensity oscillations observed for three different locations of the substrate, **a**, **b**, and **c** depicted in the bottom panel. Superlattices composed of SrTiO$_3$ (STO) and BaTiO$_3$ (BTO) were formed while mask was moved to deposit the films selectively on each area [27]

Figure 4.2c shows the scheme of temperature gradient method. As mentioned earlier, infrared laser beam is the best choice for making oxide thin films in oxygen atmosphere. The first trial was made with continuous wave Nd:YAG (Nd-doped Y$_3$Al$_5$O$_{12}$ garnet) laser [12], of which beam was brought to a view port of a chamber through optical fiber and appropriate optics with lenses as shown in Fig. 4.3. In recent years, semiconductor laser diode has become very reliable and inexpensive. The latter has a great advantage of stable laser power controlled by the current. When Inconel is used as a substrate holder, maximum temperature can be reached at 1,200°C and only 100 ∼ 150 W is needed for 1 cm^2 substrate. To obtain homogeneous temperature distribution on the substrate, four slits are given at four sides and only the corners are connected to the outside. When the three slits are continued and the center part is connected to the outer parts with one of the sides, the heat flow becomes very asymmetric. Then the laser beam is irradiated on the opposite side to create the temperature gradient. Typically, 300°C difference can be given for 1 cm long substrate [12, 28, 29]. By depositing a film on a substrate with such a temperature gradient, the optimization of growth temperature can be made by a single deposition experiment.

Figure 4.2d shows the scheme of composition spread method. Mask motion and target exchange are similar to those in Fig. 4.2b, but the mask is moved while the deposition is carried out. When target A is ablated, the mask is moved leftward and the periods of mask motion is synchronized with the completion of 1 u.c. deposition. Then the target is switched to B and the mask is moved rightward while ablating the target. By continuing these processes, one can make a film having a composition spread $A_{1-x}B_x$, where the composition x linearly corresponds to the position on the substrate [30]. If the substrate is rotated azimuthally by 120° at the target exchange of A, B, and C, one can form triangle shape film corresponding to ternary phase diagram [31]. The advantage of composition spread to discrete phase spread is the fine resolution in composition. The latter has advantages of easy accommodation of more than four components and reasonably large sample size ($100\,\mu m$–$1\,mm$) having uniform composition. When one wishes to use conventional optical characterization tools, the size of $1\,mm$ is large enough. It should be noted here that such composition spread method was already demonstrated in 1960s [32–35]. Also, there have been demonstrations of so-called natural composition spread, where two or more of material sources such as sputtering gun were faced to the substrate and inhomogeneity in composition was intentionally enlarged [36]. In this case, one does not need masks and complicated synchronization but there are two major drawbacks compared with that shown in Fig. 4.2d. One is that the coverage in composition space is not usually complete. The composition space close to the end members are missing (one cannot make pure materials). The other is that the composition has to be measured by electron probe microanalysis or other means for entire samples, which is usually very time consuming.

All functionalities presented in Fig. 4.2, including laser heating system, are accommodated in a very compact chamber as shown in Fig. 4.3. Three stepping motors for two independent masks and azimuthal rotation of the substrate, optics with manipulator, and view port and other service ports are all accommodated in a single 8 in. flange. Two stepping motors for target exchange and rotation are on the flange at the bottom, which also accommodate a view port for a pyrometer or CCD camera for local temperature measurement and monitoring the mask motion, respectively. The size of the chamber is about 40 cm diameter sphere and occupying area is $1 \times 1\,m^2$. This system was originally designed for synchrotron researches [37]. The PLD was connected to surface sensitive analysis tools such as photoelectron spectroscopy and people might need frequent attachment and detachment of PLD. This system is so compact that one student can perform these tasks in a day or two.

4.4 High Throughput Characterization

Accomplishing parallel synthesis of epitaxial thin films was rather easy. However, characterizing them in a short time has been a big challenge for many of the properties and functionalities. There are two ways of characterization: sequential and concurrent methods. Usually, sequential scanning microprobe analyses are best fit to such spread libraries. Electron probe microanalysis with energy or wavelength dispersive X-ray detectors is often used to determine the compositions. Scanning SQUID (superconducting quantum interference devices as a very sensitive magnetic field sensor) is very powerful technique to search for magnetic and superconducting new phases [30, 38]. Scanning near-field microwave microscope has been demonstrated to determine the real and imaginary parts of dielectric constants and conductivity [31, 39, 40]. Optical characterization techniques such as absorption and photo luminescence spectroscopies are readily applicable for millimeter size samples. One can use these techniques equipped with a microscope for finer resolution experiments [30]. For concurrent characterization, most beautiful example is the phosphor search. Human eyes can immediately identify the color and intensity of luminescence upon the excitation by ultraviolet light or electron beam exposure on the library [41, 42].

One of the examples for structural characterization by X-ray diffraction is depicted in Fig. 4.4a. Here, X-ray from a fine point source on a rotating cathode is focused on to a stripe pattern of $0.1 \times 12\,\mathrm{mm}^2$ in size on a substrate by a curved monochromator. The diffracted X-ray is detected by a X-ray CCD so that the sample position and diffraction angle are dispersed into two axes in CCD image [43, 44]. Three-dimensional view of intensity mapping from a superlattice library is shown in Fig. 4.4b. The superlattices composed of n u.c. of $SrTiO_3$ and n u.c. of $BaTiO_3$, with n varied from 12 to 30, were prepared in the sequence depicted in Fig. 4.5. The diffraction peak from the

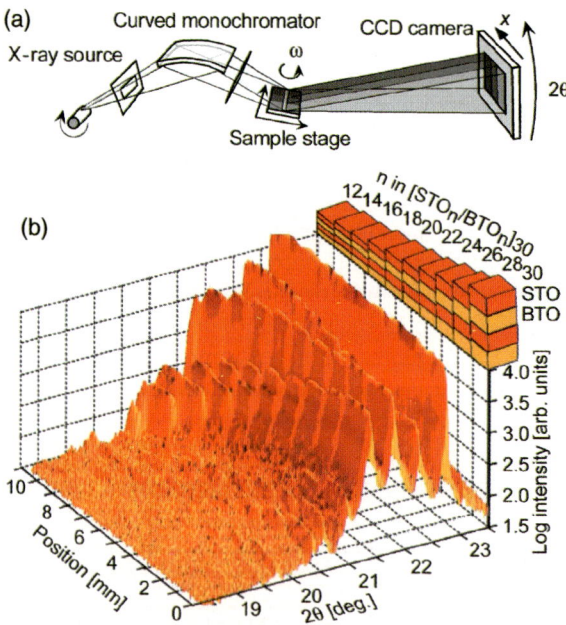

Fig. 4.5. (a) Schematics of a concurrent X-ray diffraction system. (b) Example of X-ray diffraction intensity mapping for $SrTiO_3/BaTiO_3$ superlattices [27]

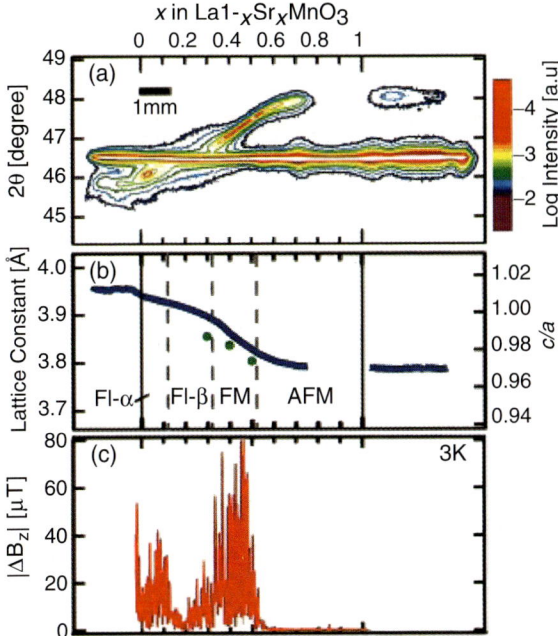

Fig. 4.6. (a) X-ray diffraction intensity mapping for a composition spread thin film of $La_{1-x}Sr_xMnO_3$. (b) Lattice constant as a function of x. (c) Signal intensity of scanning SQUID microscope [30]

substrate and main peak corresponding to the averaged lattice constant stay on the same position in 2θ, but the satellite peaks shift according to the period of superlattices. The width of superlattice is 1 mm and the resolution at the sample position is 0.1 mm or less, which is determined by the resolution of CCD pixels [27]. Another example is depicted in Fig. 4.6 for a composition spread LaMnO$_3$–SrMnO$_3$ library [30]. The sample was prepared in the sequence depicted in Fig. 4.2d. Figure 4.6a shows the gradual shift of X-ray diffraction peak according to the gradual change of the composition. In the area of $0.8 < x < 1.0$, the peak was missing due to the absence of right Perovskeite phase. The peak running horizontally is that from SrTiO$_3$ substrate. The lattice constant is plotted in Fig. 4.6b. For this library, the signal intensity of scanning SQUID microscope is also plotted in Fig. 4.2c. For the region of $0 < x < 0.5$, strong signal is observed due to ferromagnetic and canted antiferromagnetic phases. Infrared refrectance spectroscopy with microscope was also carried out for the sample. From the presence of large reflectance at low energy region due to Drude response, one can identify the metallic or insulating properties.

4.5 Oxide Semiconductors

Oxide semiconductors represented by ZnO, In_2O_3, SnO_2, and TiO_2 are all transparent with a band gap of 3.0–3.3 eV and can be easily turned into n-type conduction with oxygen deficiency or heterovalent impurity doping. Zn^{2+} has (Ar core) $3d^{10}$ and In^{3+} and Sn^{4+} have (Kr core) $4d^{10}$ electron configurations. The conduction bands are mainly formed from $3s$ (ZnO) and $4s$ (In_2O_3, SnO_2) orbitals of cations, and the valence band is from O $2p$ orbital. Ti^{4+} has (Ar core) electron configuration so that the conduction band of TiO_2 is formed from Ti $3d$ orbital. The mobilities of electrons in s-orbital compounds are very large $(100–200\,cm^2\,V\,s^{-1})$ but that in d-orbital TiO_2 is rather low $(1–10\,cm^2\,V\,s^{-1})$, depending on the band width. Degenerated s-orbital compounds are well known as transparent conducting materials and used for electrodes in solar cells and displays [4]. Recently, TiO_2 has also been shown to be a good candidate for the same purpose [45]. Besides such passive application, active devices have been intensively studied for the last decade and such devices as light emitting diodes (LED), thin film transistors (TFT), and magnetic tunnel junction (MTJ) have been already demonstrated by the group of the author. In this section, the use of parallel epitaxy is shown to be effective for such purposes.

4.5.1 Ferromagnetic Oxide Semiconductors

Semiconductors are considered to be nonmagnetic. However, by adding an impurity of magnetic elements, the charge carriers in conduction or valence bands can interact with spin degree of freedom of magnetic impurities through exchange coupling [46, 47]. Some compounds such as (In, Mn)As and (Ga,Mn)As are known to be ferromagnetic [47]. Important fact is that the charge carriers are also spin-polarized upon long-range ferromagnetic ordering of localized spins at the impurity site. By utilizing such charge carriers, one can use not only charge degree of freedom in electrons but also spin degree of freedom for the devices. This research field is called as spintronics, and one of the most famous devices are MTJ made of conventional ferromagnetic metals such as Permalloy (Ni–Fe) or Co–Fe alloy [48, 49]. By extending the spintronics to semiconductors, one can not only use this device for information technology with capabilities to write, detect, store, and change the magnetization direction of ferromagnetic components, but also can change the carrier concentration by electric field [50] and can use them for the interaction with photons such as light emission [51]. Therefore, ferromagnetic semiconductors are expected to explore new devices with unprecedented functionalities. However, there remained very crucial drawback; Currie temperature (T_C) was lower than room temperature for existing (In, Mn)As and (Ga,Mn)As. Keen demand was the discovery of ferromagnetic semiconductors with T_C higher than room temperature.

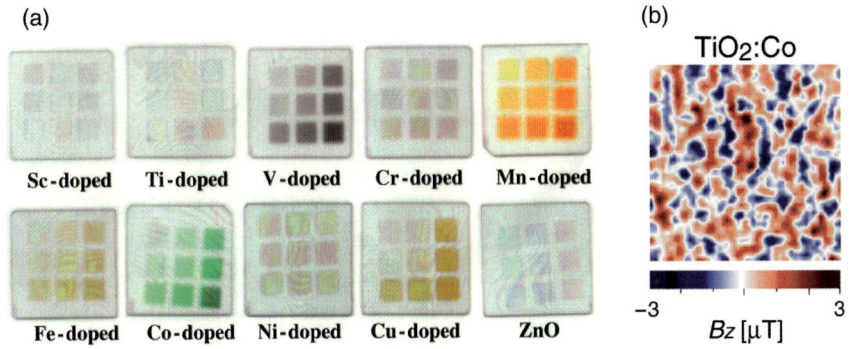

Fig. 4.7. (a) Photograph of the libraries for $Zn_{1-x}TM_xO$ samples having different x with TM extending to all the transition metal elements. The substrate and sample sizes are 16 and 3 mm, respectively [24]. (b) Scanning SQUID microscope image representing the local magnetic field intensity for a ferromagnetic TiO_2 doped with Co. The image size is $0.3 \times 0.3\,mm^2$ [38]

First attempt for making oxide semiconductors ferromagnetic was carried out by adding Mn into ZnO by Fukumura et al. [52]. The solid solution films of $Zn_{1-x}Mn_xO$ could be made into single phase up to $x = 0.35$, but these films show spin glass behavior and no long range ferromagnetism down to 2 K [53]. To rapidly screen the possibilities, parallel synthesis was made for all the host semiconductors of ZnO, In_2O_3, SnO_2, and TiO_2 by adding all the *TM* elements. Figure 4.7a shows the examples of the libraries [23, 24]. The substrate was $16 \times 16\,mm^2$ sapphire and 9 samples ($3 \times 3\,mm^2$) with different doping concentrations were integrated. The high throughput screening of ferromagnetism was carried out by scanning SQUID microscope. Among which, ferromagnetism above room temperature was observed for rutile TiO_2 doped with Co [38]. The magnetic field image observed for Co-doped sample is shown in Fig. 4.7b. The positive and negative local magnetic field sensed by SQUID pick-up loop was mapped out for $0.3 \times 0.3\,mm^2$ area. Later, another phase of TiO_2, anatase, was also confirmed to be ferromagnetic with Co doping [54]. The ferromagnetic samples were naturally n-type due to oxygen deficiency. When the samples were prepared in high oxygen pressure, the films were insulating and ferromagnetism disappeared.

Triggered by the proposal and discovery of high T_C ferromagnetic oxide semiconductors, intensive follow-up and extended studies have been made [55]. Theoretically, Mn-doped ZnO and GaN were predicted to be ferroamgentic based on Zener model description, if these semiconductors could have been made into p-type [56]. First principle calculation was also carried out for n- and p-type $Zn_{1-x}TM_xO$ and concluded that Mn-doped p-type ZnO and other TM-doped n-type ZnO would have been ferromagnetic [57]. Actually, more than 30 groups in the world postulated that $Zn_{1-x}TM_xO$ was ferromagnetic based on the experiments. On the other hand, many groups reported that ferromagnetism in $Ti_{1-x}Co_xO_2$ was not due to carrier induced ferromagnetism but due to the precipitation of ferromagnetic Co metal. The confusion was well described in one of the review papers by Fukumura [55]. The most serious problem causing the confusion was that the detection of ferromagnetism was totally relied on magnetization measurement for such a small volume samples of thin films, which were deposited on substrates with 10^4 times larger volume and therefore large contribution to magnetic signal from the impurities in high magnetic field. In this regard, scanning SQUID microscope sensing was an appropriate way because it detects remanent magnetization without applying external magnetic field.

To solve this confusion, the group of the author proposed that the reliable way of detecting ferromagnetism induced by exchange coupling between localized spin and charge carriers is to detect the sign of spin polarization of charge carries [55]. Spin polarization gives finite off-diagonal elements in real and imaginary parts of dielectric tensor. When one applies direct current method, it gives anomalous Hall effect, which can be considered as Hall effect without applying magnetic field, but net magnetization contributes as fictitious magnetic field. When one applies alternating current method such as electromagnetic wave, it gives magneto-optic effects represented by Kerr and Faraday effects from real part of off-diagonal dielectric tensor and magnetic circular dichroism (MCD) from imaginary part. Both methods detect the modification of band structure due to the emergence of ferromagnetism and rule out extrinsic effects such as precipitation.

Examples of such ferromagnetic responses were measured for both rutile and anatase phases of TiO_2 doped with Co. Figure 4.8a shows MCD spectra and their responses to the magnetic field for Co-doped anatase phase TiO_2 [58, 59] and Co-doped ZnO [60, 61]. The latter was postulated to be ferromagnetic based on the magnetization measurement [62]. Both compounds show anomaly in MCD spectra at photon energies close to the band edges. Co-doped ZnO has sharp exciton resonance at 3.3 eV both in absorption and MCD spectra. However, the MCD signal at this resonance photon energy shows paramagnetic behavior as shown in the inset. This clearly indicates that the charge carriers are not spin polarized. Co-doped anatase TiO_2 shows rather broad features both in absorption and MCD spectra, probably due to the indirect band gap nature of this compound. The magnetic field response of MCD at band edge energy clearly indicates ferromagnetic behavior of which magnetic field dependence exactly agrees with that of magnetization (not shown). Therefore, one can conclude that the magnetization and spin polarization have the same origin. Moreover, the magnetic field dependence of MCD signal turned out to be of identical shape regardless of photon energy. That indicates the ferromagnetic origin is unique and rules out the possibility of Co metal precipitation. Similar behavior was confirmed for rutile phase TiO_2 doped with Co as well [63]. In both cases, there is a phase boundary between paramagnetic and ferromagnetic ones at certain electron concentration; insulating films showed paramagnetic response although doped with Co. This fact also supports the carrier induced ferromagnetism.

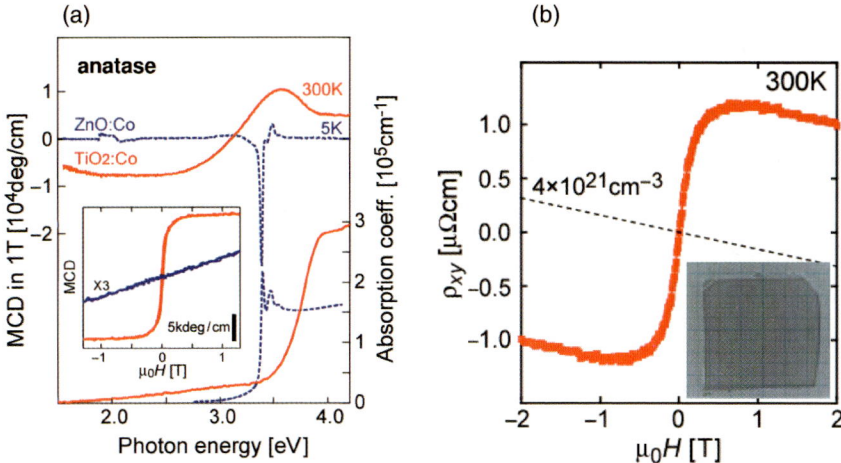

Fig. 4.8. (**a**) Magneto circular dichroism (MCD: *top*) and absorption (*bottom*) spectra for Co-doped ZnO and TiO$_2$. The inset is the magnetic field dependence of MCD signal taken at maximum responses near the band edge photon energies. (**b**) Hall resistivity as a function of magnetic field for a ferromagnetic Ti$_{1-x}$Co$_x$O$_2$. The broken line is guide to eyes that is parallel to the high field part from which carrier concentration was extracted [55]

Figure 4.8b shows anomalous Hall effect of rutile TiO$_2$ doped with Co [64]. The inset shows the picture for one of the films. The Hall resistivity of TiO$_2$ without Co doping is shown by a broken line. The straight line indicates normal response of nonmagnetic conductors, from which the carrier concentration could be extracted as 4×10^{21} cm^{-3}. Co-doped TiO$_2$ shows anomaly around zero magnetic field followed by the magnetic field dependence similar to that without Co doping at higher magnetic field. This behavior could be explained as anomalous Hall effect in these compounds. Again, clear phase boundary was identified; it is necessary for realizing ferromagnetism to have higher concentrations of electron and Co simultaneously than the boundary. Even more interesting fact is that the anomalous conductivity σ_{AF} scales with conductivity σ_{xx} as $\sigma_{AF} \sim \sigma_{xx}^{1.6}$, regardless of Co concentration, electron concentration, or temperature [64]. Very similar behavior was also observed in anatase phase [65]. It should be noted that the scaling relation holds for the samples with both metallic and semiconducting temperature dependences of resistivity. This universal scaling holds for many other ferromagnetic compounds and this fact suggests that there is underlying physics governing anomalous Hall effect [66]. This finding may open up a way to understand the microscopic mechanism of anomalous Hall effect that has been explained phenomenologically for years [67–69]. In fact, a theory based on Berry phase was proposed for more metallic compounds [70, 71]. This theory predicted similar scaling law with a power of 1.6 for less conductive regime with diffusive characteristics [71]. The details of this discussion are given by Fukumura in this book chapter [72]. Note added to this section is that a demonstration of MTJ has been already done in rutile phase TiO_2 doped with Co [73, 74]. Although preliminary, tunneling magnetoresistance was clearly shown up to 200 K in a junction of $Co/AlO_x/Ti_{1-x}Co_xO_2$.

4.5.2 Optical Devices

Recent progress in short wavelength LED and lasers based on GaN and its alloys has changed the human life in various ways. Extended research efforts include pursuing even shorter wavelength LED and highly efficient lasers. In this context, however, several drawbacks are there for nitrides. First, the abundance of mineral sources of Ga and In is rather limited. Second, high quality single crystal substrates are not yet ready in mass-production for steady supply in terms of their cost and productivity. Third, the patent situation rather limits the number of major suppliers of devices, leading to more difficult risk/cost management due to less choice of suppliers for companies making consumer products using LED and lasers.

ZnO is one of the most promising alternative wide band gap semiconductors, expanding the band gap up to 4.5 eV by employing (Mg, Zn)O alloy crystals [75]. The natural abundances of Zn and Mg are extremely rich. High quality single crystal wafers are now commercially available up to 3 in. diameter [76], where the crystal growth methods are not time-consuming vapor phase method as is the case of GaN, but hydrothermal method with very high throughput. The patent situation is not clear yet, but basic concept of device structure will be the same as those of traditional III–V LED. This is because conductive and homoepitaxial substrates are ready. In case of GaN, the employment of insulating and lattice mismatched sapphire heteroepitaxial substrates had generated several key patents, including those for the structure of electrodes and insertion of suitable buffer layers. Moreover, ZnO-based semiconductors have intrinsic physical and structural properties, which are suitable for device application. First, the exciton binding energy is as large as 60 meV, which surpasses to that of GaN (24 meV), and thermal energy of room temperature (25 meV). This can be enhanced to 110 meV in multi-quantum well structures, which again surpasses to the energy of optical phonon mode in ZnO (72 meV) [77]. Therefore, low threshold and highly efficient lasers can be expected. These expectations have been already demonstrated in optical pumping experiments for ZnO epitaxial films and superlattices [78–81]. Second, the lattice mismatch between ZnO and (Mg, Zn)O or (Zn, Cd)O can be very small (<0.5%) [82, 83]. This is due to the strong Coulomb interaction between highly ionic elements in wurtzite structure. This fact makes ZnO free from undesired internal electric field effect due to epitaxial strain of piezoelectric crystal of wurtzite.

The challenge was to produce p-type ZnO. ZnO is well known as naturally n-type semiconductor due to Zn interstitial and many other defects. Doping of Al or Ga makes ZnO highly conductive with keeping transparency, providing a chance to be used as transparent electrodes in various devices to be discussed in Sect. 4.3. Therefore, before trying to make acceptor doping, one has to eliminate any defect working as donors. The group of the author has optimized thin film crystal growth process of PLD for nondoped ZnO so as to achieve a residual donor concentration less than 10^{16} cm^{-3}, a mobility greater than 400 cm^2 V s^{-1}, and a radiative exciton recombination life time longer than 2 ns all at room temperature [84, 85]. These values were record breaking ones even if compared with bulk single crystals of ZnO. The keys are to employ lattice matched ScAlMgO$_4$ substrates [86] to insert lattice-mismatch-engineered and ultra-smooth ZnO buffer layer [87, 88], and to employ laser heating of substrate for clean and ultra-high temperature above $1{,}000°$C [12]. The growth temperature optimization was carried out by the temperature gradient method shown in Fig. 4.2c and also explained by Tsukazaki in this book chapter [89]. The RHEED oscillation observed for one of the film growth runs is depicted in Fig. 4.9. The oscillation with a period of 0.26 nm (1 u.c. of ZnO) deposition persisted for more than $4{,}000$ times for making 1 μm thick film [90].

Next challenge was to incorporate nitrogen (N) as an acceptor by keeping such high quality structure and properties of nondoped ZnO. The volatile nature of N makes it very difficult to be incorporated at high temperature [91]. Below $500°$C, one can dope 10^{19}–10^{21} cm^{-3} of nitrogen but the crystalline quality degrades substantially. To solve this dilemma, a novel concept of repeated temperature modulation (RTM) technique was developed [90]. Here, the following processes were repeated to grow micron-thick films as shown in Fig. 4.10a: low temperature growth of 15 nm ZnO/N at $400°$C, rapid jump to $1{,}000°$C and annealing, additional growth of 1 nm ZnO at high temperature to recover ultra-smooth surface, and rapid cooling to $400°$C. Surprisingly, the

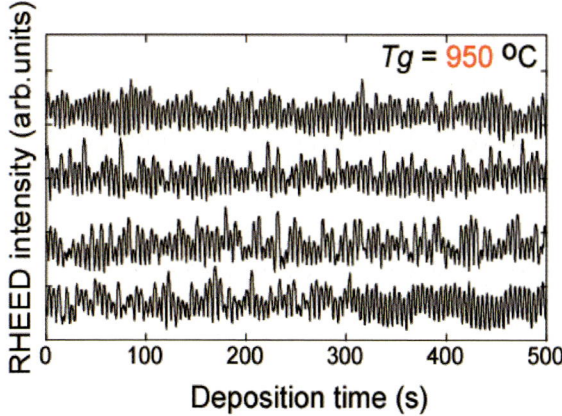

Fig. 4.9. RHEED intensity oscillation observed during ZnO growth under the optimized conditions [90]

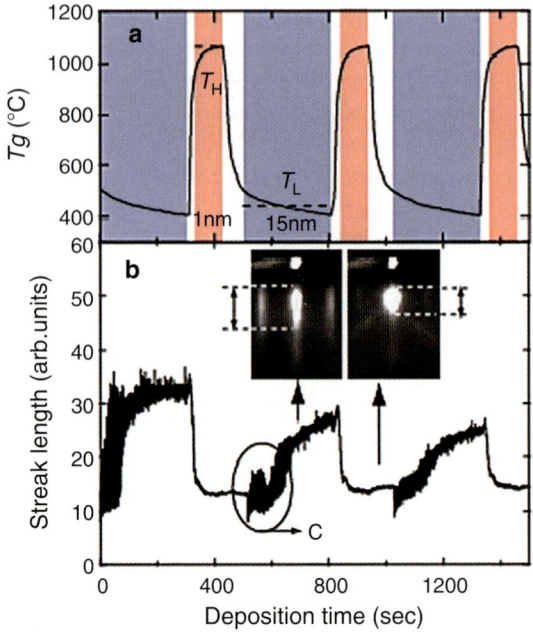

Fig. 4.10. (a) Temporal evolution of substrate temperature during the repeated temperature modulation technique. (b) The streak length indicated in the inset pictures are plotted as a function of time [90]

entire process was driven in a layer-by-layer mode, which was confirmed not only by the RHEED intensity oscillation but also by an oscillatory change in atomic scale surface roughness detected by temporal change in diffusive streak length. The length of central streak in RHEED pattern was monitored during the process and plotted in Fig. 4.10b. The oscillation has the same period with RHEED intensity oscillation but a phase shift of π. Also shown is the rapid recovery of surface roughness upon annealing. The key was to expose very flat surface to the atoms deposited at low temperature and terminate the growth before the surface becomes rough.

Through this process, p-type ZnO have been reproducibly realized, yielding in a hole concentration above 10^{16} cm^{-3} and an activation energy of about 100 meV for the films with a N concentration of 10^{20} cm^{-3}. However, the compensation ratio was still as high as 0.8. This means that there is still enough room to increase the hole concentration by reducing donor type defects by optimizing the RTM process [90]. ZnO p-i-n homoepitaxial junctions were fabricated as shown in Fig. 4.11. The devices can emit blue light centering at 420 nm [92]. This emission is attributed to the recombination in p-type ZnO due to electron injection caused by the penetration of depletion layer into p-type layer due to the imbalance of carrier concentration between n- and p-type layers [92]. Near future challenges are (1) increasing hole concentration

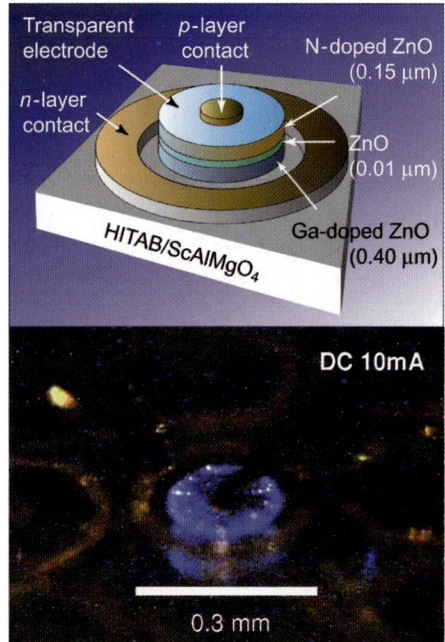

Fig. 4.11. A schematic (*top*) and a picture in dark under current bias (*bottom*) for a ZnO LED [92]

above 10^{18} cm^{-3} in ZnO:N, (2) realizing p-type (Mg Zn)O:N to give conduction band discontinuity, and (3) reproducing p-type ZnO grown on ZnO single crystal substrates by mimicking the ZnO buffer layer surface made on SaAlMgO$_4$ substrates.

4.5.3 Field Effect Devices

Making minority carrier devices such as LED is rather tough task. Majority carrier devices such as transistors are thought to be easier and intensive efforts have been made for making transparent thin film transistor (TFT) [93–99]. In this context, the group of the author started making ZnO TFT. After the patents were filed [100], experiments were carried out with an aim of replacing amorphous silicon (a-Si) TFT in liquid crystal display. The field effect mobility (μ_{fe}) of a-Si TFT is about 0.5 cm^2 V s^{-1} and the threshold voltage (V_{th}) is about 0 V. These are the two benchmarks. For replacing a-Si, one has to use amorphous substrates and resulting polycrystalline films. It is well known that polycrystalline ZnO tends to become naturally n-type and highly conducting due to crystalline defects. One can compensate the electrons by adding Li or Ni as acceptors, but this will cause degradation of mobility. However, ZnO has a big advantage of preferential texturing. Even when grown at temperatures lower than 150°C, the films tend to orient with c-axis normal to the surface.

This temperature range is significant when one considers polymer substrates. From the first stage, the substrate for device demonstration was brought from a fabrication line of liquid crystal display in Sharp Corp. The gate structures of a-SiN$_x$/TaO$_x$/Ta were prepared on a glass substrate in the fabrication line to be used for products. The typical channel size was $5\,\mu$m long and $15\,\mu$m wide. To have rather intrinsic characteristics for the films grown on a-SiN$_x$, buffer layer of polycrystalline CaHfO$_3$ was chosen by considering the sub-lattice matching of oxide ions between basal plane of ZnO and (111) plane of CaHfO$_3$. The TFT showed nice performance of $\mu_{\mathrm{fe}} = 7\,\mathrm{cm}^2\,\mathrm{V}\,\mathrm{s}^{-1}$ and V_{th} close to $0\,$V [97]. However, the drain current enhanced more than that expected from conventional semiconductor TFT model, presumably due to the presence of grain boundaries.

Extensive device simulation was carried out by modeling the device in a two-dimensional device simulator [101, 102]. The grain boundaries were treated as double Schottky barriers as usually treated in varistors and gas sensors. It turned out that the device action is far from that of single crystal transistors but it behaves as charge valves. Upon the application of gate voltage, the height of double Schottky barrier is suppressed linearly, resulting in an exponential increase of drain current. This behavior agrees well with experiments. Therefore, it is needed to evaluate the field effect mobility and Hall mobility independently, because the former has extra-term and gives overestimated values. To isolate the effects of grain boundaries, single crystalline transistors having Hall voltage probes in the channels were prepared. The performance of transistors was analyzed in detail to agree with that for the conventional semiconductor devices. The maximum μ_{fe} reached to $70\,\mathrm{cm}^2\,\mathrm{V}\,\mathrm{s}^{-1}$ [103].

Besides the electric field devices of ZnO with gate electrode, the group of the author has found that one can apply an electric field at the ZnO/Mg$_x$Zn$_{1-x}$O interfaces without gate electrode. This is because ZnO with wurtzite structure is a polar material with luck of inversion symmetry. There exists considerable spontaneous polarization and piezoelectric polarization upon strain. As a result, the mismatches in these polarizations at the ZnO/Mg$_x$Zn$_{1-x}$O interfaces induce sheet charge or two-dimensional electron gas (2 DEG) as shown in Fig. 4.12 [104]. The carrier concentration and two-dimensionality could be tuned by using temperature gradient method as shown in Fig. 4.2c and also explained by Tsukazaki in this book chapter [89]. Optimized sample had a carrier concentration as low as $10^{12}\,\mathrm{cm}^{-2}$ and exhibited both Shubnikov-de Haas (SdH) oscillations in magneto-resistance and Landau plateaus in Hall resistance (quantum Hall effect: QHE) as shown in Fig. 4.13. SdH oscillation was observed also in SrTiO$_3$/LaAlO$_3$ by Ohtomo and Hwang [105] previously. But the QHE was first observed among oxides in this system. Demonstration of the QHE in an oxide heterostructure presents the possibility to combine quantum Hall physics with the versatile functionality of metal oxides in complex heterostructures.

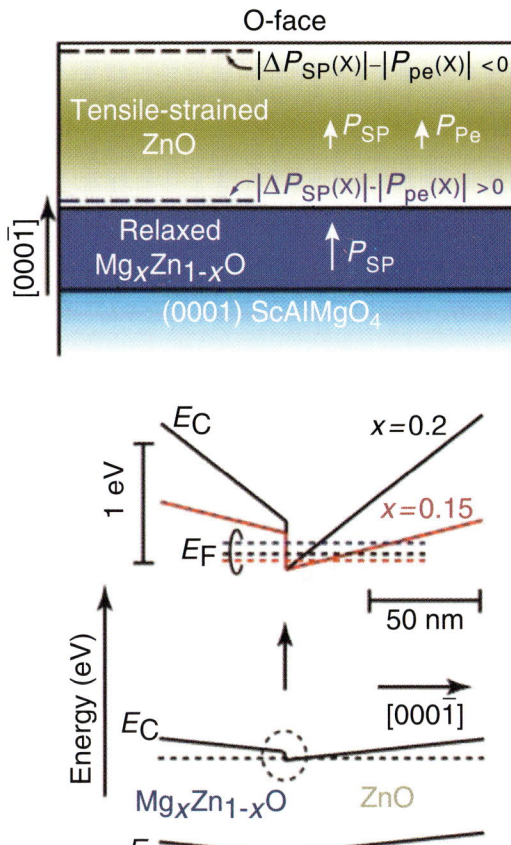

Fig. 4.12. A schematic representation of ZnO/Mg$_x$Zn$_{1-x}$O heterostrucutre that exhibits two-dimensional electron gas at the interface. P_{sp} and P_{pe} stand for spontaneous and piezoelectric polarizations. The bottom diagrams are the band structures for the heterostructure [104]

4.6 Conclusion and Future Prospects

As described in this chapter, epitaxial technology has substantially advanced in the last two decades, both in atomic scale controllability and parallel integration of samples. The advantage of the latter has been clearly demonstrated by the discovery of transparent ferromagnetic semiconductors as well as systematic optimization of p-type ZnO and 2 DEG devices. What has to be developed further is the high throughput characterization techniques. It has not mentioned above but what is very crucial is to develop methods to characterize electrical properties of the thin film samples in a high throughput fashion. The measurement that needs many electrical lead wires is not so compatible with integrated libraries. Also important is the handling of the

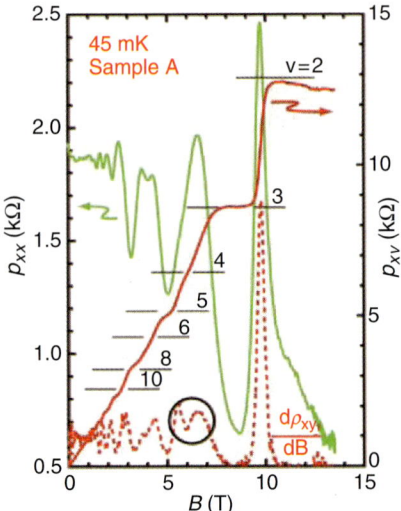

Fig. 4.13. Magnetic field dependences of resistivity ρ_{xx} and Hall resistivity ρ_{xy} for a ZnO/Mg$_x$Zn$_{1-x}$O heterostrucutre. Shubnikov-de Haas oscillations and quantum Hall effect are clearly seen [104]

huge amount of the data produced by automatic measurements for integrated libraries. There has been an attempt to solve this problem and this approach was named as "materials informatics."

In terms of further progress of oxide semiconductor devices, a real application of these devices has to be accomplished. ZnO LED will be very promising but there have been identified many problems to be solved. However, the present author believes that these are the technical ones and will be solved by the injection of enough resources. For the ferromagnetic semiconductors, one needs to demonstrate cleared semiconducting device action. MTJ alone cannot motivate industry people because conventional MTJ has behaved already very excellent. Therefore, electric field effect modulation of magnetic states will be the important milestone. The ZnO TFT has many advantages compared with a-Si TFT but the resources already injected to liquid crystal display industry has been so huge that it will be difficult to replace by having better performance by a factor or two. One really needs to demonstrate performances surpassing existing devices by orders of magnitude. Therefore, replacement of a-Si may not be a good choice. For science, the present author is confident that oxides will provide many of new and exciting discoveries for many years [106].

Acknowledgement. The work was supported by Grant-in-Aid for Creative Research MEXT-14GS0204, Grant-in-Aid for Priority Area MEXT-436, Combinatorial Materials Exploration and Technology Project of MEXT, NEDO Grant 02BR3, and CREST-JST. The author thanks T. Fukumura, A. Ohtomo, A. Tsukazaki, K. Ueno, and all the graduate students in his group for the collaboration.

References

1. J.G. Bednorz, K.A. Müller, Z. Phys. B **64**, 189 (1986)
2. M. Imada, A. Fujimori, Y. Tokura, Rev. Mod. Phys. **70**, 1039 (1998)
3. H. Koinuma, Thin Solid Films **486**, 2 (2005)
4. T. Minami, Semicond. Sci. Technol. **20**, S35 (2005)
5. M. Kawasaki, M. Nantoh, MRS Bull. **19(9)**, 33 (1994)
6. R. Tsuchiya, M. Kawasaki, H. Kubota, J. Nishino, H. Satoh, H. Akoh, H. Koinuma, Appl. Phys. Lett. **71**, 1570 (1997)
7. M. Lebl, J. Comb. Chem. **1**, 3 (1999)
8. W.F. Maier, Angew. Chem. Int. Ed. **38**, 1216 (1999)
9. H. Koinuma, I. Takeuchi, Nat. Mater. **3**, 429 (2004)
10. X.D. Xiang, I. Takeuchi (eds.), *Combinatorial Materials Synthesis* (Marcel Dekker, New York, 2003)
11. J.H. Haeni, S.D. Theis, D.G. Schlom, J. Electroceramics **4**, 385 (2000)
12. S. Ohashi, M. Lippmaa, N. Nakagawa, H. Nagasawa, H. Koinuma, M. Kawasaki, Rev. Sci. Instrum. **70**, 178 (1999)
13. M. Lippmaa, T. Furumochi, S. Ohashi, M. Kawasaki, H. Koinuma, T. Satoh, T. Ishida, H. Nagasawa, Rev. Sci. Instrum. **72**, 1755 (2001)
14. G.J.H.M. Rijnders, G. Koster, D.H.A. Blank, H. Rogalla, Appl. Phys. Lett. **70**, 1888 (1997)
15. T. Terashima, Y. Bando, K. Iijima, K. Yamamoto, K. Hirata, K. Hayashi, K. Kamigaki, H. Terauchi, Phys. Rev. Lett. **65**, 2684 (1990)
16. M. Yoshimoto, H. Nagata, T. Tsukahara, H. Koinuma, Jpn. J. Appl. Phys. **29**, L1199 (1990)
17. M. Kawasaki, K. Takahashi, T. Maeda, R. Tsuchiya, M. Shinohara, O. Ishiyama, T. Yonezawa, M. Yoshimoto, H. Koinuma, Science **266**, 1540 (1994)
18. M. Izumi, Y. Murrakami, Y. Konishi, T. Manako, M. Kawasaki, Y. Tokura, Phys. Rev. B **60**, 1211 (1999)
19. Y. Ogawa, H. Yamada, T. Ogasawara, T. Arima, H. Okamoto, M. Kawasaki, Y. Tokura, Phys. Rev. Lett. **90**, 217403 (2003)
20. H. Yamada, Y. Ogawa, Y. Ishii, H. Sato, M. Kawasaki, H. Akoh, Y. Tokura, Science **305**, 646 (2004)
21. K. Ueda, H. Tabata, T. Kawai, Science **280**, 1064 (1998)
22. Y. Matsumoto, M. Murakami, Z.W. Jin, A. Ohtomo, M. Lippmaa, M. Kawasaki, H. Koinuma, Jpn. J. Appl. Phys. **38**, L603 (1999)
23. Z.W. Jin, M. Murakami, Y. Matsumoto, A. Ohtomo, M. Kawasaki, H. Koinuma, J. Cryst. Growth **214/215**, 55 (2000)
24. Z.W. Jin, T. Fukumura, M. Kawasaki, K. Ando, H. Saito, T. Sekiguchi, Y.Z. Yoo, M. Murakami, Y. Matsumoto, T. Hasegawa, H. Koinuma, Appl. Phys. Lett. **78**, 3824 (2001)
25. X.D. Xiang, Annu. Rev. Mater. Sci. **29**, 149 (1999)
26. T. Ohnishi, D. Komiyama, T. Koida, S. Ohashi, C. Stauter, H. Koinuma, A. Ohtomo, M. Lippmaa, N. Nakagawa, M. Kawasaki, T. Kikuchi, K. Omote, Appl. Phys. Lett. **79**, 536 (2001)
27. M. Ohtani, M. Lippmaa, T. Ohnishi, M. Kawasaki, Rev. Sci. Instrum. **76**, 62218 (2005)

28. T. Koida, D. Komiyama, M. Ohtani, M. Lippmaa, M. Kawasaki, H. Koinuma, Appl. Phys. Lett. **80**, 565 (2002)
29. A. Tsukazaki, H. Saito, K. Tamura, M. Ohtani, H. Koinuma, M. Sumiya, S. Fuke, T. Fukumura, M. Kawasaki, Appl. Phys. Lett. **81**, 235 (2002)
30. T. Fukumura, Y. Okimoto, M. Ohtani, T. Kageeeyama, T. Koidda, M. Kawasaki, T. Hasegawa, Y. Tokura, H. Koinuma, Appl. Phys. Lett. **77**, 3426 (2000)
31. K. Hasegawa, P. Ahmet, N. Okazaki, T. Hasegawa, K. Fujimoto, M. Watanabe, T. Chikyow, H. Koinuma, Appl. Surf. Sci. **223**, 229 (2004)
32. K. Kennedy, T. Stefansky, G. Davy, V.F. Zacky, E. Parker, J. Appl. Phys. **36**, 3808 (1965)
33. N.C. Miller, G. Shirn, Appl. Phys. Lett. **10**, 86 (1967)
34. J.J. Hanak, J.I. Gittleman, J.P. Pellicane, S. Bozowski, Phys. Lett. **A 30**, 201 (1969)
35. J.J. Hanak, J. Mater. Sci. **5**, 964 (1970)
36. R.B. van Dover, L.F. Schneemeyer, R.M. Fleming, Nature **392**, 162 (1998)
37. K. Horiba, H. Ohguchi, H. Kumigashira, M. Oshima, K. Ono, N. Nakagawa, M. Lippmaa, M. Kawasaki, H. Koinuma, Rev. Sci. Instrum. **74**, 3406 (2003)
38. Y. Matsumoto, M. Murakami, T. Shono, T. Hasegawa, T. Fukumura, M. Kawasaki, P. Ahmet, T. Chikyow, S. Koshihara, H. Koinuma, Science **291**, 854 (2001)
39. T. Wei, X.D. Xiang, W.G. Wallace-Freedman, P.G. Schultz, Appl. Phys. Lett. **68**, 3506 (1996)
40. H. Chang, I. Takeuchi, X.-D. Xiang, Appl. Phys. Lett. **74**, 1165 (1999)
41. J. Wang, Y. Yoo, C. Gao, I. Takeuchi, X. Sun, H. Chang, X.D. Xiang, P.G. Schultz, Science **279**, 1712 (1998)
42. V.Z. Mordkovich, H. Hayashi, M. Haemori, T. Fukumura, M. Kawasaki, Adv. Funct. Mater. **13**, 519 (2003)
43. M. Ohtani, T. Fukumura, M. Kawasaki, K. Omote, T. Kikuchi, J. Harada, A. Ohtomo, M. Lippmaa, T. Ohnishi, D. Komiyama, R. Takahashi, Y. Matsumoto, H. Koinuma, Appl. Phys. Lett. **79**, 3594 (2001)
44. M. Ohtani, T. Fukumura, M. Kawasaki, K. Omote, T. Kikuchi, J. Harada, H. Koinuma, Appl. Phys. Lett. **80**, 2066 (2002)
45. Y. Furubayashi, T. Hitotsugi, Y. Yamamoto, K. Inaba, G. Kinoda, Y. Hirose, T. Shimada, T. Hasegawa, Appl. Phys. Lett. **86**, 252101 (2005)
46. W. Giriat, J.K. Furdyna, in *Semiconductors Semimetals*, vol. 25, ed. by R.K. Willardson, A.C. Beer (Academic Press, Boston, MA, 1988)
47. H. Ohno, Science **281**, 951 (1998)
48. S.S.P. Parkin, C. Kaiser, A. Panchula, P.M. Rice, B. Hughes, M. Samant, S.H. Yang, Nat. Mater. **3**, 862 (2004)
49. S. Yuasa, T. Nagahama, A. Fukushima, Y. Suzuki, K. Ando, Nat. Mater. **3**, 868 (2004)
50. H. Ohno, D. Chiba, F. Matsukura, T. Omiya, E. Abe, T. Dietl, Y. Ohno, K. Ohtani, Nature **408**, 944 (2000)
51. Y. Ohno, D.K. Young, B. Beschoten, F. Matsukara, H. Ohno, D.D. Awschalom, Nature **402**, 790 (1999)
52. T. Fukumura, Z.W. Jin, A. Ohtomo, H. Koinuma, M. Kawasaki, Appl. Phys. Lett. **75**, 3366 (1999)

53. T. Fukumura, Z.W. Jin, M. Kawasaki, T. Shono, T. Hasegawa, S. Koshihara, H. Koinuma, Appl. Phys. Lett. **78**, 958 (2001)
54. Y. Matsumoto, R. Takahashi, M. Murakami, T. Koida, X.J. Fan, T. Hasegawa, T. Fukumura, M. Kawasaki, S. Koshihara, H. Koinuma, Jpn. J. Appl. Phys. **40**, L1204 (2001)
55. T. Fukumura, H. Toyosaki, Y. Yamada, Semicond. Sci. Technol. **20**, S103 (2005)
56. T. Dietl, H. Ohno, F. Matsukura, J. Cibert, D. Ferrand, Science **287**, 1019 (2000)
57. K. Sato, H.K. Yoshida, Jpn. J. Appl. Phys. **39**, L555 (2000)
58. T. Fukumura, Y. Yamada, K. Tamura, K. Nakajima, T. Aoyama, A. Tsukazaki, M. Sumiya, S. Fuke, Y. Segawa, T. Chikyow, T. Hasegawa, H. Koinuma, M. Kawasaki, Jpn. J. Appl. Phys. **42**, L105 (2003)
59. Y. Yamada, H. Toyosaki, A. Tsukazaki, T. Fukumura, K. Tamura, Y. Segawa, K. Nakajima, T. Aoyama, T. Chikyow, T. Hasegawa, H. Koinuma, M. Kawasaki, J. Appl. Phys. **96**, 5097 (2004)
60. K. Ando, H. Saito, Z.W. Jin, T. Fukumura, M. Kawasaki, Y. Matsumoto, H. Koinuma, Appl. Phys. Lett. **78**, 2700 (2001)
61. K. Ando, H. Saito, Z.W. Jin, T. Fukumura, M. Kawasaki, Y. Matsumoto, H. Koinuma, J. Appl. Phys. **89**, 7284 (2001)
62. K. Ueda, H. Tabata, T. Kawai, Appl. Phys. Lett. **79**, 988 (2001)
63. H. Toyosaki, T. Fukumura, Y. Yamada, M. Kawasaki, Appl. Phys. Lett. **86**, 182503 (2005)
64. H. Toyosaki, T. Fukumura, Y. Yamada, K. Nakajima, T. Chikyow, T. Hasegawa, H. Koinuma, M. Kawasaki, Nat. Mater. **3**, 221 (2004)
65. K. Ueno, T. Fukumura, H. Toyosaki, M. Nakano, M. Kawasaki, Appl. Phys. Lett. **90**, 72103 (2007)
66. T. Fukumura, H. Toyosaki, K. Ueno, M. Nakano, M. Kawasaki, Jpn. J. Appl. Phys. **46**, L642 (2007)
67. C.L. Chien, C.R. Westgate, *The Hall Effect and Its Applications* (Plenum, New York, 1979)
68. J. Smit, Physica (Amsterdam) **24**, 39 (1958)
69. L. Berger, Phys. Rev. B **2**, 4559 (1970)
70. N. Nagaosa, J. Phys. Soc. Jpn **75**, 042001 (2006)
71. S. Onoda, N. Sugimoto, N. Nagaosa, Phys. Rev. Lett. **97**, 126602 (2006)
72. T. Fukumura, H. Toyosaki, K. Ueno, M. Nakano, M. Kawasaki, Advances in Materials Research, **10**, 87–92
73. H. Toyosaki, T. Fukumura, K. Ueno, M. Nakano, M. Kawasaki, Jpn. J. Appl. Phys. **44**, L896 (2005)
74. H. Toyosaki, T. Fukumura, K. Ueno, M. Nakano, M. Kawasaki, J. Appl. Phys. **99**, 08M102 (2006)
75. A. Ohtomo, A. Tsukazaki, Semicond. Sci. Technol. **20**, S1 (2005)
76. K. Maeda, M. Sato, I. Niikura, T. Fukuda, Semicond. Sci. Technol. **20**, S49 (2005)
77. T. Makino, Y. Segawa, M. Kawasaki, H. Koinuma, Semicond. Sci. Technol. **20**, S78 (2005)
78. P. Yu, Z.K. Tang, G.K.L. Wong, M. Kawasaki, A. Ohtomo, H. Koinuma, Y. Segawa, 23rd *International Conference on the Physics of Semiconductors*, ed. by M. Scheffler, R. Zimmermann (World Scientific, Singapore, 1996), p. 1453

74   M. Kawasaki

79. P. Yu, Z.K. Tang, G.K.L. Wong, M. Kawasaki, A. Ohtomo, H. Koinuma, Y. Segawa, Solid State Commun. **103**, 459 (1997)
80. Z.K. Tang, G.K.L. Wong, P. Yu, M. Kawasaki, A. Ohtomo, H. Koinuma, Y. Segawa, Appl. Phys. Lett. **72**, 3270 (1998)
81. A. Ohtomo, K. Tamura, M. Kawasaki, T. Makino, Y. Segawa, Z.K. Tang, G.K.L. Wong, Y. Matsumoto, H. Koinuma, Appl. Phys. Lett. **77**, 2204 (2000)
82. A. Ohtomo, M. Kawasaki, T. Koida, K. Masubuchi, H. Koinuma, Y. Sakurai, Y. Yoshida, T. Yasuda, Y. Segawa, Appl. Phys. Lett. **72**, 2466 (1998)
83. T. Makino, Y. Segawa, M. Kawasaki, A. Ohtomo, R. Shiroki, K. Tamura, T. Yasuda, H. Koinuma, Appl. Phys. Lett. **78**, 1237 (2001)
84. A. Tsukazaki, A. Ohtomo, M. Kawasaki, Appl. Phys. Lett. **88**, 152106 (2006)
85. S.F. Chichibu, A. Uedono, A. Tsukazaki, T. Onuma, M. Zamfirescu, A. Ohtomo, A. Kavokin, G. Cantwell, C.W. Litton, T. Sota, M. Kawasaki, Semicond. Sci. Technol. **20**, S67 (2005)
86. A. Ohtomo, K. Tamura, K. Saikusa, K. Takahashi, T. Makino, Y. Segawa, H. Koinuma, M. Kawasaki, Appl. Phys. Lett. **75**, 2635 (1999)
87. A. Tsukazaki, A. Ohtomo, S. Yoshida, M. Kawasaki, C.H. Chia, T. Makino, Y. Segawa, T. Koida, S.F. Chichibu, H. Koinuma, Appl. Phys. Lett. **83**, 2784 (2003)
88. A. Tsukazaki, A. Ohtomo, M. Kawasaki, T. Makino, C.H. Chia, Y. Segawa, H. Koinuma, Appl. Phys. Lett. **84**, 3858 (2004)
89. A. Tsukazaki, A. Ohtomo, M. Kawasaki, Advances in Materials Research, **10**, 77–86
90. A. Tsukazaki, T. Onuma, M. Ohtani, T. Makino, M. Sumiya, K. Ohtani, S.F. Chichibu, S. Fuke, Y. Segawa, H. Ohno, H. Koinuma, M. Kawasaki, Nat. Mater. **4**, 42 (2005)
91. K. Tamura, A. Ohtomo, K. Sakurai, Y. Osaka, T. Segawa, H. Koinuma, M. Kawasaki, J. Cryst. Growth **214/215**, 59 (2000)
92. A. Tsukazaki, M. Kubota, A. Ohtomo, T. Onuma, S.F. Chichibu, K. Ohtani, H. Ohno, M. Kawasaki, Jpn. J. Appl. Phys. **44**, L643 (2005)
93. S. Arulkumaran, M. Sakai, T. Egawa, H. Ishikawa, T. Jimbo, T. Shibata, K. Asai, S. Sumiya, Y. Kuraoka, M. Tanaka, O. Oda, Appl. Phys. Lett. **81**, 1131 (2002)
94. S. Harada, S. Suzuki, J. Senzaki, R. Kosugi, K. Adachi, K. Fukuda, K. Arai, Mater. Sci. Forum **389–393**, 1069 (2002)
95. M. Kubovic, M. Kasu, I. Kallfass, M. Neuburger, A. Aleksov, G. Koley, M.G. Spencer, E. Kohn, Diamond Relat. Mater. **13**, 802 (2004)
96. R.L. Hoffman, B.J. Norris, J.F. Wager, Appl. Phys. Lett. **82**, 733 (2003)
97. J. Nishii, F.M. Hossain, S. Takagi, T. Aita, K. Saikusa, Y. Ohmaki, I. Ohkubo, S. Kishimoto, A. Ohtomo, T. Fukumura, F. Matsukura, Y. Ohno, H. Koinuma, H. Ohno, M. Kawasaki, Jpn. J. Appl. Phys. **42**, L347 (2003)
98. K. Nomura, H. Ohta, K. Ueda, T. Kamiya, M. Hirano, H. Hosono, Science **300**, 1269 (2003)
99. K. Nomura, H. Ohta, A. Takagi, T. Kamiya, M. Hirano, H. Hosono, Nature **432**, 488 (2004)
100. M. Kawasaki, H. Ohno, Jpn Patent 3,276,930, 2002 [filed as H10-326,889 (1998) and disclosed as 2000–150,900 (2000)]; US Patent 6,727,522, 2004; 7,064,346, 2006

101. F.M. Hossain, J. Nishii, S. Takagi, A. Ohtomo, T. Fukumura, H. Fujioka, H. Ohno, H. Koinuma, M. Kawasaki, J. Appl. Phys. **94**, 7768 (2003)
102. F.M. Hossain, J. Nishii, S. Takagi, T. Sugihara, A. Ohtomo, T. Fukumura, H. Koinuma, H. Ohno, M. Kawasaki, Physica E **21**, 911–915 (2004)
103. T.I. Suzuki, A. Ohtomo, A. Tsukazaki, F. Sato, J. Nishii, H. Ohno, M. Kawasaki, Adv. Mat. **16**, 1887 (2004)
104. A. Tsukazaki, A. Ohtomo, T. Kita, Y. Ohno, H. Ohno, M. Kawasaki, Science **315**, 1388 (2007)
105. A. Ohtomo, H.Y. Hwang, Nature **427**, 423 (2004)
106. A.P. Ramirez, Science **315**, 1377 (2007)

5

Epitaxial Growth and Transport Properties of High-Mobility ZnO-Based Heterostructures

A. Tsukazaki, A. Ohtomo, and M. Kawasaki

Summary: Optimization of growth conditions is most important for extracting desired material properties, but it always requires time-consuming experiments. The temperature gradient method was applied for high-throughput optimization of the growth temperature by pulsed laser deposition to ZnO-based heterostructures. Surface morphology, photoluminescence, and electrical transport properties depend systematically on the growth temperature. Consequently, enhancement of two-dimensional growth, as detected from atomic force microscope images, can elucidate good physical properties, e.g., the observation of higher-order exciton emissions and highest electron mobility. By further optimizing the structure of heterojunction, we found growth conditions enabling the quantum Hall effect at the $ZnO/Mg_xZn_{1-x}O$ heterointerface.

As an oxide semiconductor, ZnO has attracted much attention for optoelectronic device applications. Large exciton binding energy of 60 meV and a wide direct band gap of 3.37 eV are advantages for optical devices such as light-emitting diodes and lasers [1, 2]. In addition, large bulk single crystals grown using chemical vapor transport, hydrothermal method, and pressured melt growth are available, which are useful for suitable substrates of high-quality devices [3–5]. However, the difficulty in synthesis of p-type ZnO prevents us from developing such devices [6]. The n-type conductivity in ZnO is well known to be attributable to the low formation energy of donor-like defects such as Zn interstitial or oxygen vacancy. Our approaches to materialize p-type conduction are the reduction of those crystalline defects through optimization of growth processes using high-throughput techniques and development of an effective acceptor doping method. We have made efforts to reduce crystalline defects through adoption of a lattice-matched $ScAlMgO_4$ (SCAM) substrate [7], insertion of high-temperature annealed buffer (HITAB) layer [8], and optimization of growth conditions. In fact, improvement of physical properties is obtainable as evidence of reduction of crystalline defects in undoped ZnO films [9–11]. Additionally, a repeated temperature modulation method has been developed to realize effective acceptor doping with repressing self-compensation effect, attaining p-type conduction. As a consequence, blue electroluminescence with current injection was observed from a p-n homojunction diode [12].

Band-gap engineering is an important technique for applying the quantum size effect to improve the device performance of semiconductor devices. The band gap of ZnO can be controlled from about 2 to 4.5 eV while maintaining the wurtzite crystal structure by doping of Cd or Mg [13, 14]. Systematic studies by combinatorial techniques for observation of the quantum size effect have been reported in optical properties with single-quantum and multi-quantum well structures [15]. However, quantum transport properties such as the quantum Hall effect (QHE) or a resonant tunneling diode had not been reported in ZnO heterostructures because of the lack of ability for controlling interfaces on an atomic scale. On the other hand, many researchers have studied electron transport properties in transparent electrodes and field effect transistors with n-type and undoped ZnO for realization of transparent electronic circuits [16–18]. Recently, $Mg_xZn_{1-x}O/ZnO$ heterostructures have been applied to high-electron-mobility transistors [19, 20]. Moreover, the abrupt interface realized high electron mobility, which can reach the observation of QHE [21].

In this review paper, we report on the optimization of growth temperature for ZnO on HITAB using temperature gradient method. Comparison with surface morphology, photoluminescence (PL) spectra, and electrical transport properties showed a consistent optimum growth temperature region to our pulsed laser deposition systems. Through this optimization, we were able to detect the QHE in $ZnO/Mg_xZn_{1-x}O$ heterointerface at the higher growth temperature.

The crystal structure of ZnO is wurtzite; it is depicted in Fig. 5.1 as black spheres of Zn and white spheres of oxygen. It has inversion asymmetry along the [000–1] direction. The ZnO films and ZnO-HITAB were prepared on SCAM substrates using pulsed laser deposition. A small 0.09% lattice

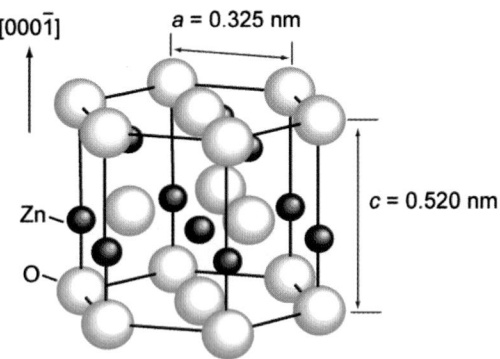

Fig. 5.1. Crystal structure of ZnO drawn with white sphere of O^{2-} and black sphere of Zn^{2+}. The lattice constants of the a-axis and c-axis are, respectively, 0.325 and 0.520 nm. The perpendicular direction is the [000–1]

mismatch between ZnO and the SCAM substrate gave us a good chance for high-quality thin film growth. ZnO single crystal target is ablated by KrF excimer laser pulses ($\lambda = 248\,\mathrm{nm}$, $5\,\mathrm{Hz}$, $1\,\mathrm{J\ cm^{-2}}$) in a vacuum chamber to supply film-growth precursors. We inserted ZnO-HITAB or $\mathrm{Mg}_x\mathrm{Zn}_{1-x}\mathrm{O}$-HITAB on SCAM to relax the strain from the substrate and to enhance surface migration. The HITAB was annealed in a vacuum chamber with $1\,\mathrm{mTorr}$ oxygen flow at $1{,}273\,\mathrm{K}$ following $100\,\mathrm{nm}$ deposition with oxygen flow of 1×10^{-6} Torr at $923\,\mathrm{K}$. The temperature gradient method is a combinatorial technique for parallel synthesis of thin films and high-throughput optimization of growth temperature. Figure 5.2a,b, respectively, depict schematic cross-sectional views of heating systems at the temperature gradient configuration and top view of the susceptor. One side of the susceptor was heated using focused near-infrared semiconductor laser ($\lambda = 808\,\mathrm{nm}$). The growth temperature was measured using a pyrometer, which can scan the substrate. In this configuration, the growth temperature on a substrate changed linearly at each position along one direction, as shown in the temperature gradation image in Fig. 5.2, with gray scale.

During $1\,\mathrm{\mu m}$-thickness ZnO film growth on HITAB, we were able to observe the intensity oscillation of reflection high-energy electron diffraction because of the atomically flat surfaces of HITAB [8]. Atomic force microscope (AFM) topographies in Fig. 5.3 show surface morphologies for ZnO film grown on

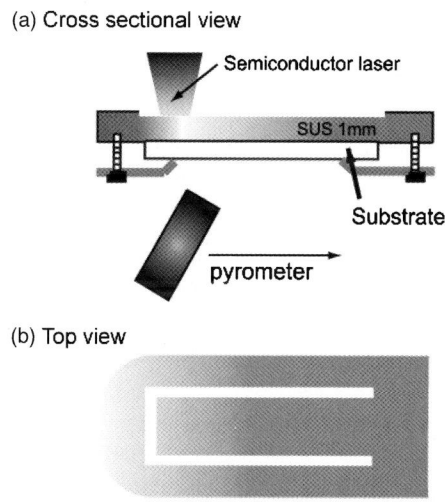

(a) Cross sectional view

Semiconductor laser

SUS 1mm

Substrate

pyrometer

(b) Top view

Fig. 5.2. The schematic illustration of laser heating system used in temperature gradient method. (**a**) Cross-sectional view of susceptor holding substrate. The focused semiconductor laser illuminates the left side of the susceptor. The pyrometer can move along the long direction of the substrate for measuring the growth temperature at each position. (**b**) Top view of the susceptor. The color gradation indicates the temperature gradient on the substrate

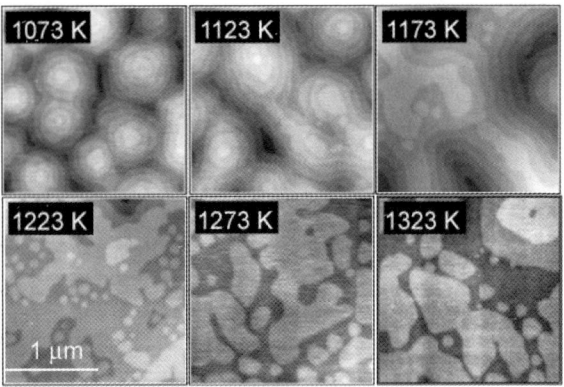

Fig. 5.3. Surface morphologies are shown by $2 \times 2\,\mu m^2$ AFM images obtained from a ZnO film grown on ZnO-HITAB using temperature gradient method. Each growth temperature is denoted in the images. The number of steps is reduced with increasing growth temperature; also, the terrace width widens

ZnO-HITAB. These images were obtained from each position on a film grown using temperature gradient method. All AFM images show a step and terrace structure, representing the half unit cell (0.26 nm) of ZnO crystal structure. At a low temperature of 1,073 K, a hexagonal column surface structure having steps and terraces appeared, which is similar to that of ZnO films grown on sapphire substrates [1]. With increasing growth temperature, the terraces widen and the number of steps decreases with connecting neighboring hexagons. The enhancement of surface migration length by high-temperature growth yielded these surface morphology changes. Therefore, the optimum growth temperature regarding surface flatness was judged at temperatures greater than 1,223 K because of its fewest step and widest terrace structure.

Figure 5.4 presents the PL spectra at 5 K of ZnO films grown on ZnO-HITAB at 1,023, 1,123, and 1,223 K, including the donor bound exciton emission lines (D^0X) at 3.361 and 3.367 eV, ground state free A-exciton line $(FE_A, n = 1)$ at 3.377 eV, and higher-order exciton lines $(n = 2)$ at 3.418 eV (D^0X) and at 3.425 eV (FE_A) [22]. The excitation source for PL measurement was the 325 nm line from a He-Cd laser. Clear observation of higher-order exciton lines indicated the enhancement of crystalline coherency, judging from the facts that the Bohr radius of $n = 1$ exciton is 18 Å and that of $n = 2$ is 72 Å. The peak intensities of free A-exciton and higher-order excitons gradually weaken with decreasing growth temperature. Therefore, the high-temperature grown ZnO films have wide crystalline coherency and good optical properties involving not only PL spectra but also reflection and transmission spectra (not shown) [9]. In fact, the nonradiative PL lifetime at room temperature has growth temperature dependence, which decreased with decreasing growth temperature, indicating that the generation of nonradiative crystalline defects

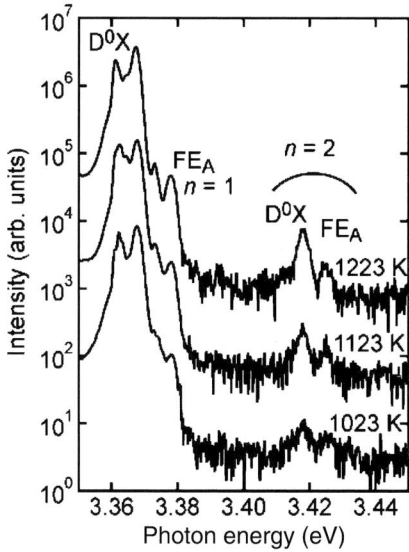

Fig. 5.4. Photoluminescence spectra at 5 K of ZnO film grown on ZnO-HITAB at 1,223, 1,123, and 1,023 K. D^0X indicates the donor-bound exciton emission lines and FE_A indicates the free A-exciton emission lines of ground state ($n = 1$) and higher-order exciton state ($n = 2$)

depends on the growth temperature [10]. The longest lifetime of 3.8 ns was indicated for films grown at around 1,223 K.

Next, we performed Hall effect measurement for ZnO films grown on $Mg_{0.15}Zn_{0.85}O$-HITAB, whose layer structure is shown in the inset of Fig. 5.5a. The $Mg_{0.15}Zn_{0.85}O$-HITAB has atomically flat surfaces and is semi-insulating ($\rho > 10^2 \, \Omega cm$) so that we can characterize those transport properties only of the ZnO layers. The layered samples were processed using conventional photolithography and Ar ion milling. The Au/Ti electrode was formed by electron beam evaporation, giving good ohmic contact even at low temperatures. The electron concentration (n) was evaluated by Hall effect measurements at low magnetic fields ($\pm 1 \, T$) and mobility (μ) was calculated from $\mu = 1/\rho en$, where ρ is resistivity at a zero magnetic field and e is the elementary charge. Figure 5.5a,b, respectively, shows growth temperature dependences of n and μ at 300 K. The value of n decreases with decreasing growth temperature below about 1,250 K because of the generation of an acceptor like defect of Zn vacancy, which is consistent with the positron annihilation measurement [23]. On the other hand, at temperatures greater than 1,250 K, n drastically increases to greater than the order of $10^{17} \, cm^{-3}$ because of the increment of donor-like Zn interstitial defects. In this experiment, we can obtain an optimum growth temperature of around 1,250 K at the intersection for two-temperature dependence of the formation of crystalline defects because

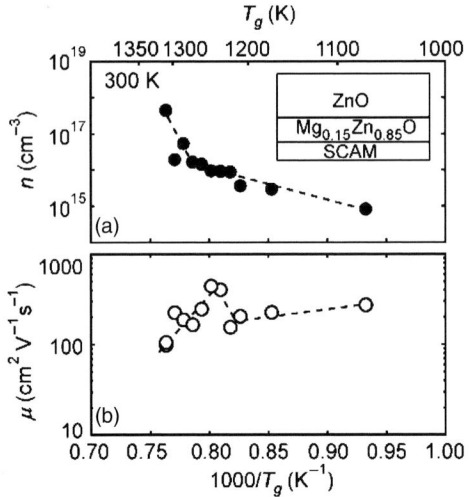

Fig. 5.5. Growth temperature dependence of n and μ at 300 K are shown, respectively, in (**a**) and (**b**). The sample structure is depicted in the inset of (**a**)

the highest μ of $440 \, \mathrm{cm^2 \, V^{-1} \, s^{-1}}$ was observed at that growth temperature. It is noteworthy that the temperature gradient method enabled us to find the optimum growth temperature using only one film to eliminate crystalline defects.

We found metallic conduction at temperature dependence of resistivity from layered structure samples grown at temperatures greater than $1,300 \, \mathrm{K}$ with slightly different ablation excimer laser conditions. The n and μ values are shown, respectively, as functions of temperature in Fig. 5.6 by open and closed squares. Generally, the n in semiconductor ZnO is frozen out at low temperature because of the low activation energy [11, 24]. However, n of $7 \times 10^{11} \, \mathrm{cm^{-2}}$ was observed even at temperatures lower than 0.1 K in this sample. The μ exhibited metallic behavior, as evidenced by the monotonic increment with decreasing temperature. The highest μ of $5,500 \, \mathrm{cm^2 \, V^{-1} \, s^{-1}}$ was obtained at 1 K. We concluded from these temperature dependences of n and μ that the two-dimensional electron gases (2DEGs) were formed at the heterointerface between ZnO and $\mathrm{Mg_{0.15}Zn_{0.85}O}$. The electron concentrations of 2DEGs were tunable according to the Mg content in barrier layer and electron concentration in top ZnO layer with growth temperature [21].

Magnetotransport properties were examined using a standard lock-in technique with AC excitation of 10 nA in diluted refrigerator at 45 mK. Figure 5.7a,b, respectively, shows ρ_{xy} and ρ_{xx} as functions of the magnetic field. The Shubnikov-de Haas oscillation and quantum Hall plateau started appearing at a magnetic field above 1.5 T. It is somewhat surprising that clear QHE is observable despite the heavy electron effective mass ($m^* = 0.32 \pm 0.03 \, m_0$, where m_0 is the free electron mass) [21]. Zeeman

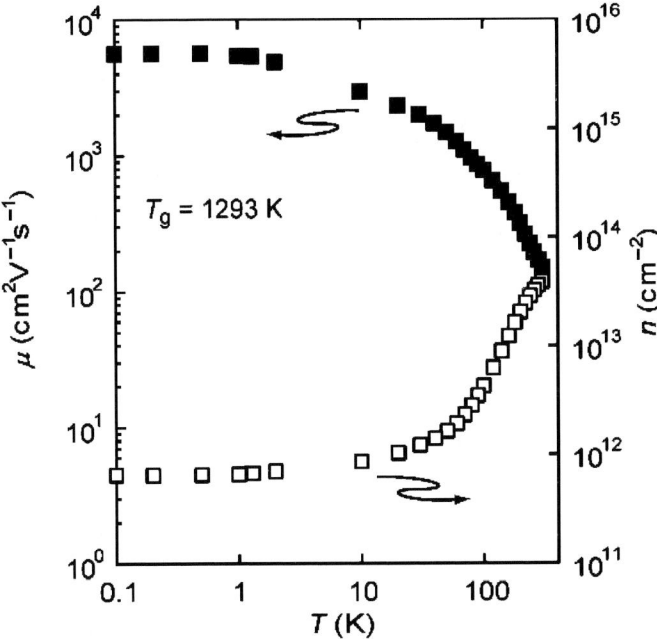

Fig. 5.6. The n (*open square*) and μ (*filled square*) for ZnO film grown on Mg$_{0.15}$Zn$_{0.85}$O-HITAB at high temperature of 1,293 K are shown as a function of temperature

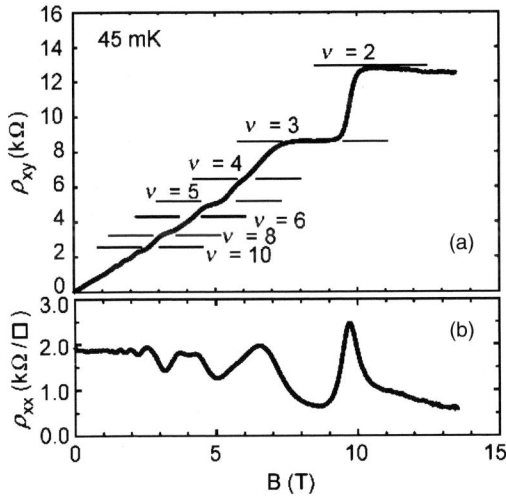

Fig. 5.7. Magnetotransport properties of ρ_{xy} and ρ_{xx} for ZnO/Mg$_{0.15}$Zn$_{0.85}$O heterostructures at 45 mK are shown, respectively, in (**a**) and (**b**). The indices of the filling factor (ν) are indicated for the quantum plateaus in (**a**)

splitting, the emergence of Landau levels with odd filling indices, is clearly resolved because the Zeeman energy ΔE becomes comparable to the level width Γ at $B \sim 5\,\mathrm{T}$, where $\Delta E = g^* \mu_B B$, g^* is the effective g factor of ZnO being equal to -1.93 [25], μ_B is Bohr magneton, and $\Gamma = [(2/\pi)(\hbar\omega_c)(\hbar/\tau)]^{1/2}$ [26]. We do not expect, however, that the odd states have much wider Hall plateaus and larger amplitudes of ρ_{xx} minima compared to those of the even states as $\nu = 3$ and 5 states. Consequently, the even states such as $\nu = 4$ are barely identified. Exchange enhancement of the g factor might account for this anomaly, but that remains unclear at the moment [27].

In conclusion, we performed high-throughput optimization of growth conditions for pulsed laser deposition using temperature gradient method on ZnO films and $\mathrm{ZnO/Mg_{0.15}Zn_{0.85}O}$ heterostructures. Experimental results obtained for surface flatness, the higher-order exciton emission at PL measurement, and electrical transport properties at $300\,\mathrm{K}$ revealed the optimum growth temperature around $1{,}223$–$1{,}273\,\mathrm{K}$. We were also able to find the 2DEGs at heterointerfaces exhibiting QHE from $\mathrm{ZnO/Mg_{0.15}Zn_{0.85}O}$ heterostructures grown at higher temperature. Results show that the temperature gradient method is a powerful technique to optimize the growth temperature and find interesting physical properties.

Acknowledgements. We are grateful to T. Makino, Y. Segawa, T. Onuma, S.F. Chichibu, T. Kita, Y. Ohno, and H. Ohno for technical assistance and fruitful discussion.

References

1. P. Yu et al., Solid State Commun. **103**, 459 (1997); Z.K. Tang et al., Appl. Phys. Lett. **72**, 3270 (1998)
2. D.M. Bagnall et al., Appl. Phys. Lett. **70**, 2230 (1997)
3. D.C. Look et al., Solid State Commun. **105**, 399 (1998)
4. E. Ohshima et al., J. Cryst. Growth **260**, 166 (2004); K. Maeda, M. Sato, I. Niikura, T. Fukuda, Semicond. Sci. Technol. **20**, S49 (2005)
5. J. Nause, B. Nemeth, Semicond. Sci. Technol. **20**, S45 (2005)
6. D.C. Look, B. Claflin, Y.I. Alivov, S.J. Park, Phys. Stat. Sol. A **201**, 2203 (2004)
7. A. Ohtomo et al., Appl. Phys. Lett. **75**, 2635 (1999)
8. A. Tsukazaki et al., Appl. Phys. Lett. **83**, 2784 (2003)
9. A. Tsukazaki et al., Appl. Phys. Lett. **84**, 3858 (2004)
10. S.F. Chichibu et al., J. Appl. Phys. **99**, 093505 (2006)
11. A. Tsukazaki, A. Ohtomo, M. Kawasaki, Appl. Phys. Lett. **88**, 152106 (2006)
12. A. Tsukazaki et al., Nat. Mater. **4**, 42 (2005); A. Tsukazaki et al., Jpn. J. Appl. Phys. **44**, L643 (2005)
13. A. Ohtomo, A. Tsukazaki, Semicond. Sci. Technol. **20**, S1 (2005)
14. S. Shigemori, A. Nakamura, J. Ishihara, T. Aoki, J. Temmyo, Jpn. J. Appl. Phys. **43**, L1088 (2004)

15. T. Makino, Y. Segawa, M. Kawasaki, H. Koinuma, Semicond. Sci. Technol. **20**, S78 (2005)
16. T. Minami, Semicond. Sci. Technol. **20**, S35 (2005)
17. R.L. Hoffman, B.J. Norris, J.F. Wager, Appl. Phys. Lett. **82**, 733 (2003)
18. J. Nishii et al., Jpn. J. Appl. Phys. **42**, L347 (2003)
19. K. Koike et al., Jpn. J. Appl. Phys. **43**, L1372 (2004)
20. H. Tampo et al., Appl. Phys. Lett. **89**, 132113 (2006)
21. A. Tsukazaki, A. Ohtomo, T. Kita, Y. Ohno, H. Ohno, M. Kawasaki, Science **315**, 1388 (2007)
22. E. Mollwo, in *Semiconductors: Physics of II-VI and I-VII Compounds, Semimagnetic Semiconductors*, vol. 17, ed. by O. Madelung, M. Schulz, H. Weiss (Springer, Berlin Heidelberg New York, 1982), p. 35
23. A. Uedono et al., J. Appl. Phys. **93**, 2481 (2003)
24. D.C. Look, J.W. Hemsky, J.R. Sizelove, Phys. Rev. Lett. **82**, 2552 (1999)
25. D.C. Reynolds, C.W. Litton, T.C. Collins, Phys. Rev. **140**, A1726 (1965)
26. T. Ando, Y. Uemura, J. Phys. Soc. Jpn. **36**, 959 (1974)
27. T. Ando, A.B. Fowler, F. Stern, Rev. Mod. Phys. **54**, 437 (1982)

6

A Scaling Behavior of Anomalous Hall Effect in Cobalt Doped TiO_2

T. Fukumura, H. Toyosaki, K. Ueno, M. Nakano, T. Yamasaki, and M. Kawasaki

Summary: Anomalous Hall effect (AHE) is a generally observed phenomenon in ferromagnetic metals representing spin polarized nature of itinerant carriers. However, the microscopic mechanism has not been clarified for long debates. Recent advances in the theory are to unveil the mechanism. Here, we present an AHE in a room temperature ferromagnetic semiconductor cobalt doped TiO_2. This compound shows a scaling behavior of the AHE: the anomalous Hall conductivity σ_{AH} approximately follows the relation $\sigma_{AH} \propto \sigma_{xx}^{1.6}$ (σ_{xx}, conductivity) over five decades of σ_{xx}, irrespective of the electronic state, i.e., metallic or insulating conduction.

Anomalous Hall effect (AHE) is generally observed for ferromagnetic metals. Empirical expression of Hall resistivity including AHE is $\rho_H = R_0 H + R_S M$ (ρ_H, Hall resistivity; R_0, normal Hall coefficient; H, magnetic field; R_S, anomalous Hall coefficient; M, magnetization): the first and the second terms are the normal and the anomalous Hall terms, respectively [1]. Microscopic theory of AHE has been of long debate for half a century, and is being developed quite recently to explain AHE in various ferromagnetic metals quantitatively, in which the Berry phase of Bloch wave function plays a crucial role [2]. Previously, the skew scattering [3] and side jump mechanisms [4] have been often employed for the interpretation of AHE, where the former and the latter lead, respectively, to the relations $\rho_H \propto \rho_{xx}$ and $\rho_H \propto \rho_{xx}^2$ (ρ, resistivity). For general consideration of AHE, more comprehensive treatment is needed than those two theories.

In a recent theory considering multiband ferromagnetic metals with dilute impurities, the power law dependence of the anomalous Hall conductivity σ_{AH} on the conductivity σ_{xx} changes in various regimes of the conductivity, representing the extrinsic-to-intrinsic crossover [5]. The extrinsic skew scattering mechanism ($\sigma_{AH} \propto \sigma_{xx}$) appears in the clean limit, whereas the intrinsic contribution is dominant with the lowering conductivity. In the dirty limit, the intrinsic contribution is subject to the damping due to the impurities, leading to the relation $\sigma_{AH} \propto \sigma_{xx}^{1.6}$, where the exponent is close to that in normal Hall effect of quantum Hall insulator [6]. Recent experiment on ferromagnetic metals supported the theory in the regime of high conductivity [7]. The above theory is based on the use of Bloch wave function assuming the metallic conduction [5]; hence the result is valid only for ferromagnetic metals. However, the AHE in cobalt-doped TiO_2 suggests that the relation $\sigma_{AH} \propto \sigma_{xx}^{1.6}$ is plausible even in the regime of nearly insulating conduction as described below.

Cobalt-doped TiO_2, $(Ti, Co)O_2$, is a high temperature ferromagnetic semiconductor [8–10], and one of the promising compounds for semiconductor spintronics. Recently, the tunneling magnetoresistance effect can be observed up to 200 K [11, 12]. This compound has both rutile and anatase phases, and these electronic states are significantly different from each other [13, 14]. Figure 6.1 shows the dependence of resistivity on temperature for rutile $(Ti, Co)O_2$. The growth oxygen pressures were varied to control the contents of oxygen vacancies as electron donors. The resistivity shows the systematic variation with contents of oxygen vacancies: the lower resistivity with increasing oxygen vacancies. The resistivity exponentially increases with decreasing temperature as a result of freezing out of the charge carriers. The carrier density can be varied for several decades, whereas the mobility is typically an order of $0.1 \, \mathrm{cm^2 \, V^{-1} \, s^{-1}}$ without significant temperature dependence.

On the other hand, the anatase $(Ti, Co)O_2$ shows a degenerate conduction behavior with the lower carrier density partly because of the much higher mobility in comparison with the rutile $(Ti, Co)O_2$. Figure 6.2 shows the dependence of resistivity on temperature for the anatase $(Ti, Co)O_2$. The resistivity shows a degenerate behavior with slight increase below 100 K, and the carrier density is nearly constant for the wide range of temperature in contrast

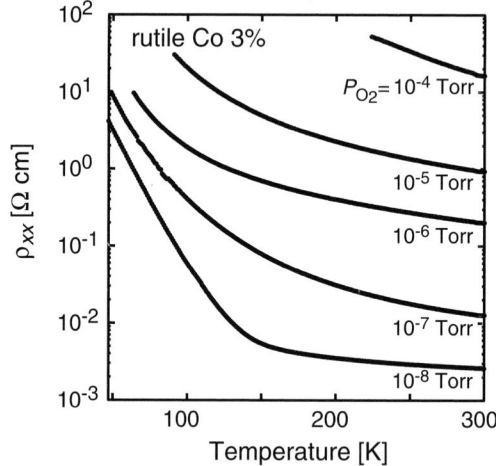

Fig. 6.1. Temperature dependence of resistivity (ρ_{xx}) for rutile $Ti_{0.97}Co_{0.03}O_{2-\delta}$ grown under different growth oxygen pressures (P_{O2})

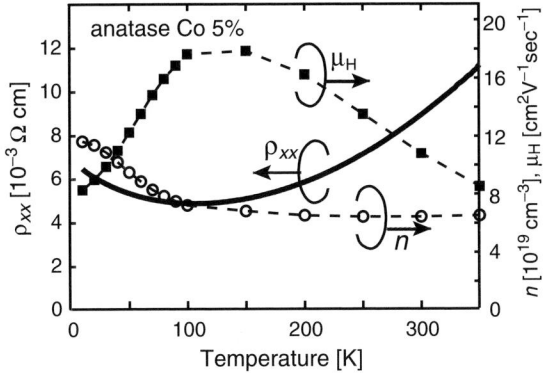

Fig. 6.2. Temperature dependence of resistivity (ρ_{xx}: *solid line*), carrier density (n: *open circles*) and Hall mobility (μ_H: *solid squares*) for anatase Ti$_{0.95}$Co$_{0.05}$O$_{2-\delta}$

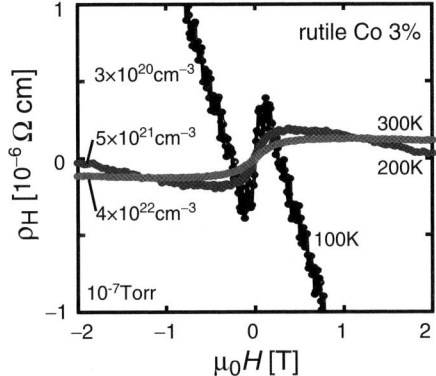

Fig. 6.3. Magnetic field dependence of Hall resistivity (ρ_H) at different temperatures for rutile Ti$_{0.97}$Co$_{0.03}$O$_{2-\delta}$ grown under $P_{O2} = 10^{-7}$ Torr. The carrier density evaluated from ordinary part of ρ_H is labeled for each curve

with the rutile (Ti, Co)O$_2$. The mobility is two decades higher than that of the rutile (Ti, Co)O$_2$ and shows a sizable temperature dependence having the maximum around 100 K.

The AHE in the rutile and anatase (Ti, Co)O$_2$ shows different behavior, too [13, 14]. Figure 6.3 shows the dependence of Hall resistivity on magnetic field at different temperatures for the rutile (Ti, Co)O$_2$. The anomalous Hall term is dominant over the normal Hall term at 300 K, whereas the normal Hall term becomes dominant at lower temperature. This is because the freezing out of the charge carriers with decreasing temperature leads to the increase in the normal Hall term, which is inversely proportional to the carrier density. For the anatase (Ti, Co)O$_2$ as shown in the inset of Fig. 6.4, the normal Hall term is almost dominant even at 300 K mainly due to the lower carrier density

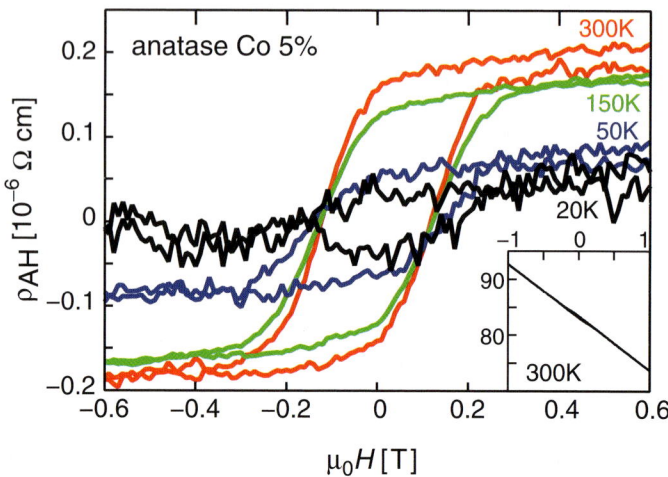

Fig. 6.4. Magnetic field dependence of anomalous Hall resistivity (ρ_{AH}) at different temperatures for anatase $\mathrm{Ti}_{0.95}\mathrm{Co}_{0.05}\mathrm{O}_{2-\delta}$. Inset shows Hall resistivity at $300\,\mathrm{K}$, including normal Hall term

than the rutile $(\mathrm{Ti},\mathrm{Co})\mathrm{O}_2$, and the anomalous Hall term is hardly seen. By subtracting the normal Hall term that is proportional to magnetic field, the anomalous Hall term can be observed as shown in Fig. 6.4. The anomalous Hall term decreases with decreasing temperature, while the normal Hall term shows little dependence on the temperature, as a result of the degenerate conduction. The hysteretic behavior is different between the rutile and anatase $(\mathrm{Ti},\mathrm{Co})\mathrm{O}_2$, representing the difference in the coercive force, which is consistent with the measurements of magnetization and magnetic circular dichroism.

Previously, the rutile $(\mathrm{Ti},\mathrm{Co})\mathrm{O}_2$ was reported to show a scaling behavior of AHE: the anomalous Hall conductivity approximately scales with the conductivity as $\sigma_{\mathrm{AH}} \propto \sigma_{xx}^{1.5-1.7}$, irrespective of the Co content, the measurement temperature, and the value of conductivity [13]. Figure 6.5 shows the relation between the anomalous Hall conductivity and the conductivity for the rutile and anatase $(\mathrm{Ti},\mathrm{Co})\mathrm{O}_2$. The anomalous Hall conductivity was approximately deduced to be $\rho_{\mathrm{H}}/\rho_{xx}^2$ since $\rho_{\mathrm{H}} \ll \rho_{xx}$, where the contribution of the normal Hall term was eliminated [13, 14]. The rutile and anatase $(\mathrm{Ti},\mathrm{Co})\mathrm{O}_2$ follows the same scaling line for five decades of the conductivity in spite of their dissimilar electronic states. The lower inset of Fig. 6.5 shows a magnification of data of the anatase $(\mathrm{Ti},\mathrm{Co})\mathrm{O}_2$ and the small hysteretic temperature dependence is seen. Such dependence disappears in the plot of the anomalous Hall conductivity vs. the mobility, representing $\sigma_{\mathrm{AH}} \propto \mu^2$ relation as shown in the upper inset of Fig. 6.5, although the underlying mechanism of its dependence is unclear at present.

Figure 6.5 includes the data of rutile and anatase $(\mathrm{Ti},\mathrm{Co})\mathrm{O}_2$ from the different research groups [15–18]. All data of $(\mathrm{Ti},\mathrm{Co})\mathrm{O}_2$ nearly follow the same

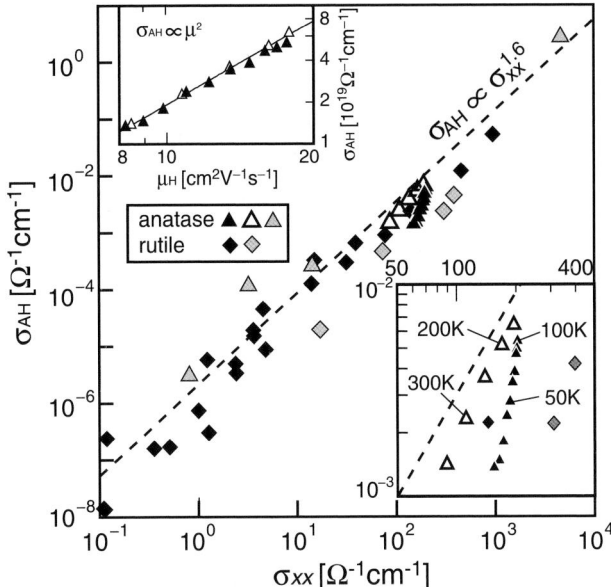

Fig. 6.5. Relation between anomalous Hall conductivity (σ_{AH}) and conductivity (σ_{xx}) for rutile Ti$_{1-x}$Co$_x$O$_{2-\delta}$ (diamonds [13, 15]) and anatase Ti$_{0.95}$Co$_{0.05}$O$_{2-\delta}$ (triangles [14, 16–18]). The solid gray symbols correspond to data taken from other groups. The lower inset shows a magnification around the data for the anatase Ti$_{0.95}$Co$_{0.05}$O$_{2-\delta}$, where the open and solid triangles represent the data for higher ($\geq$150 K) and lower ($\leq$100 K) temperatures. The upper inset shows log–log plot of σ_{AH} vs. μ_H curve for the anatase Ti$_{0.95}$Co$_{0.05}$O$_{2-\delta}$

scaling line $\sigma_{AH} \propto \sigma_{xx}^{1.6}$. Higgins et al. attributed the presence of the AHE to the precipitation of ferromagnetic Co metal in the TiO$_2$ matrix, claiming that the ferromagnetism in (Ti, Co)O$_2$ was extrinsic [15]. However, the Co content (2%) was much below the percolation limit, so that the dilute Co metal could not yield the observable AHE. Instead, their result can be interpreted as the appearance of AHE in (Ti, Co)O$_2$, where the small amounts of precipitation observed might not contribute to AHE.

The scaling relation observed in (Ti, Co)O$_2$ agrees with the theory in the regime of low conductivity [5], although the conductivity of (Ti, Co)O$_2$ extends to much lower values. The above mentioned theory is based on the use of Bloch wave function assuming the metallic conduction. As shown in this study, however, the scaling relation is evidenced also in the regime of insulating conduction.

In summary, a ferromagnetic semiconductor cobalt doped TiO$_2$, both rutile and anatase phases, shows a scaling relation of the AHE, $\sigma_{AH} \propto \sigma_{xx}^{1.6}$, irrespective of their different electronic state. This relation is consistent with the recent theory by Onoda et al. [5]. Comparison with the AHE in various ferromagnetic metals is described elsewhere [19].

Acknowledgements. The authors acknowledge N. Nagaosa, S. Onoda, H. Ohno, and F. Matsukura for fruitful discussions. This work was supported by MEXT for Young Scientists (A19686021) and for Scientific Research on Priority Areas (16076205), New Energy and Industrial Technology Development Organization, Industrial Technology Research Grant Program (05A24020d), and Tokyo Ohka Foundation for the Promotion of Science and Technology.

References

1. C.L. Chien, C.R. Westgate, *The Hall Effect and Its Applications* (Plenum, New York, 1979)
2. N. Nagaosa, J. Phys. Soc. Jpn, **75**, 042001 (2006)
3. J. Smit, Physica (Amsterdam) **24**, 39 (1958)
4. L. Berger, Phys. Rev. B **2**, 4559 (1970)
5. S. Onoda, N. Sugimoto, N. Nagaosa, Phys. Rev. Lett. **97**, 126602 (2006)
6. L.P. Pryadko, A. Auerbach, Phys. Rev. Lett. **82**, 1253 (1999)
7. T. Miyasato, N. Abe, T. Fujii, A. Asamitsu, S. Onoda, Y. Onose, N. Nagaosa, Y. Tokura, Phys. Rev. Lett. **99**, 086602 (2007)
8. T. Fukumura, H. Toyosaki, Y. Yamada, Semicond. Sci. Technol. **20**, S103 (2005)
9. Y. Matsumoto, M. Murakami, T. Shono, T. Hasegawa, T. Fukumura, M. Kawasaki, P. Ahmet, T. Chikyow, S. Koshihara, H. Koinuma, Science **291**, 854 (2001)
10. Y. Matsumoto, R. Takahashi, M. Murakami, T. Koida, X.J. Fan, T. Hasegawa, T. Fukumura, M. Kawasaki, S. Koshihara, H. Koinuma, Jpn. J. Appl. Phys. **40**, L1204 (2001)
11. H. Toyosaki, T. Fukumura, K. Ueno, M. Nakano, M. Kawasaki, Jpn. J. Appl. Phys. **44**, L896 (2005)
12. H. Toyosaki, T. Fukumura, K. Ueno, M. Nakano, M. Kawasaki, J. Appl. Phys. **99**, 08M102 (2006)
13. H. Toyosaki, T. Fukumura, Y. Yamada, K. Nakajima, T. Chikyow, T. Hasegawa, H. Koinuma, M. Kawasaki, Nat. Mater. **3**, 221 (2004)
14. K. Ueno, T. Fukumura, H. Toyosaki, M. Nakano, M. Kawasaki, Appl. Phys. Lett. **90**, 072103 (2007)
15. J.S. Higgins, S.R. Shinde, S.B. Ogale, T. Venkatesan, R.L. Greene, Phys. Rev. B **69**, 073201 (2004)
16. T. Hitosugi, G. Kinoda, Y. Yamamoto, Y. Furubayashi, K. Inaba, Y. Hirose, K. Nakajima, T. Chikyow, T. Shimada, T. Hasegawa, J. Appl. Phys. **99**, 08M121 (2006)
17. J.H. Cho, T.J. Hwang, D.H. Kim, Y.G. Joh, E.C. Kim, D.H. Kim, W.S. Yoon, H.C. Ri, J. Korean Phys. Soc. **48**, 1400 (2006)
18. R. Ramaneti, J.C. Lodder, R. Jansen, Appl. Phys. Lett. **91**, 012502 (2007)
19. T. Fukumura, H. Toyosaki, K. Ueno, M. Nakano, T. Yamasaki, M. Kawasaki, Jpn. J. Appl. Phys. **46**, L642 (2007)

7

Synthesis, Phase Diagram, and Evolution of Electronic Properties in Li_xZrNCl Superconductors

Y. Taguchi, A. Kitora, T. Takano, T. Kawabata, M. Hisakabe, and Y. Iwasa

Summary: We succeed in synthesizing a series of single phase Li_xZrNCl samples with controlled doping levels ($0 \leq x \leq 0.31$) by adopting an appropriate annealing procedure for lightly doped samples. An insulator-to-superconductor transition was found to occur at approximately $x = 0.05$, and the superconducting transition temperature (T_c) anomalously increases rapidly below $x = 0.12$ as the insulating phase is approached from the superconducting side. Doping- and temperature-dependent Raman scattering measurements indicated that electron–phonon interaction strength rather decreases upon reducing carrier concentration, suggesting that charge fluctuation plays an important role in the enhancement of T_c in the reduced carrier density regime.

ZrNCl- and HfNCl-based superconductors have recently attracted much interest because of their much higher T_c values (=15 and 25 K, respectively) than are expected from the several physical parameters that usually determine T_c in conventional phonon-mediated superconductors. Undoped β-ZrNCl and β-HfNCl are band insulators with layered structure, and have been found by Yamanaka et al. [1, 2] to become superconductors upon electron doping by means of alkali-metal intercalation. According to recent band calculations [3–7], doped electrons are accommodated into a highly two-dimensional band consisting of Zr 4d (Hf 5d) orbitals strongly hybridized with N2p states. The two-dimensionality of the electronic state has theoretically been predicted by the band calculations [3–7], and has experimentally been confirmed by magnetic measurements as well as X-ray absorption measurements [8, 9]. Reflecting the two-dimensionality of the electronic states, the density of states at the Fermi level is predicted to be almost filling-independent. One of the most remarkable features of these materials is that the constant value of the density of states is considerably small as clearly demonstrated by a recent magnetic susceptibility measurement for Li_x(THF)$_y$HfNCl [10] and a specific heat measurement for $Li_{0.12}$ZrNCl [11]: Pauli paramagnetic susceptibility χ of Li_x(THF)$_y$HfNCl was reported to be 1.7×10^{-5} emu mol^{-1} [10] and the electronic specific heat coefficient γ was determined to be 1.0 ± 0.1 mJ mol^{-1} K^{-2}

for $Li_{0.12}ZrNCl$ [11]. These values of χ and γ are approximately one order of magnitude smaller than those of typical superconductors with similar T_c values. From the comparison of the experimentally observed value of γ and calculated density of states at the Fermi level, the maximum value of dimensionless electron–phonon coupling constant λ is estimated to be 0.22, which is also much smaller than that expected for a superconductor with higher T_c than 10 K. These observations are in accord with the theoretical prediction that the density of states at the Fermi level is small and the electron–phonon interaction is also very weak. By contrast, very large values of superconducting gap ratio $2\Delta/k_BT_c$ have been reported for $Li_x(THF)_yHfNCl$ by a tunneling experiment [12] and for the $Li_{0.12}ZrNCl$ by a specific heat measurement [11]. μSR measurements have also indicated [13, 14] that these superconductors belongs to a very strong coupling regime. To understand these apparently contradicting observations, it has been proposed theoretically [15, 16] that other bosonic fluctuation than phonon, most probably charge fluctuation, also additionally contributes to and reinforces the pairing interaction. The importance of charge fluctuation contribution to the pairing interaction is emphasized especially for small carrier-density systems by Takada [17] and also by Kohno et al. [18]. Unfortunately, however, systematic investigations on samples with a wide range of doping concentration have never been performed thus far in this system because of the difficulty in synthesizing single phase samples, especially in the lightly doped regime, and also the extreme sensitivity of the Li-intercalated samples to the air. Therefore, we have attempted to synthesize a series of Li_xZrNCl samples with systematically controlled doping levels, and to clarify the evolution of electronic properties upon Li intercalation, with particular focus on the very lightly doped regime in the vicinity of insulator–superconductor transition.

β-ZrNCl was prepared by chemical vapor transport technique, following the procedures described in [19]. Obtained β-ZrNCl powder was dispersed into n-BuLi/hexane solutions ($\sim$5 mL) of appropriate molarity for the intended doping level x ($0 \leq x \leq 0.31$) for 1 day in an Ar-filled glove box. In the cases of $x > 0.16$, the solutions were kept at 70°C for 2 h after the 1 day of soaking at ambient temperature to promote Li intercalation. The powder was collected by filtration, and quick examination of the samples was performed by taking powder X-ray diffraction pattern with Cu Kα radiation. Since the samples with $x \leq 0.10$ were found to be easily phase-separated, we subjected them to annealing procedure in sealed quartz tubes at 600°C for 30 min to achieve uniform distribution of Li ions. This annealing temperature for lightly doped samples was determined on the basis of the result of high temperature X-ray diffraction experiment performed at BL02B2, SPring-8, which is shown in Fig. 7.1. In the pristine β-ZrNCl, Cl-[ZrN]$_2$-Cl layers stack along c-axis to form a so-called SmSI structure. Upon Li intercalation, the stacking sequence along the c-axis changes from the SmSI-type to YOF-type [20, 21]. Associated with the change in stacking sequence along the c-axis, X-ray diffraction intensity of (104) and (015) peak exhibits significant variation: In the pristine β-ZrNCl

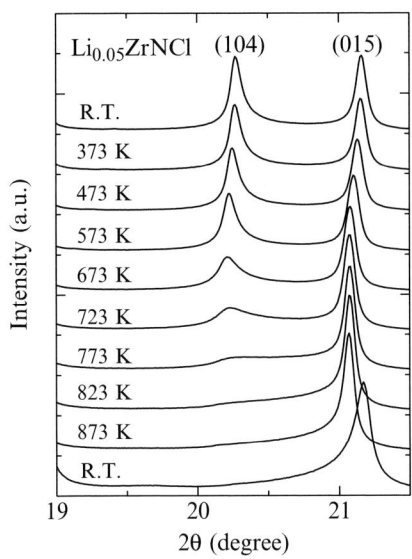

Fig. 7.1. X-ray diffraction patterns of Li$_{0.05}$ZrNCl sample at room temperature (before and after heating) and at elevated temperatures. In the pristine β-ZrNCl with SmSI-type structure, (104) peak is strong and (015) peak is almost invisible, and vice verse for Li-intercalated ZrNCl phase with YOF-type structure

with SmSI-type structure, the intensity of (104) peak is strong and (015) peak is almost invisible, and vice verse for Li-intercalated sample with YOF-type structure. Therefore, a diffraction pattern shown at the top of Fig. 7.1, which is the pattern of as-intercalated sample with $x = 0.05$, indicates that the sample contains two phases of pristine and intercalated phase. However, when we heat the sample, the (104) peak gradually loses the intensity and eventually disappears above $T = 823$ K. After the sample is cooled to room temperature, the (104) peak is still invisible (the pattern shown at the bottom of Fig. 7.1), and the whole sample is of the YOF-type structure. Therefore, we reached a conclusion that we can obtain single phase samples in the very lightly doped regime by annealing at 600°C after the intercalation process. For thus prepared samples, we collected powder X-ray diffraction patterns at BL02B2, SPring-8, and confirmed that all the samples are of single phase. Li concentration (x) in the products was determined within an accuracy of ±0.01 by inductively coupled plasma (ICP) spectroscopy. Thus determined x values are used throughout the paper. For resistivity measurement, we prepared c-axis oriented compressed pellets and adopted the conventional four-probe method with current flowing parallel to the conduction plane. Magnetization measurements were performed using a Quantum Design SQUID magnetometer. Highly oriented compressed pellets for low temperature Raman scattering measurements were generated and mounted in a MicrostatHe (Oxford instruments) with a quartz window. These procedures were done in an Ar-filled glove box to

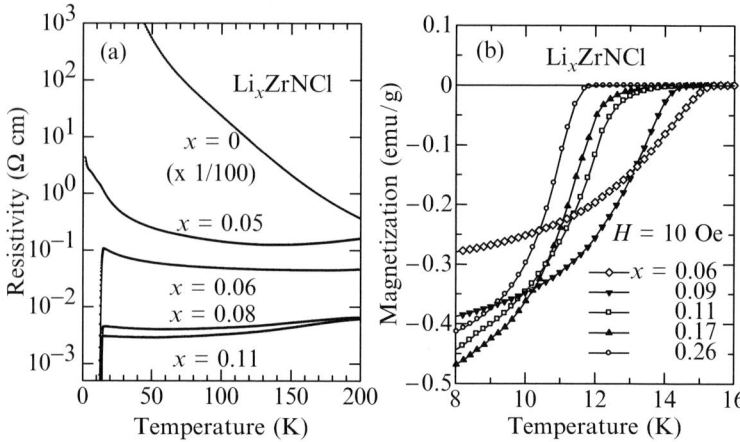

Fig. 7.2. (a) Temperature dependence of resistivity for pristine and lightly Li-doped ZrNCl. The resistivity of pristine sample is multiplied by 0.01. (b) Magnetization in a field of 10 Oe is plotted against temperature without correction of demagnetizing field for selected samples of $Li_x ZrNCl$

avoid sample degradation due to the oxygen and/or moisture. Raman spectra of the samples were recorded at several temperatures between 4 and 250 K, using a micro-Raman spectrometer with an excitation wavelength of 532 nm. We have measured the temperature (T) dependence of Raman spectra for six samples in total, whose $x(T_c)$ values are 0 (non-superconducting), 0.07 (15.2 K), 0.09 (14.2 K), 0.14 (12.1 K), 0.24 (11.7 K), and 0.31 (11.3 K).

In Fig. 7.2a, we show temperature dependence of the resistivity for un-doped ZrNCl and Li-intercalated samples [22]. Pristine sample exhibits insulating temperature dependence of resistivity below room temperature, and the resistivity below 70 K is in accord with activation type behavior with an activation energy of 21 meV, which is much smaller than optical gap that well exceeds 3 eV. Upon the Li-intercalation, the resistivity becomes systematically smaller. Although the temperature dependence above 140 K of the sample with $x = 0.05$ is metallic and the absolute value is much smaller than that of the undoped material, the resistivity shows an upturn at approximately 140 K and does not show superconducting transition. This material is an Anderson insulator due to the disorder arising from the random distribution of the intercalated Li ions. Further increase in doping level reduces the absolute value of resistivity, but even the sample with $x = 0.11$ shows an upturn at about 50 K. Such upturns that are observed robustly at low temperatures are also ascribed to the Anderson localization effect due to the disorder, which is probably reinforced by the nature of the present specimen, namely, a compressed pellet. Despite such upturn of the resistivity, the samples with x larger than 0.05 show superconducting transition at low temperatures. Therefore, insulator-to-superconductor transition in this system occurs at approximately $x = 0.05$.

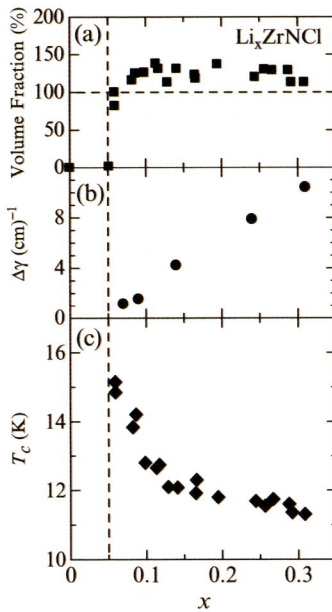

Fig. 7.3. Li concentration (x) dependence of (**a**) superconducting volume fraction at 2 or 5 K, (**b**) the difference of $\gamma_x - \gamma_{\mathrm{pris}}$ between low temperature and high temperature, and (**c**) superconducting transition temperature

In Fig. 7.2b, we show the temperature dependence of magnetization for several selected samples [22]. In this plot, correction for the demagnetizing field is not performed. In all the samples with $x \geq 0.06$, clear diamagnetic signal is observed associated with the superconducting transition. The most important observation in this figure is that the T_c increases as the doping level is reduced, and the most lightly doped sample with $x = 0.06$ exhibits the highest T_c($=15.2$ K) among others.

In Fig. 7.3a, we show the superconducting volume fraction as a function of Li concentration x. The volume fraction was estimated from the zero-field cooled *MH* measurement ($-20\,\mathrm{Oe} \leq H \leq 20\,\mathrm{Oe}$) at 2 or 5 K. Here, we performed correction for demagnetizing field, assuming that each powder sample consists of small plates with a completely random orientation, despite the fact that there is actually some preferred orientation. Because of this imperfect correction of demagnetizing field, the volume fraction takes the value exceeding 100%, but an important point to be noted is that the volume fraction is sufficiently high to ensure that the observed diamagnetic signal represents the bulk property. In the two samples that are located near the superconductor-to-insulator transition, the volume fraction is slightly smaller than the other samples, but is still high enough to rule out the possibility that superconducting signal comes from some minority phases.

Having confirmed that the samples are of high quality in terms of super-conducting volume fraction, we next discuss the doping dependence of T_c, which was determined by MT measurements shown in Fig. 7.2b and is plotted in Fig. 7.3c as a function of Li concentration x. For the doping region of $0.12 \leq x \leq 0.31$, T_c slightly increases as x is reduced in accordance with the previous result [23]. However, when x is further reduced below 0.12, the trend changes and T_c rapidly increases to reach 15.2 K at $x = 0.06$, and then suddenly disappears at $x = 0.05$, which is an Anderson insulator as discussed earlier. Such an increase in T_c upon approaching an Anderson-localization phase seems to be quite anomalous since T_c usually decreases as the Anderson-insulator phase is approached, as well established in a thin-film experiment [24]. There may be a possibility that charge ordering or Wigner crystal phase prevails in the insulating side of the phase diagram, although no evidence has been obtained thus far. According to the band calculations [3–7], the density of states at the Fermi level is almost constant, reflecting the two-dimensional nature of the present system, or rather decreases as the carrier density is reduced. Therefore, the doping dependence of T_c is opposite to that expected from the doping-variation of the density of the states. The only possibility that explains the observed x dependence of T_c within a simple framework that takes into account the phonon contribution alone is that the electron–phonon interaction strength is enhanced as the carrier concentration is reduced.

To obtain information about the doping dependence of electron–phonon interaction strength, we have performed T-dependent Raman scattering measurements and focused on the doping dependence of thermal broadening behavior of the phonon line width. In the inset of Fig. 7.4a, we show the Raman

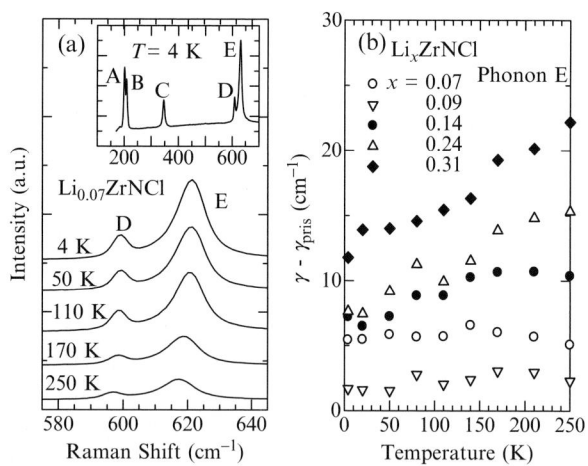

Fig. 7.4. (a) Raman scattering intensity of phonons D and E for $Li_{0.07}ZrNCl$ sample at several temperatures. The inset shows the whole spectrum of the $x = 0.07$ sample at $T = 4$ K. (b) Temperature dependence of the line width (γ), subtracted by that of pristine material (γ_{pris}), of the phonon E for several samples with different doping levels

scattering spectra of $x = 0.07$ sample at $T = 4$ K [25]. In this range of Raman shift, we observed five phonon lines (labeled as A–E, hereafter) in accord with previous results [26, 27]. Phonons B, C, and D are assigned to vibrations along c-axis, and phonons A and E are ascribed to those within the ab-plane according to the lattice dynamics calculation [27]. As the doping proceeds, all the phonon modes exhibit slight softening and broadening. It should be noted, however, that the observed changes are far less drastic than those in BaPb$_{1-x}$Bi$_x$O$_3$ [28] and MgB$_2$ [29, 30]. Another important point is that asymmetry of the line shape due to the Fano resonance is hardly observed even in the $x = 0.31$ sample with the highest doping in the present study. These observations would imply that the electron–phonon interaction is weak in this system.

In the main panel of Fig. 7.4a, we show the enlarged view of the phonon D and E of $x = 0.07$ sample at several temperatures. Clearly, both phonon lines exhibit broadening as well as softening as the temperature is increased. Actually, the phonons C and D show the strongest temperature dependence, but the detailed analysis indicated that the T-variation of the phonon lines of these two modes is mostly due to the phonon anharmonicity. Therefore, we focus on the phonon E, hereafter, which is predicted to interact most strongly with electronic system among all the phonon modes [7].

In Fig. 7.4b, $\gamma - \gamma_{\mathrm{pris}}$ is plotted against temperature for various samples. Here, γ and γ_{pris} are the line width of the phonon E of the doped and pristine samples, respectively. By subtracting γ_{pris}, the contribution from the phonon anharmonicity, or phonon–phonon scattering, is eliminated, and hence the plotted quantity represents the contributions from disorder scattering and electron–phonon scattering. The width due to the disorder scattering should be constant, while the broadening due to the electron–phonon scattering should depend on the temperature. Therefore, the T-dependent part of the $\gamma - \gamma_{\mathrm{pris}}(\equiv \Delta\gamma)$ represents the contribution from the electron–phonon scattering, and is plotted against doping in Fig. 7.3b. It is readily noticed that $\Delta\gamma$ increases almost linearly as a function of doping. Taking into account the relationship [31, 32] among the line-width, electron–phonon coupling constant, and the density of states at the Fermi level as well as the fact that the density of states at the Fermi level is filling-independent due to the two-dimensionality, the increase in $\Delta\gamma$ upon increasing doping is interpreted as enhancement of the electron–phonon coupling constant. This doping dependence of the electron–phonon interaction is opposite to that expected from the doping-variation of T_c if we assume a simple framework of the superconductivity that takes into account the phonon contribution alone to the pairing-interaction. Therefore, the present observation would be another supporting evidence for the relevance of other fluctuation than phonon to the pairing-interaction. In view of the enhanced T_c upon reducing carrier density, charge fluctuation would play important roles in this system.

In summary, we succeeded in synthesizing single phase samples of Li_xZrNCl with controlled doping levels ($0 \leq x \leq 0.31$) by adopting an appropriate annealing procedure for lightly doped samples. An insulator-to-superconductor transition was observed to occur at approximately $x = 0.05$. The superconducting transition temperature gradually increases upon reducing the carrier density, and increases rapidly below $x = 0.12$, and takes a maximum value of $15.2\,K$ at $x = 0.06$. Raman scattering experiments at low temperatures revealed that electron–phonon interaction strength rather decreases as the doping concentration is reduced. The doping dependence of T_c and electron–phonon coupling strength are not in accord with each other within a simple framework of superconductivity that takes into account the phonon contribution alone to the pairing-interaction. These observations may be indicative of the importance of charge fluctuation in this small carrier-density system.

Acknowledgements. We acknowledge S. Ishihra and H. Matsueda for useful discussion, and K. Kanoda for pointing out the possibility of Wigner crystal phase in the insulating side of the phase diagram. This work was in part supported by Grant-in-Aid for Scientific Research from MEXT, Japan, and by the Nano-Material Developing Project of IMR, Tohoku University. Magnetization measurements were in part carried out at the Center for Low-Temperature Science, Tohoku University. Chemical analysis of the sample was performed at analytical research core for advanced materials, IMR, Tohoku University. Synchrotron X-ray diffraction measurements were performed at BL02B2, SPring-8, with the approval of JASRI.

References

1. S. Yamanaka, H. Kawaji, K. Hotehama, M. Ohashi, Adv. Mater. **8**, 771 (1996)
2. S. Yamanaka, K. Hotehama, H. Kawaji, Nature (London) **392**, 580 (1998)
3. R. Weht, A. Filippetti, W.E. Pickett, Europhys. Lett. **48**, 320 (1999)
4. I. Hase, Y. Nishihara, Phys. Rev. B **60**, 1573 (1999)
5. C. Felser, R. Seshadri, J. Mater. Chem. **9**, 459 (1999)
6. H. Sugimoto, T. Oguchi, J. Phys. Soc. Jpn. **3**, 2771 (2004)
7. R. Heid, K.-P. Bohnen, Phys. Rev. B **72**, 134527 (2005)
8. H. Tou, Y. Maniwa, T. Koiwasaki, S. Yamanaka, Phys. Rev. B **63**, 020508(R)(2000)
9. T. Yokoya, Y. Ishiwata, S. Shin, S. Shamoto, K. Iizawa, T. Kajitani, I. Hase, T. Takahashi, Phys. Rev. B **64**, 153107 (2001)
10. H. Tou, Y. Maniwa, T. Koiwasaki, S. Yamanaka, Phys. Rev. Lett. **86**, 5775 (2001)
11. Y. Taguchi, M. Hisakabe, Y. Iwasa, Phys. Rev. Lett. **94**, 217002 (2005)
12. T. Ekino, T. Takasaki, T. Muranaka, H. Fujii, J. Akimitsu, S. Yamanaka, Phys. B **328**, 23 (2003)
13. Y.J. Uemura, Y. Fudamoto, I.M. Gat, M.I. Larkin, G.M. Luke, J. Merrin, K.M. Kojima, K. Itoh, S. Yamanaka, R.H. Heffner, D.E. MacLaughlin, Phys. B **289–290**, 389 (2000)

14. T. Ito, Y. Fudamoto, A. Fukaya, I.M. Gat-Malureanu, M.I. Larkin, P.L. Russo, A. Savici, Y.J. Uemura, K. Groves, R. Breslow, K. Hotehama, S. Yamanaka, P. Kyriakou, M. Rovers, G.M. Luke, K.M. Kojima, Phys. Rev. B **69**, 134522 (2004)
15. A. Bill, H. Morawitz, V.Z. Kresin, Phys Rev. B **66**, 100501(R) (2002)
16. A. Bill, H. Morawitz, V.Z. Kresin, Phys Rev. B **68**, 144519 (2003)
17. Y. Takada, Phys. Rev. B **47**, 5202 (1993)
18. H. Kohno, K. Miyake, H. Harima, Phys. B **312–313**, 148 (2002)
19. M. Ohashi, S. Yamanaka, M. Hattori, J. Solid State Chem. **77**, 342 (1988)
20. S. Shamoto, T. Kato, Y. Ono, Y. Miyazaki, K. Ohoyama, M. Ohashi, Y. Yamaguchi, T. Kajitani, Phys. C **306**, 7 (1998)
21. X. Chen, L. Zhu, S. Yamanaka, J. Solid State Chem. **169**, 149 (2002)
22. Y. Taguchi, A. Kitora, Y. Iwasa, Phys. Rev. Lett. **97**, 107001 (2006)
23. H. Kawaji, K. Hotehama, S. Yamanaka, Chem. Mater. **9**, 2127 (1997)
24. D.B. Haviland, Y. Liu, A.M. Goldman, Phys. Rev. Lett. **62**, 2180 (1989)
25. A. Kitora, Y. Taguchi, Y. Iwasa, J. Phys. Soc. Jpn. **76**, 023706 (2007)
26. P. Adelmann, B. Renker, H. Schober, M. Braden, F. Fernandez-Diaz, J. Low Temp. Phys. **117**, 449 (1999)
27. A. Cros, A. Cantarero, D. Beltran-Porter, J. Oro-Sole, A. Fuertes, Phys. Rev. B **67**, 104502 (2003)
28. S. Sugai, S. Uchida, K. Kitazawa, S. Tanaka, A. Katsui, Phys. Rev. Lett. **55**, 426 (1985)
29. K.-P. Bohnen, R. Heid, B. Renker, Phys. Rev. Lett. **86**, 5771 (2001)
30. T. Masui, S. Lee, S. Tajima, Phys. Rev. B **70**, 024504 (2004)
31. P.B. Allen, Phys. Rev. B **6**, 2577 (1972)
32. E. Cappelluti, Phys. Rev. B **73**, 140505(R) (2006)

8

Ambipolar Tetraphenylpyrene (TPPy) Single-Crystal Field-Effect Transistor with Symmetric and Asymmetric Electrodes

S.Z. Bisri, T. Takahashi, T. Takenobu, M. Yahiro, C. Adachi, and Y. Iwasa

Summary: An ambipolar field-effect transistor (FET) based on a 1,3,6, 8-tetraphenylpyrene (TPPy) single-crystal, a high photoluminescent material, has been successfully fabricated using symmetric and asymmetric electrodes. Several kinds of metal electrodes have been employed to investigate the charge injection characteristics in the single-crystal FET. Hole and electron mobilities of 0.34 and $7.7 \times 10^{-2}\,\mathrm{cm^2\,V^{-1}\,s^{-1}}$ were achieved by using Au and Ca electrodes, respectively. The ambipolar characteristic of this device gives a prospect for further development in light-emitting FET operation.

Research in organic semiconductors has become a highlight in materials science and technology, since they offer chances for low cost, environmental-friendly, and wide-range electronic applications. On the other hand, organic material research has provided opportunities to explore new device physics, such as ambipolar transport in organic semiconductors [1]. Ambipolar organic field-effect transistor (FET) has become a great interest in recent years due to its uniqueness and prospect. Both electrons and holes are equally mobile inside an ambipolar organic transistor, which leads to an opportunity for novel device applications, such as light-emitting field-effect transistors (LEFETs) [2–4] as well as photo-voltaic devices [5].

The progress in LEFETs has provided possibilities in developing organic lasers based on FET structures [1, 6]. The existing ambipolar LEFET devices are based on polymer thin film [2, 3, 7], carbon nanotubes [8], organic thin-film multilayer structures [9, 10], and small molecules thin film [4, 11–14]. Previously, we have succeeded in fabricating an ambipolar rubrene single-crystal FET [15], followed by an ambipolar LEFET based on tetracene single crystal, in order to utilize the structural ordering of single crystals [16]. However, the light emission efficiency was still remaining low. This might be due to the relatively low photoluminescence efficiency of tetracene single crystal, the active material itself. In contrary, the polymer thin film based LEFET [2, 3, 7] has relatively high photoluminescence efficiency, but their charge carrier mobilities are extremely low, less than $10^{-3}\,\mathrm{cm^2\,V^{-1}\,s^{-1}}$. For further improvements, FET device based on highly photoluminescent materials with high carrier mobilities are crucially demanded.

1,3,6,8-Tetraphenylpyprene (TPPy) is one of the promising materials for light-emitting FETs, due to its high photoluminescence efficiency of 68% [17]. TPPy itself has been used as an active material of organic light-emitting diodes (OLEDs) [18] and thin-film LEFETs [19], which are doped with rubrene molecules, and the resulting external electroluminescence efficiency was 0.5% with relatively low carrier mobility, $10^{-5}\,cm^2\,V^{-1}\,s^{-1}$. Therefore, in this letter, we report fabrication of TPPy single-crystal FETs for the purpose of attaining higher carrier mobilities of this high photoluminescent material as well as acquainting its intrinsic nature. We successfully observed an ambipolar characteristic in the TPPy single-crystal device with high carrier mobilities. We tried several kinds of metal electrodes, which involve Au, Ca, Ag-paste, and Mg, in the device fabrications. However, in this report, we focus on Au and Ca electrodes, which exhibited the highest hole and electron mobilities.

The device was fabricated as a bottom-gated FET, with a top-contact configuration. A SiO_2 layer of 400 nm in thickness, which was thermally oxidized on an n-type highly doped silicon wafer, acts as a gate dielectric. To diminish the electron traps at the interface between the SiO_2 and the single crystal [20], a 4 nm thick polymethylmetacrylate (PMMA) buffer layer was spin-coated on the SiO_2 surface from toluene solution. Subsequently, substrate was heated overnight on a hot plate at 70°C and annealed at 100°C for 3 h in nitrogen atmosphere.

Single crystal of TPPy was prepared by using a physical vapor transport method with TPPy powder [17] as a starting material. The heating temperature of the starting material was set in between 305 and 310°C. Ar gas stream of $25\,cm^3\,s^{-1}$ was used as a transporting gas. A vast amount of single-crystals were obtained at the growth zone, where most of large single crystals were transparent and in the form of platelets, with a thickness of 750 nm – measured by a profilometer. After the single-crystal growth was finished, the glass-tube was sealed at both ends, with the transporting gas filled inside. Then, the sealed glass-tube was transported into a N_2-filled glove box ($< 1\,ppm\,O_2$, $< 1\,ppm\,H_2O$), where the single-crystal devices were fabricated. Therefore, single crystals were not exposed to air along the crystal growth, device fabrication, and measurement processes. A TPPy single crystal was placed on the SiO_2 substrate following the report by Takeya et al. [21].

Au and Ca electrodes, which we focus on in this letter, were fabricated by a vacuum evaporation technique, at a deposition pressure of less than $10^{-3}\,Pa$. The evaporation chamber was directly connected to the N_2-filled glove box. Metals were deposited onto the single-crystal samples through a patterned shadow mask. Since Ca metal is easily oxidized, which may reduce the device durability, thick electrodes were made with a very high deposition rate. Even though all of the processes are performed in an air-free condition, without any oxygen exposure, Ca electrode on the device might remain vulnerable due to some oxidation during the evaporation process – from the degassed oxygen of the metal source – or even from the small fraction of oxygen ($< 1\,\text{ppm}$) inside the glove box. The deposition rate was kept at 25–50 Å s^{-1} and the obtained thickness was 300 nm. By adopting such higher deposition rates and thicker electrodes, device reproducibility and durability was notably improved. In making FETs with Au electrodes, on the other hand, evaporation process was done in a relatively moderate deposition rate and thickness, which are 0.3 Å s^{-1} and 30 nm, respectively.

The three-terminal measurements in order to examine the FET transfer characteristics of the devices were carried out in a dark condition inside the N$_2$-filled glove box, using an Agilent E5207 semiconductor parameter analyzer. Therefore, we expected that there is almost no degradation of the devices due to interaction with the surroundings, within a measurement period. The FET measurements were done in both p-channel and n-channel operations with drain and gate biases up to 200 V.

Figure 8.1a shows a schematic diagram of the FET device and its optical micrograph picture. Figure 8.1b shows an energy diagram of the HOMO and LUMO levels of TPPy as well as the work function of metal electrodes that were used in the device fabrication. The HOMO and LUMO of TPPy are 5.7 and 2.7 eV, respectively, as determined by UV-photoelectron spectroscopy [17]. Au was used as hole injecting electrodes because its work function is in the vicinity of the TPPy HOMO level, whereas Ca was appropriate for the electron injection into LUMO level.

In the case of Au electrodes, the TPPy single-crystal FET exhibited an ambipolar characteristic with dominating hole injection, as shown in Fig. 8.2a. The hole mobility, extracted from the saturation region, was as high as 0.27 cm^2 V^{-1} s^{-1} in the device, with channel length and width of 60 and

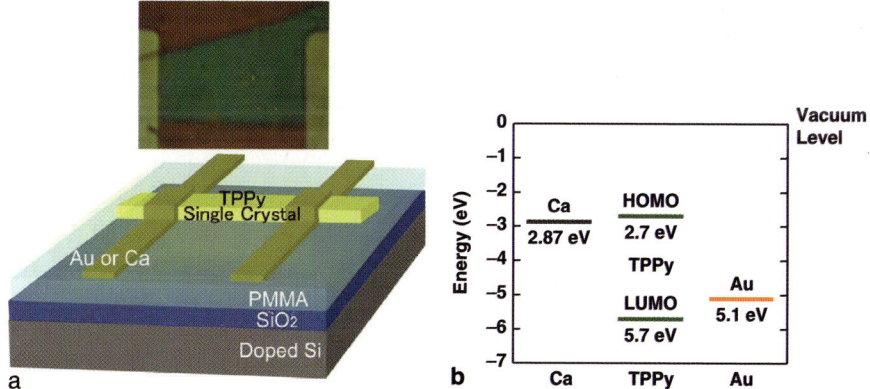

Fig. 8.1. (a) Optical microscope image and schematic illustration of a TPPy single crystal FET with symmetric electrodes. (b) Energy diagram of TPPy and the utilized metal electrodes

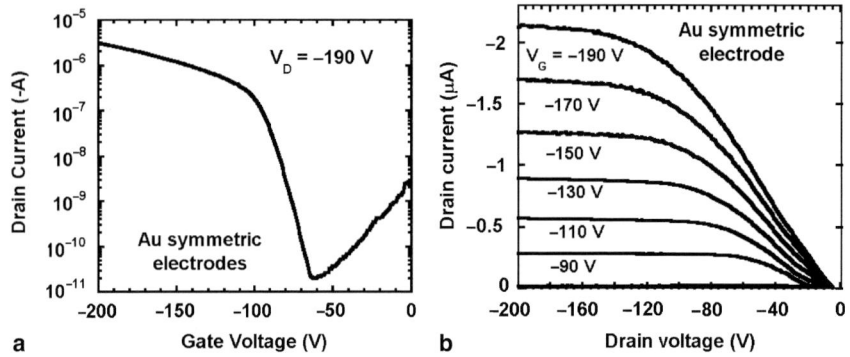

Fig. 8.2. FET characteristics of a TPPy single-crystal FET device with Au symmetric electrodes. (**a**) Transfer curves showing an ambipolar characteristic with dominant hole injection. (**b**) Output characteristics for various gate voltages

110 µm, respectively. The obtained hole mobility is much higher than that in TPPy thin film FET, in the order of 10^{-5} cm^2 V^{-1} s^{-1} [19]. Meanwhile, the electron mobility obtained was as 2.9×10^{-3} cm^2 V^{-1} s^{-1}, despite a large energy-band mismatch between Au work function and TPPy LUMO that reach 2.4 eV. From the view point of thermal injection, this electron injection should be impossible. This phenomenon itself is not well-explained yet; therefore, we propose that the electron injection is possibly assisted by in-gap states at the Au/TPPy interfaces, caused by defects that were formed due to thermal evaporation of Au on top of organic single-crystal [22]. However, the nature and the energy distribution of the in-gap states are quantitatively unknown. In addition to what has been obtained by symmetric Au electrodes, the highest hole mobility of 0.34 cm^2 V^{-1} s^{-1} was observed in the asymmetric electrodes configuration, where Au and Mg were used at the opposite sides of the single crystal, with Au as hole injection electrode.

In the devices with Ca electrodes, on the other hand, electron injection was dominantly observed, with also an ambipolar characteristic, as shown in Fig. 8.3a. Electron mobility of 7.7×10^{-2} cm^2 V^{-1} s^{-1} and hole mobility of 9.4×10^{-3} cm^2 V^{-1} s^{-1} were achieved from the device, with channel length and width of 600 and 240 µm, respectively. The electron injection is enhanced by a small band-mismatch between the Ca electrode and TPPy LUMO level, whereas the hole injection is suppressed due to a large band-mismatch of the TPPy HOMO level. This electron transport is also supported by reduced number of electron traps at the interface of single-crystal and gate dielectric, due to the insertion of PMMA buffer layer, which was 4 nm thick. The subthreshold swing of this device was found to be as low as 6 V/decade for electron injection. By using a well-known relation between S and shallow-trap density, the electron-trap density was obtained as 3.1×10^{13} cm^2 eV. This value is only one or two order higher than the hole-trap density in pentacene and rubrene

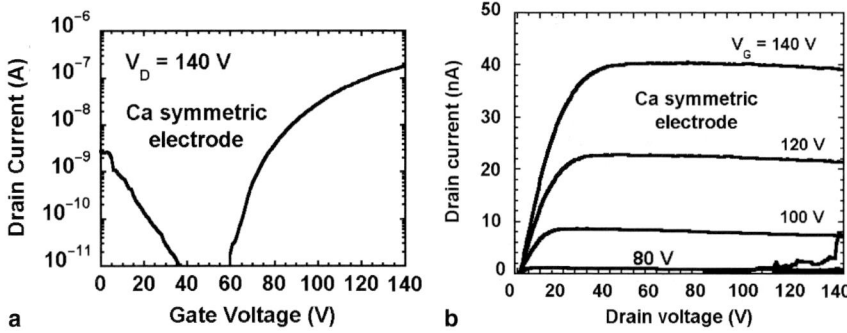

Fig. 8.3. FET characteristics of a TPPy single-crystal FET device with Ca symmetric electrodes. (**a**) Transfer curves showing an ambipolar characteristic with dominant electron injection, and (**b**) output characteristics, which shows ohmic-like behavior at the origin, compared with its counterpart (Fig. 8.2(**b**))

FET, respectively [21]. However, the obtained value of electron-trap density might be responsible for the large threshold-voltage for electron injection in this device. By also concerning the other successful ambipolar single-crystal FETs [15, 16], it might be perceived that the insertion of PMMA buffer layer and air-free device fabrication process are crucial factors for realizing ambipolar characteristics.

From comparison of the $I_D - V_D$ characteristics, as seen in Figs. 8.2b and 8.3b, one realizes that the linearity in the low V_D region for electron injection by using Ca electrodes is much better than that for hole injection from Au electrodes. This suggests that the Schottky barrier for hole injection from Au electrodes is much larger. Relatively large band-mismatch between the work function of Au, 5.1 eV, and TPPy HOMO level, 5.7 eV, might be responsible for the larger Schottky barrier for holes. The electron injection band-mismatch between the work function of Ca, 2.87 eV, and TPPy LUMO level, 2.7 eV, is smaller compared with the hole band-mismatch.

The above results show that there is only a weak Fermi level pinning effect in the TPPy single crystal. Therefore, the Schottky barrier height in this single crystal is mostly determined by the metal electrode work function. Both holes and electrons are nicely injected into the single-crystal by using different kinds of electrode, work functions of which are not so different from either LUMO level or HOMO level of TPPy. However, for ambipolar LEFET operation, both electron and hole should be simultaneously injected into the single crystal from different electrodes. Therefore, FET devices with asymmetric electrodes are still to be investigated. In comparison with a tetracene single-crystal FET [16], which has already been successful as a LEFET, the electron and hole mobility of TPPy is one order smaller. However, the photoluminescence efficiency of TPPy single crystal is about two orders of magnitude higher than that of tetracene. On considering polymer thin-film based LEFET that has comparable photoluminescence efficiency [3], this TPPy single-crystal FET

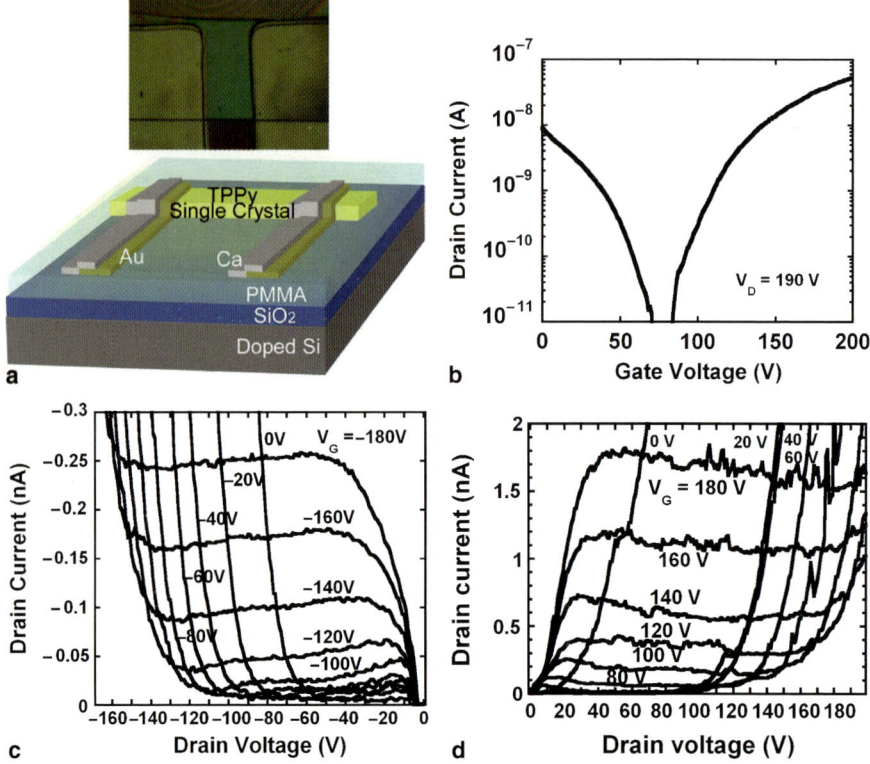

Fig. 8.4. (a) Optical microscope image and schematic illustration of a TPPy single crystal FET with asymmetric electrodes. FET characteristics of a TPPy single-crystal FET device with Au/Ca asymmetric electrodes, (**b**) transfer curves showing an ambipolar characteristic, (**c**) output characteristics for hole channel, and (**d**) output characteristics for electron channel

possessed more than one order higher charge carrier mobilities. With nearly equal electron and hole mobility as well as good charge carrier injection, an ambipolar LEFET operation of this device is highly expected by using a Au electrode in one side for hole injection and a Ca electrode in the other side for electron injection. Actually, we have preliminary success in the fabrication of TPPy single-crystal FETs with Au/Ca asymmetric electrodes (Fig. 8.4), although the obtained mobilities were much smaller than the maximum values. Improvement in fabrication process is now in progress. In addition, some other efforts are still being made in order to increase charge carrier densities for bright light emission intention, especially by lowering threshold voltage through utilization of different organic buffer layer, as well as by making a shorter channel transistor.

In conclusion, we have successfully fabricated an ambipolar FET of TPPy single crystal by using symmetric and asymmetric electrodes. The utilization of various electrodes allowed estimating the highest hole and electron field-induced mobilities of this single-crystal. In addition to the good injection of holes and electrons in this ambipolar FET, the high photoluminescence efficiency of the single-crystal may lead to a good prospect for ambipolar LEFET operation.

Acknowledgements. This study has been partially supported by a Grant-in-Aid from the Ministry of Education, Culture, Sport, Science, and Technology of Japan (17069003, 17204022, and 18710091).

References

1. M. Muccini, Nat. Mater. **5**, 605 (2006)
2. J. Zaumseil, R.H. Friend, H. Sirringhaus, Nat. Mater. **5**, 69 (2006)
3. J. Zaumseil, C.L. Donley, J.S. Kim, R.H. Friend, H. Sirringhaus, Adv. Mater. **18**, 2708 (2006)
4. C. Rost, S. Karg, W. Riess, M.A. Loi, M. Murgia, M. Muccini, Appl. Phys. Lett. **85**, 1613 (2004)
5. S. Cho, J. Yuen, J.Y. Kim, K. Lee, A.J. Heeger, Appl. Phys. Lett. **90**, 063511 (2007)
6. M.A. Baldo, R.J. Holmes, S.R. Forrest, Phys. Rev. B **66**, 035321 (2002)
7. J.S. Swensen, C. Soci, A.J. Heeger, Appl. Phys. Lett. **87**, 253511 (2005)
8. J.A. Misewich, R. Martel, P. Avouris, J.C. Tsang, S. Heinze, J. Tersoff, Science **300**, 783 (2003)
9. S. de Vusser, S. Schols, S. Steudel, S. Verlaak, J. Genoe, W.D. Oosterbaan, L. Lutsen, D. Vanderzande, P. Heremans, Appl. Phys. Lett. **89**, 223504 (2006)
10. F. Dinelli, R. Capelli, M.A. Loi, M. Murgia, M. Muccini, A. Facchetti, T.J. Marks, Adv. Mater. **18**, 1416 (2006)
11. C. Santato, R. Capelli, M.A. Loi, M. Murgia, F. Cicoira, V.A.L. Roy, P. Stallinga, R. Zamboni, C. Rost, S.F. Karg, M. Muccini, Synth. Met. **146**, 329 (2004)
12. J. Reynaert, D. Cheyns, D. Janssen, R. Müller, V.I. Arkhipov, J. Genoe, G. Borghs, P. Heremans, J. Appl. Phys. **97**, 114501 (2005)
13. A. Hepp, H. Heil, W. Weise, M. Ahles, R. Schmechel, H. von Seggern, Phys. Rev. Lett. **91**, 157406 (2003)
14. E.C.P. Smits, S. Setayesh, T.D. Anthopoulos, M. Buechel, W. Nijssen, R. Coehoorn, P.W.M. Blom, B. de Boer, D.M. de Leeuw, Adv. Mater. **19**, 734 (2007)
15. T. Takahashi, T. Takenobu, J. Takeya, Y. Iwasa, Appl. Phys. Lett. **88**, 033505 (2006)
16. T. Takahashi, T. Takenobu, J. Takeya, Y. Iwasa, Adv. Funct. Mater. (in press)
17. T. Oyamada, S. Akiyama, M. Yahiro, M. Saigou, M. Shiro, H. Sasabe, C. Adachi, Chem. Phys. Lett. **421**, 295 (2006)
18. T. Oyamada, H. Sasabe, Y. Oku, N. Shimoji, C. Adachi, Appl. Phys. Lett. **88**, 093514 (2006)
19. T. Oyamada, H. Uchiuzou, S. Akiyama, Y. Oku, N. Shimoji, K. Matsushige, H. Sasabe, C. Adachi, J. Appl. Phys. **98**, 074516 (2005)

20. L.L. Chua, J. Zaumseil, J.F. Chang, E.C.W. Ou, P.K.H. Ho, H. Sirringhaus, R.H. Friend, Nature **434**, 194 (2005)
21. J. Takeya, T. Nishikawa, T. Takenobu, S. Kobayashi, Y. Iwasa, T. Mitani, C. Goldmann, C. Krellner, B. Batlogg, Appl. Phys. Lett. **85**, 5078 (2004)
22. T. Takenobu, T. Takahashi, J. Takeya, Y. Iwasa, Appl. Phys. Lett. **90**, 013507 (2007)

9

Bulk Zinc Oxide and Gallium Nitride Crystals by Solvothermal Techniques

D. Ehrentraut and T. Fukuda

Summary: We report on recent achievements from the growth of hydrothermal zinc oxide (ZnO) and ammonothermal gallium nitride (GaN). A thin-film deposition technique under conditions near the thermodynamic equilibrium, liquid phase epitaxy (LPE) is applied for fast screening of dopants and their effects on physical properties of ZnO. In particular, super fast luminescent decay will be reported from some donor–acceptor co-doped ZnO films prepared by LPE.

9.1 Introduction

Based on the unique combination of electrical and optical properties, the wide-band-gap semiconductors zinc oxide (ZnO, $E_g = 3.37$ eV at 300 K) and gallium nitride (GaN, $E_g = 3.42$ eV at 300 K) are going to play a crucial role in modern optoelectronics, high-frequency high-power electronics, fast-speed communication, etc., due to high electron mobility ($\mu = 250$–280 and $5,300$ cm^{-2} V s for ZnO and GaN, respectively), saturation velocity (for GaN $v \leq 19 \times 10^6$ cm s^{-1}), and breakdown voltage ($V_{BR} = 35$ and 26–33×10^6 cm s^{-1} for ZnO and GaN, respectively) [1]. Device application in harsh environment is substantial as no thermodynamic phase transition occurs in ZnO and GaN upon decomposition. Both compounds are characterized by high radiation hardness.

Because of thermodynamic limitation, for example, GaN decomposes at 1,150 K under atmospheric pressure [2], the III-nitrides cannot be grown from the stoichiometric melt without employment of extreme pressure and temperature of > 6 GPa at $2,220°$C [3]. Therefore, only vapor phase and solution techniques are applicable to grow the III-nitrides in an economic way. The development of lattice-matched substrates for III-nitride (AlN, GaN, InN, and their alloys) device technology is currently being made by hydride vapor phase epitaxy (HVPE), Na-flux growth, and high-pressure solution growth technique in the case of GaN. More recently, the ammonothermal growth of GaN is considered as a potentially very powerful route of mass-production [4].

Like GaN, free-standing $\geq$ 2-in. size ZnO substrates with low defect and impurity concentration are still available. Besides the growth from the vapor phase and from the melt [5], the hydrothermal growth of ZnO is now well established and already 3 and 4-in. size crystal has been shown [6, 7].

Both, the ammonothermal and hydrothermal growth techniques are so-called solvothermal techniques, which encompass a separate class of technologies capable to fabricate single crystals under elevated pressure and temperature regimes. Basic principle is the use of a liquid polar solvent (H_2O in hydrothermal and ammonia, NH_3, in ammonothermal growth), which forms metastable products with the solute (nutrient). Mineralizers are essential to amplify the solubility of the nutrient. If a closed system is utilized, where the exchange of matter with ambient is impossible, the solvent often takes over a supercritical state, which also improves the solubility of the nutrient.

Merits of the solvothermal growth technology comprise the operation near the thermodynamic equilibrium, with ability to generate a high crystallinity; control of large crystal quantities over long process time, thus enabling a high throughput; no need for vacuum technology; environmentally benign conditions for production and capability for recycling of the solution.

The hydrothermal growth of the low temperature modification of quartz (SiO_2) is now already over 60 years in use for mass-production. ZnO in fact is the first semiconducting crystal, grown on an industrial scale by a solvothermal route for the purpose of wafer production. Figure 9.1 gives an overview over the present state-of-the-art in the solvothermal growth of hydrothermal SiO_2, hydrothermal ZnO, and ammonothermal GaN with respect to the estimated time spent for research (dashed line). Diameter (inch) and thickness (mm) for a (0001) GaN crystal are calculated values (open squares).

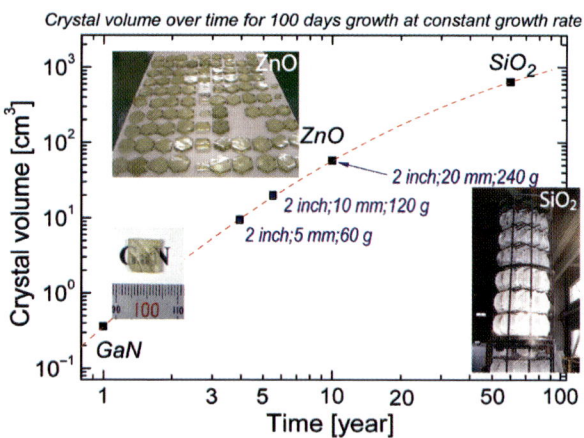

Fig. 9.1. Comparison of crystal volume of hydrothermal SiO_2, hydrothermal ZnO, and ammonothermal GaN with respect to the estimated time spent for research (*dashed line*). Diameter (inch) and thickness (mm) for a (0001) GaN crystal are calculated values

In this paper, efforts in the ammonothermal growth of hexagonal GaN are analyzed and compared with the developments of the hydrothermal growth of ZnO. To shorten time for development, the doping of ZnO under near-thermodynamic equilibrium conditions is being screened by our recently demonstrated liquid phase epitaxy (LPE) [8].

9.2 Hydrothermal Growth of ZnO

The hydrothermal growth of ZnO was much inspired by the successful hydrothermal growth of the low-temperature phase of quartz (α-SiO_2). Basic to the technology is using an autoclave (lined with Pt) as shown in Fig. 9.2, which contains precursor (pressed and sintered ZnO powder) and mineralizer (LiOH and KOH) in the supercritical water. The seed crystals serve for controlled nucleation and successive growth of a new ZnO crystal. A proper temperature gradient between zones for dissolution of the precursor (lower part in Fig. 9.2) and growth (upper part with seed crystals in Fig. 9.2) is arranged by a certain heater arrangement [6].

The development of the hydrothermal growth technology for ZnO again has seen tremendous improvements over the past few years. In Fig. 9.3 are shown ZnO crystals grown in the early 2000s in our group from autoclaves with 20 mm inner diameter (ID). The length of the crystals was typically about 10 mm along the c-axis. Since industrialization of the growth technique went parallel and in collaboration with our work, soon 3 in. ZnO crystal and wafer

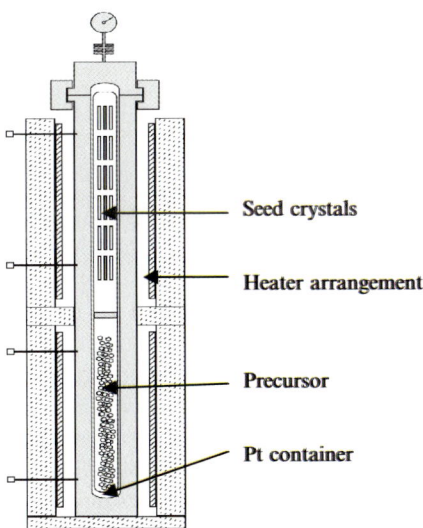

Seed crystals

Heater arrangement

Precursor

Pt container

Fig. 9.2. Schematic of an autoclave for the growth of SiO_2, ZnO, and future GaN crystals

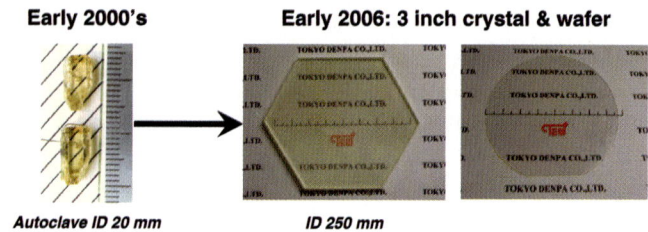

Fig. 9.3. ZnO crystals of 1 cm size along $\langle 0001 \rangle$ direction from the early 2000s (*left*) and a 3-in. size ZnO crystal and wafer from early 2006 (*right*)

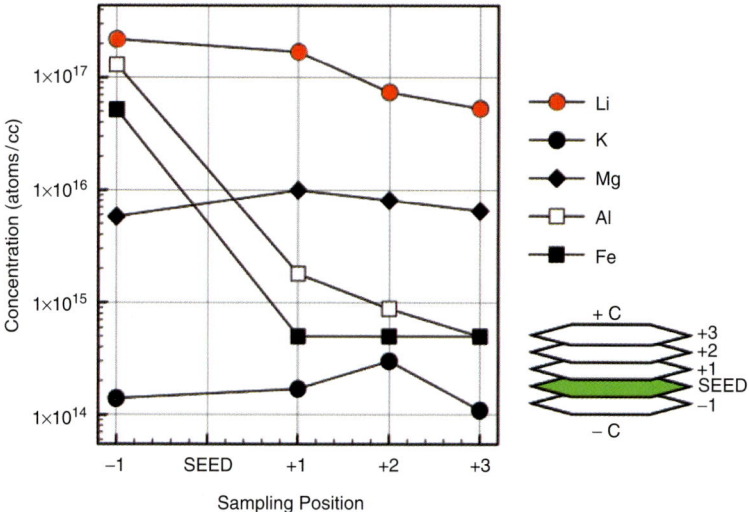

Fig. 9.4. SIMS profile of some hydrothermal ZnO wafers cut from the c+ and c− face of the same crystal

has been demonstrated (Fig. 9.3). The later is the result from developments at Tokyo Denpa, who use large size autoclaves with >200 mm ID and several meters of inner length [9].

Since basic mineralizers are most effective to use, the Zn-species in the solution, $Zn(OH)_4^{2-}$, and after decomposition on the crystal surface, ZnO_2^{2-}, are negatively charged, which yields the higher growth rate for the (0001) face [6]. Latter is Zn terminated, and by contrast the (000$\bar{1}$) face is O terminated. Not only the growth rate, but also the crystal quality is superior for growth on the (0001) face. This gets clearer from Fig. 9.4, where the concentration of Li, K, Mg, Al, and Fe is shown for some wafers cut from a single ZnO crystal. Secondary-ion mass spectroscopy (SIMS; Cs^+ at 5 kV, 350 nA; sputter speed around 100 nm min^{-1}) was applied. Particularly striking is the effect of Al and Fe, which both can be found in wafers cut from (000$\bar{1}$) in concentrations about two orders of magnitude higher than in (0001) wafers. Magnesium is basically not a threat if replacing for Zn in the lattice. In this case, stable $Mg_xZn_{1-x}O$ solid solution is formed causing a slight UV-shift in the luminescence. This has been demonstrated for solution-grown LPE film of $Mg_xZn_{1-x}O$, with maximum $x = 0.06$ [10].

Lithium can be removed upon proper annealing regime at about $1,100°C$; thus lowering the concentration down to about $10^{16}\,cm^{-3}$ [11]. Likewise, the Al content can be decreased about 2–3 times to levels around $3 \times 10^{15}\,cm^{-3}$ [6]. However, further improvement is necessary to reduce impurity concentration in the hydrothermal ZnO wafers, in particular, for the surface-near region which is in direct contact with the device structure deposited above. A route could be to get the impurities in the center of the wafer or to deposit a thin and highly pure ZnO film atop the wafer to produce epi-ready ZnO wafer.

Positron annihilation spectroscopy on hydrothermal ZnO from Tokyo Denpa has revealed about $10^{16}\,cm^{-3}$ zinc vacancies (V_{Zn}) as neutral defect complexes and about $10^{17}\,cm^{-3}$ oxygen vacancies (V_O) as neutral oxygen [12]. Clear effects of V_{Zn} and V_O and Li-related defects can be seen in low temperature (10 K) photoluminescence (PL) spectrum obtained from excitation with a 325 nm laser ($P_{out} = 1.6\,mW$). Figure 9.3 shows PL from a bulk sample, with the surface prepared by chemical–mechanical polishing (CMP). A broad band peaking around 2.3 eV appears in the visible region and the bands at 2.53, 2.35, and 2.17 eV have been assigned to V_{Zn} and V_{Zn} and Li-related defects, respectively [13]. More dominant is the strong emission around the band edge, with pronounced peaks around 3.233, 3.307, 3.356, and 3.37 eV, which are assigned due to donor–acceptor pair (DAP) transition, two-electron satellite (TES) transitions, neutral acceptor bound exciton recombination, and the free A-exciton recombination, respectively [6].

Liquid phase epitaxy (LPE) of ZnO has been developed in our group around 2005, and is now employed as a tool for fast screening of doping of ZnO under thermodynamic equilibrium conditions. ZnO films are grown by LPE at a constant temperature of 640°C under air atmosphere and pressure. The growth solution contains LiCl as a solvent and $ZnCl_2$ as a source for zinc. K_2CO_3 was chosen as the oxygen source activated through thermal decomposition into metastable K_2O and CO_2, the latter simply escaping as gas. Best film qualities have been fabricated from the ZnO concentration of 13 mmol per mol LiCl. Details on film preparation and doping have been given elsewhere [6, 8]. Doped and undoped ZnO films in the range of 0.5–5 µm thickness have been grown on ZnO substrates prepared from hydrothermally grown crystals. Figure 9.5 shows a Li, Sb co-doped (0001) ZnO film. The structural quality is very high as can be seen from the parallel aligned steps on the surface. The X-ray crystallinity remains unchanged from the substrate crystal, that is, a FHWM from (0002) reflection is typically of the order 25–40 arcsec. A closer look at the surface morphology of undoped LPE films has been done previously and atomic steps have been revealed [8]. However, the nature of the dopant strongly affects the morphology.

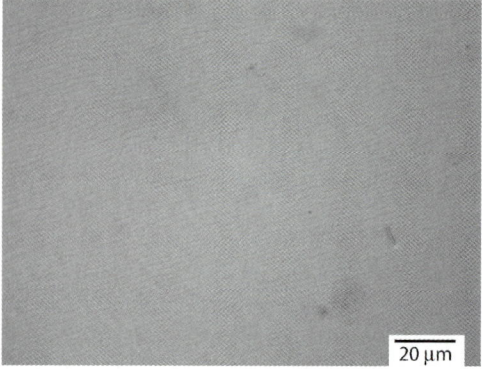

Fig. 9.5. Nomarski interference micrograph from a Li, Sb co-doped ZnO film grown by LPE. A very regular step pattern is on display

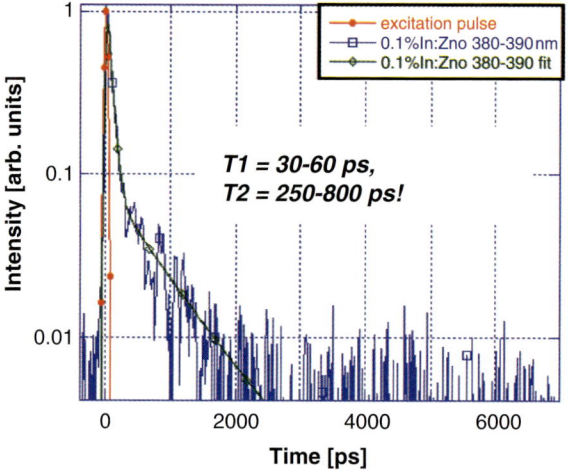

Fig. 9.6. Room-temperature photoluminescence decay from the 0.1 mol% In-doped ZnO film by LPE

Detailed measurements of PL at low temperatures have shown interesting effects for Li, Sb and Li, In co-doped samples. Broad emission in the 400–420 nm spectral region was obtained in the Li^+, In^{3+}-doped films at RT. The latter was ascribed to DAP recombination enabled by high content of Li^+ ions in the films [14]. The RT PL decay of 0.1 mol% In-doped ZnO film upon femtosecond laser excitation is shown in Fig. 9.6. The wavelength region of 380–390 nm was extracted. The decay follows a two-exponential course, which is found, with varying mutual intensity of two involved decay components, at all the LPE films studied. The faster component shows typically 30–60 ps decay time, while that of the slower component ranged from 250 to

800 ps. Intensity of the latter component was increasing towards longer emission wavelengths [14]. ZnO seems promising to obtain super fast scintillation time response, which can be applied for devices where high 2D resolution is required.

9.3 Ammonothermal Growth of Gallium Nitride

The synthesis of small-size GaN crystals from a solution of supercritical NH_3 and basic mineralizers like lithium and potassium amides at $\geq 550°C$ temperature and 400–500 MPa pressure was published in 1997 by Dwiliński et al. [15]. Recently, the uniform growth of GaN on an over-1-in. HVPE seed was reported by Nakamura's group [16]. Applied temperature and pressure was 625–675°C and about 214 MPa, respectively. More recently, Wang et al. [17] have demonstrated single crystals of $\leq 10 \times 10 \times 1\,mm^3$ in size grown under similar growth conditions of 475–625°C and 100–300 MPa. A growth rate of up to $2\,\mu m\,h^{-1}$ was achieved. In both cases, basic mineralizers ($NaNH_2$ + NaI and KNH_2, respectively) were employed in a Ni-based superalloy autoclave. When using basic mineralizers, the Ni contributes to impurities at the interface of the grown GaN crystal to the seed [18]. Also, Ni seems to act as catalyst for GaN [19] and therefore direct contact with the solution should better be avoided.

The use of basic mineralizers in supercritical ammonia causes a retrograde solubility of GaN [17, 20]. This means that a lightly increased temperature in the GaN growth zone is required to initiate nucleation on the GaN seed crystal. What more, growth temperature and pressure is very high, which is a real challenge for the material of the autoclave. Knowingly, the choice of alloys suited as autoclave material under such harsh conditions is limited. The general trend in the growth technology from autoclaves is to reduce temperature and pressure of a process.

We use the acidic mineralizer NH_4Cl to enhance the solubility of GaN precursor (small HVPE-GaN crystals or pressed GaN powder) [21]. A Pt inner liner is necessary to prevent the autoclave from corrosion. In Fig. 9.1 is shown an about $100\,\mu m$ thick, virtually colorless GaN crystal grown on a HVPE seed under the conditions of 550°C and ≤ 150 MPa from an autoclave of 16 mm ID. A relatively high growth rate of about $1\,\mu m\,h^{-1}$ was achieved over a growth period of 10 days [21]. In the same paper we reported the tuning of the average growth rate by modifying the precursor. A mixed Ga/GaN precursor yielded abovementioned growth rate.

Our recent research aimed at optimizing the mineralizer to achieve a high yield of hexagonal GaN in order to vary the speed of the entire process from dissolving the GaN precursor until the re-crystallization on the GaN seed [22]. The employment of the mixed mineralizers like $\geq 80\,mol\%$ $NH_4Cl/\leq 20\,mol\%$ NH_4Br and $\geq 80\,mol\%$ $NH_4Cl/\leq 20\,mol\%$ NH_4I has been proven successful.

Knowledge of the solubility behavior of the solute is a key figure in any solvothermal technology. Using a high-pressure cell experimental setup, we have recently reported the solubility of GaN under acidic conditions for different molar ratios of NH_4Cl to NH_3 [4]. The effect of mineralizer on the solubility was clearly obtained. Increasing the concentration of mineralizer in the solvent NH_3 improves the solubility of GaN. Data fitting by using a simple first exponential function (Arrhenius plot: Log solubility of GaN over reciprocal temperature) shows a strictly linear behavior of the solubility with temperature. This indicates that only one chemical process is dominating the dissolution of GaN under given conditions, which is a basic requirement to control the growth from solution. The energy of formation is $15.9 \, kcal \, mol^{-1}$ for the temperature range 673–820 K.

Using NH_4Cl as mineralizer and a one-in large HVPE seed yielded about 0.5 mm thick single-crystalline GaN crystal for the first experiment at a growth temperature of about 500°C and <130 MPa system pressure [4]. Figure 9.7 shows that the crystal appears in a translucent brownish coloration due to defects. Nitrogen deficiency has been claimed to cause the dark color [17]. Cracks from the seed crystal continued into the grown crystals. Yet, the feasibility to produce large GaN crystals from acidic ammonothermal conditions under significantly lower pressures and temperatures than from basic mineralizers is evidenced.

Figure 9.8 shows the low-temperature (10 K) Pl signal from the near band edge region for a recently grown (exp 106) and a former (exp 101) crystal of ammonothermal GaN in comparison to the HVPE-GaN substrate. There are clear peaks at 3.27 and 3.466 eV, which have been ascribed as DAP and acceptor bound exciton (ABE) emissions, respectively [23]. The first ABO low phonon replica (LO) is found at 3.378 eV. The main peak at 3.472 eV is due to donor bound exciton (DBE) and was related to the O and Si donors [23]. Elastic scattering of free exciton at Si donor is seen at 3.456 eV and TES at 3.451 eV. What gets clear from Fig. 9.8 is the striking similarity between the best sample of ammonothermal GaN and the HVPE grown GaN. In other words, the crystal quality and defects are obviously very comparable. This was found for the first time in a GaN crystal grown by ammonothermal technique and demonstrates the recently great improvement in the technology.

HVPE GaN seed   Ammonothermal GaN

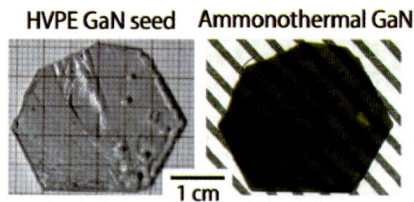

Fig. 9.7. A one-inch size HVPE seed crystal (*left*) and the subsequently grown ammonothermal GaN crystal (*right*)

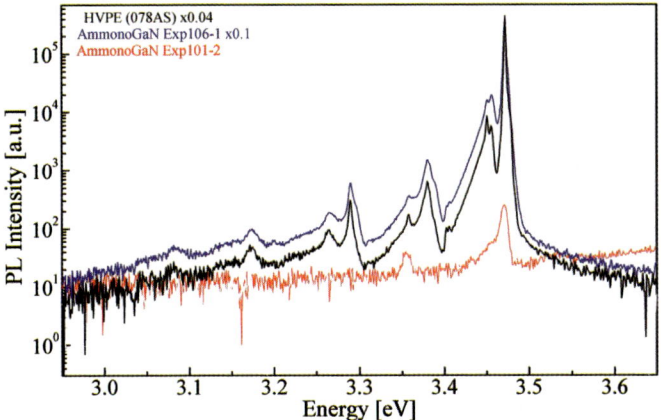

Fig. 9.8. Photoluminescence spectra from HVPE GaN in comparison to a recently grown (exp 106) and formerly grown (exp 101) ammonothermal GaN crystal. The N-face was subject for measurement

Generally, from our experiments the crystal quality is higher for GaN nucleated on the $(000\bar{1})$ or N-terminated face than for GaN nucleated on the (0001) or Ga-terminated face. This is coherent to the growth from basic mineralizers [16]. The reason for this variation, similar to the hydrothermal ZnO system that uses basic mineralizers, might be that the main growth species must be positively charged like $[Ga(NH_3)_5Cl]^{2+}$, etc. [24]. This however is still under debate and further work is currently going on in our group. In contrast to ZnO, better crystal quality is grown on the $(000\bar{1})$ face. Quite opposite to the hydrothermal growth of ZnO, the growth rate in $\langle 000\bar{1}\rangle$ is faster than that in $\langle 0001\rangle$.

9.4 Conclusion

The solvothermal growth of large size crystals of ZnO and GaN was reviewed in the light of our own activities. Liquid phase epitaxy of ZnO has been developed and is used for fast screening of doping under the conditions of the thermodynamic equilibrium. It was shown that the ammonothermal growth of GaN from acidic mineralizers holds great potential to mass-produce GaN bulk crystal in an economic fashion. A regular solubility of GaN under acidic ammonothermal conditions enables rather easy control of the growth process.

References

1. S. Adachi, *Properties of Group-IV, III-V and II-VI Semiconductors* (Wiley, Chichester, 2005)
2. J. Karpinski, J. Jun, S. Porowski, J. Cryst. Growth **66**, 1 (1984)

3. W. Utsumi, H. Saitoh, H. Kaneko, T. Watanuki, K. Aoki, O. Shimomura, Nat. Mater. **2**, 735 (2003)
4. T. Fukuda, D. Ehrentraut, J. Cryst. Growth **305**, 304 (2007)
5. R. Helbig, J. Cryst. Growth **15**, 25 (1972); D.C. Look, D.C. Reynolds, J.R. Sizelove, R.L. Jones, C.W. Litton, G. Cantwell, W.C. Harsch, Solid State Comm. **105**, 399 (1998); J. Nause, B. Nemeth, Semicond. Sci. Technol. **20**, S45 (2005)
6. D. Ehrentraut, H. Sato, Y. Kagamitani, H. Sato, A. Yoshikawa, T. Fukuda, Prog. Cryst. Growth Char. Mater. **52**, 280 (2006)
7. IMR Workshop, IMR, Tohoku University, Sendai, March 1, 2007
8. D. Ehrentraut, H. Sato, M. Miyamoto, T. Fukuda, M. Nikl, K. Maeda, I. Niikura, J. Cryst. Growth **287**, 367 (2006)
9. K. Maeda, M. Sato, I. Niikura, T. Fukuda, Semicond. Sci. Technol. **20**, S49 (2005)
10. H. Sato, D. Ehrentraut, T. Fukuda, Jpn. J. Appl. Phys. **45**, 190 (2006)
11. T. Fukuda, M. Mikawa, D. Ehrentraut, Proc. SPIE **6474**, 647412 (2007)
12. F. Tuomisto, Proc. SPIE **6474**, 647413 (2007)
13. T. Moe Børseth, B.G. Svensson, A.Yu. Kuznetzov, P. Klason, Q.X. Zhao, M. Willander, Appl. Phys. Lett. **89**, 262112 (2006)
14. D. Ehrentraut, H. Sato, Y. Kagamitani, A. Yoshikawa, T. Fukuda, J. Pejchal, K. Polak, M. Nikl, H. Odaka, K. Hatanaka, H. Fukumura, J. Mater. Chem. **16**, 3369 (2006)
15. R. Dwiliński, J. Doradziński, J. Garczyński, L. Sierzputowski, J.M. Baranowski, M. Kamińska, Mater. Sci. Eng. B **50**, 46 (1997)
16. T. Hashimoto, K. Fujito, M. Saito, J.S. Speck, S. Nakamura, Jpn. J. Appl. Phys. **44**, L1570 (2005)
17. B. Wang, M.J. Callahan, K.D. Rakes, L.O. Bouthillette, S.Q. Wang, D.F. Bliss, J.W. Kolis, J. Cryst. Growth **287**, 376 (2006)
18. B. Raghothamachar, J. Bai, M. Dudley, R. Dalmau, D. Zhuang, Z. Herro,R. Schlesser, Z. Sitar, B. Wang, M. Callahan, K. Rakes, P. Konkapaka, M. Spencer, J. Cryst. Growth **287**, 349 (2006)
19. M.J. Callahan, *Solvothermal Growth of GaN and ZnO Crystals*. Paper presented at the Symposium on Solvothermal Growth of Wide-bandgap Materials, Sendai, Miyagi, Japan (2004)
20. T. Hashimoto, K. Fujito, F. Wu, B.A. Haskell, P.T. Fini, J.S. Speck, S. Nakamura, Mater. Res. Soc. Symp. Proc. **831**, 281 (2005)
21. Y. Kagamitani, D. Ehrentraut, A. Yoshikawa, N. Hoshino, T. Fukuda, S. Kawabata, K. Inaba, Jpn. J. Appl. Phys. **45**, 4018 (2006)
22. D. Ehrentraut, N. Hoshino, Y. Kagamitani, A. Yoshikawa, T. Fukuda, H. Itoh, S. Kawabata, J. Mater. Chem. **17**, 886 (2007)
23. B. Monemar, P.P. Paskov, F. Tuomisto, K. Saarinen, M. Iwaya, S. Kamiyama, H. Amano, I. Akasaki, S. Kimura, Phys. B **376–377**, 440 (2006)
24. H. Yamane et al., Acta Cryst. E **63**, i59 (2007)

Part II

Materials for Ecological and Biological Systems

High-Quality Si Multicrystals with Same Grain Orientation and Large Grain Size by the Newly Developed Dendritic Casting Method for High-Efficiency Solar Cell Applications

K. Nakajima, K. Fujiwara, and N. Usami

Summary: Si multicrystals have many grains with different orientations and sizes, resulting in many grain boundaries with different characteristics. To obtain the high conversion efficiency of the solar cells prepared by Si multicrystals that is very close to that of Si single crystals, Si multicrystals that have grains with a same orientation and proper sizes and have electrically inactive grain boundaries are required. This concept was tried and the dendritic casting method was newly developed to obtain extremely high-quality Si multicrystal ingots and to largely increase the yield of high-quality Si multicrystal ingots. To develop such a new growth technology by understanding the crystal growth mechanism to control the grain orientation and the grain size, the in-situ observation system was newly developed to directly observe the growing interface of Si crystals at temperatures higher than 1,400°C. Using the in-situ observation system, it was found that the new growth mode of Si dendrite crystals appeared along the bottom of the crucible at the initial stage of the growth. Such dendrite crystals were very effective to control the grain orientation and the grain size of Si multicrystals. Using the developed dendritic casting method, extremely high-quality Si multicrystal ingots with the same grain orientation and very large grain size were obtained. The conversion efficiency of the solar cells prepared by such Si multicrystals was much higher than that of the solar cells with different grain orientations, especially for the upper parts of the Si ingots.

In face of the destruction of the global environment and the depletion of world-wide natural resources and energy sources in the twenty-first century, we should take materials research forward into the problems of the global environment and energy to aim at sustaining human development and providing high living standards, and to rapidly develop the new environment-friendly clean energy conversion system. These studies, especially the development of future solar cells to establish the clear-energy based world or the green world, will offer our next generations a brighter future.

Within the next 100 years, most of natural resources and energy sources will be exhausted. In particular, the consumption of oil, natural gas, and uranium is a serious problem. In comparison with the traditional energy sources, solar energy is the only ultimate natural energy source. Although

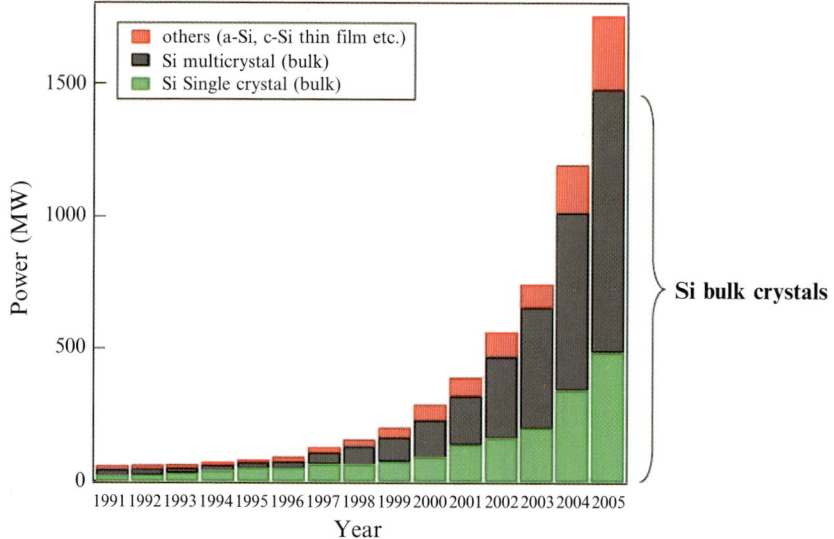

Fig. 10.1. Shipped total power of solar cells in the world

~30% of total solar energy is reflected from the earth surface, the most part of ~70% is utilized by human being. The world's energy consumption is about 10,000 Mtoe (M ton oil equivalent) at 2005. However, it is only 0.01% of the available solar energy. If we can use the solar energy for 10% of the world energy consumption, we must manufacture 40 GW solar cells every year for 40 years and the required Si feedstock is about 400,000 ton per year. To attain this target, the world production of Si feedstock needs to be increased to about 27 times as much as the present world production.

Figure 10.1 shows the shipped total power of solar cells in the world, which is rapidly increasing in the recent years. Si single and multi-bulk crystals contribute to 85% of the annual shipment of solar cells. The share of the Si multicrystal solar cells is the largest one in all types of solar cells. Thus, the development of high-quality and low-cost Si feed stock, including both multicrystal and single crystal Si, is the most important problem needed to be solved. To largely promote the applications of solar cells as an energy source, the present energy cost of the solar cells should be halved. For example, the cost of Si multicrystal solar cells is mainly determined by the cost of Si feed stock, the yield of Si ingots, and the conversion efficiency of solar cells. If the cost of Si feed stock is reduced from $70 per kg to $35 per kg, the yield of Si ingots increases from 60% to 90% and the conversion efficiency increase from 15% to 20%, and the total cost will be reduced to half. So, the main technologies to be developed are the production technology to obtain low-cost and high-quality Si feed stock, the growth method to largely increase the yield of high-quality Si ingots, and the growth method to obtain extremely high-quality Si ingots.

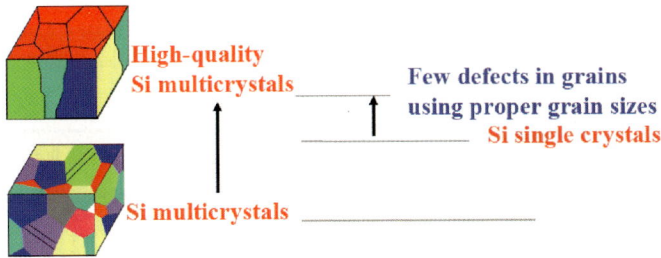

Fig. 10.2. Si multicrystals that have grains with the same orientation and proper sizes and have grain boundaries with electrically inactive characteristics

On the view point of crystal growth, the target is to develop a method that can effectively increase the yield of high-quality Si multicrystals from 60% to 90% and to obtain high-quality Si multicrystals with conversion efficiency higher than 20%. To obtain high-quality Si multicrystals, many efforts have been attempted on the view point of purification for a long time. The best conversion efficiency of 20% has been attained from small size cells using texture structure and hydrogen passivation. However, Si multicrystals have many grains with different orientations and sizes, resulting in many grain boundaries with different characteristics. If we can obtain Si multicrystals that have grains with same orientation and proper sizes as shown in Fig. 10.2 and grain boundaries are electrically inactive, such as $\Sigma 3$, the conversion efficiency of the Si multicrystals should be higher than that of Si single crystals because there are fewer defects such as vacancy within grains. This concept has not been tried yet because there are no breakthrough ideas to obtain such high-quality Si multicrystals by a practical method. We have recently developed a reliable method for both purification and structure control to obtain high-quality Si multicrystals with the same grain orientation and large grain sizes by the newly developed dendritic casting method.

10.1 In-Situ Observation System to Directly Observe the Growing Interface of Si Crystals

To develop the high-quality Si multicrystals, the growth mechanism should be well known to control the grain orientations and the grain sizes. An in-situ observation system was newly developed to directly characterize the growing interface of Si crystals at a temperature higher than 1,400°C [1] as shown in Fig. 10.3. A small amount of Si growth melt was placed in the silica crucible, and they were set inside a small furnace with a temperature gradient to obtain the unidirectional growth as shown in Fig. 10.3. The growing process with the motion of solid/liquid interfaces can be directly observed by the microscope through a silica window.

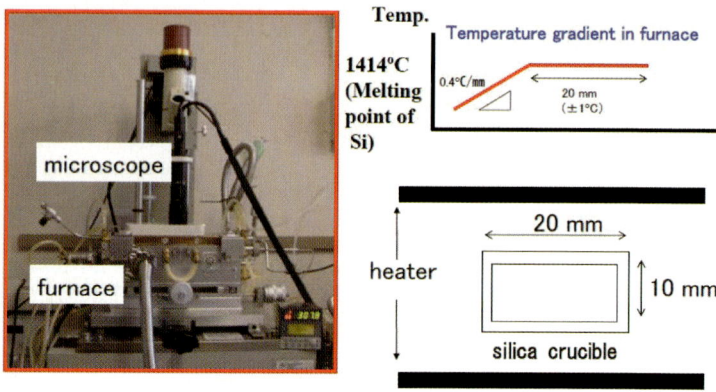

Fig. 10.3. Newly developed in situ observation system to directly observe the growing interfaces of Si crystals

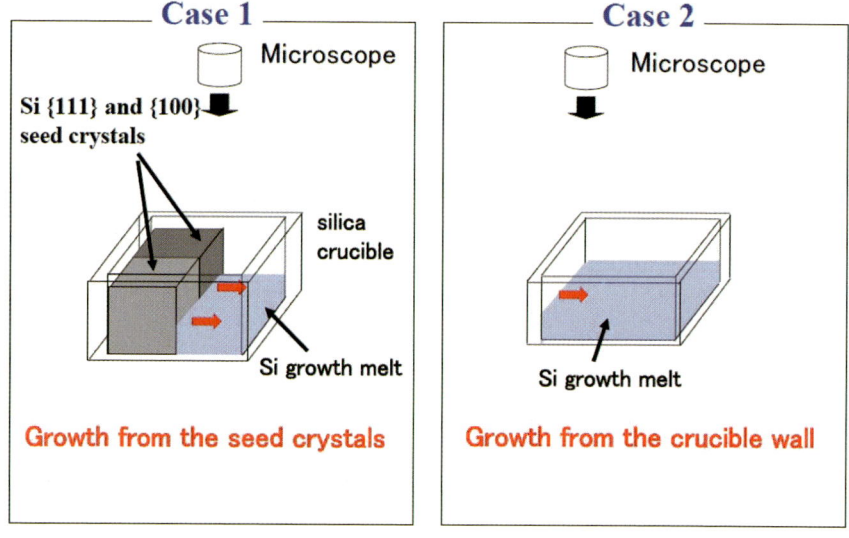

Fig. 10.4. Arrangement of the Si growth melt for the in situ observation system

Figure 10.4 shows examples of the arrangement of Si growth melts. In Case 1, Si {111} and {100} seed crystals are set side by side to contact with the Si growth melt. In Case 2, the Si growth melt directly contacts with the crucible. Before setting the growth melt in the furnace, wafer tips of Si (p-type, $1 \sim 5 \ \Omega \, cm$) were dipped in $HF : H_2O = 2.5 : 97.5$ solution for 120 s to remove the oxide layer. After melting the wafer tips in the furnace, the furnace was cooled down at a rate between 1 and $50°C \ min^{-1}$. Figure 10.5 is a microscopic image of the Si growth in Case 1. The Si melt was cooled down at $30°C \ min^{-1}$. The crystal grown on the {111} seed has a {111} flat facet, whereas the crystal grown on the {100} seed has many steps with {111}

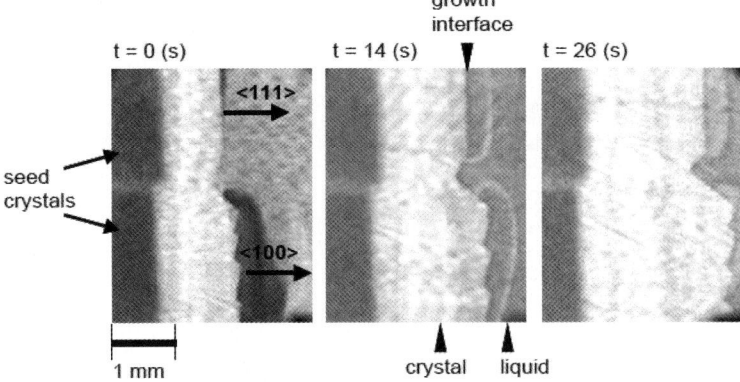

Fig. 10.5. Microscopic image of the Si melt growth in the case 1

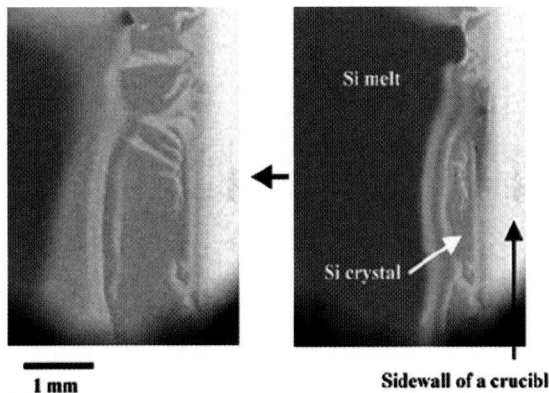

Fig. 10.6. Microscopic image of the Si melt growth in the case 2

facet. The growth rate of the crystal on the {100} seed is faster than that on the {111} seed because the surface energy of the closed packed {111} face is the smallest one of Si and the {111} face is smoother than that of the {100} face. Figure 10.6 is a microscopic image of the Si growth in Case 2. The Si melt was cooled down at 30°C min^{-1}. With the proper growth conditions, it was found that a Si thin layer grows along the bottom of the crucible at the initial stage, and then Si multicrystals grow on the upper surface of the Si layer. The surface of the very thin Si growth melt was directly observed to know the state of this Si thin layer. The thickness of the Si growth melt was about 1 mm. It was found that a dendritic crystal [2] grows along the bottom of the crucible as shown in Fig. 10.7. The details of the dendrite crystals can be observed by changing the temperature gradient in the Si growth melt as shown in Fig. 10.8. A sharp and long needle-like crystal grows at the top of a dendrite crystal, and then crystals grow on both sides of the facets of the long needle-like crystal.

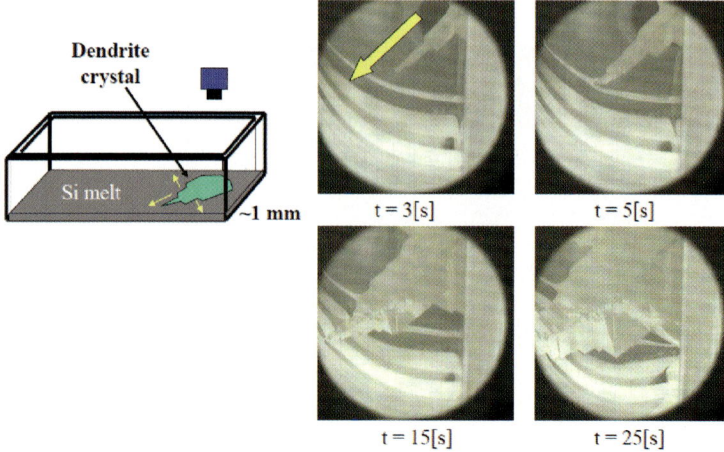

Fig. 10.7. A dendrite crystal grown along the bottom of the crucible

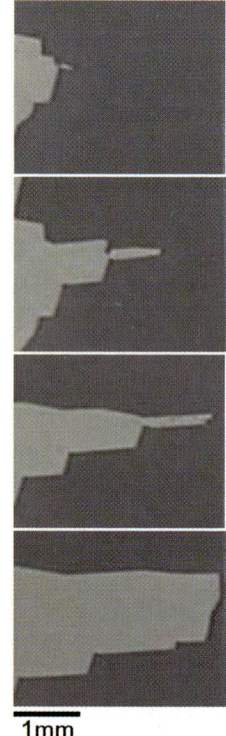

Fig. 10.8. Growth of dendrite crystals by changing the temperature gradient in a Si growth melt

10.2 Formation Mechanism of Parallel Twins Related to Si-Faceted Dendrite Growth

The fundamental study of growth of Si or Ge dendritic crystals that are so-called faceted dendrites has had a long history since the first report by Billig [2]. It has been shown that more than two parallel twins exist at the center of a faceted dendrite [3–7]. So the parallel twins are known to play an important role for the growth of a dendrite [3, 4]. Although many researchers have studied the growth mechanism, it has not been well known how the parallel twins were formed in a dendrite crystal during melting growth. In other words, it has not been well known why the twins were generated parallel to each other, and why they were a prerequisite factor for the formation of a faceted dendrite.

To answer the question, we carefully observed the growth process of Si crystals in Si melts using an in-situ observation system to monitor how the parallel twins were formed in a crystal [8]. Figure 10.9 shows microscopic images of the solid/liquid growing interface during Si melt growth. The shape of the interface changed with decrease in the temperatures of the melt as shown in Fig. 10.9a–d. At first, a faceted interface was observed as shown in Fig. 10.9a. At this stage of the growth, a faceted dendrite did not grow because the temperature of the melt was not low enough to produce large supercooling, although parallel twins might have already been formed in the crystal. When the temperature of the melt became much lower, a faceted dendrite grew in a constant direction from a part of the faceted interface as shown in Fig. 10.9b–d. It should be noted that the growth direction of the dendrite was parallel to the facet plane as shown in Fig. 10.9c. This indicates that the parallel twins

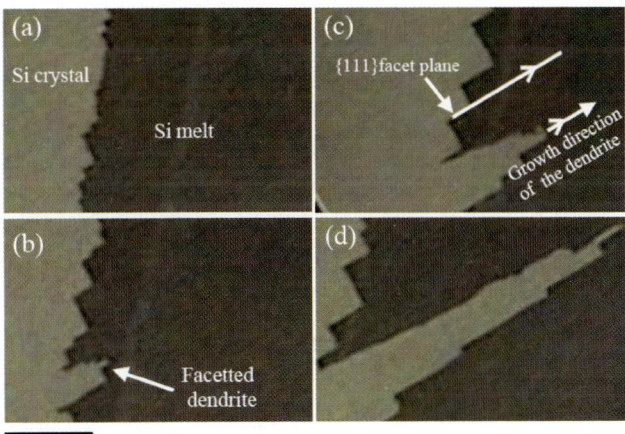

1 mm

Fig. 10.9. Microscopic images of the solid/liquid growing interface during Si melt growth. The shape of the interface changed from (**a**) to (**d**), with decreasing temperatures of the Si melt [8]

at the center of the faceted dendrite are formed parallel to the facet plane. We confirmed the existence of the {111} parallel twins at the center of the faceted dendrite after crystallization using electron backscattering diffraction pattern (EBSP) as shown in Fig. 10.9. Then, we found that the facet planes on the growth interface, parallel to the growth direction of the faceted dendrite, were {111} planes.

As shown in Fig. 10.9, the parallel twins related to the faceted dendrite growth were found to be formed parallel to the {111} facet plane on the growth interface. Based on this fact, we should consider how parallel twins are formed although there have been several reports concerning when and/or where a twin boundary is formed, that is, whether a nucleus originally contains a twin boundary [9] or a twin is formed after nucleation [10].

Figure 10.10 shows our schematic diagrams for the generation of the parallel twins during Si melt growth [8]. Figure 10.10a shows an initial solid/liquid growing interface of a Si crystal. It has been theoretically explained that the growing interface of Si crystals is faceted and {111} planes appear on the surface [11]. Such a faceted interface of Si crystals was directly observed using the in-situ observation system by us. The shape of the faceted interface strongly depends on the orientation of grown crystals as shown in Fig. 10.10. Since the growing interface is always faceted as mentioned earlier, Si crystals grow always on {111} facet planes. If Si atoms attach on a facet plane to generate a twin, a thin layer that maintains the twin relationship to the

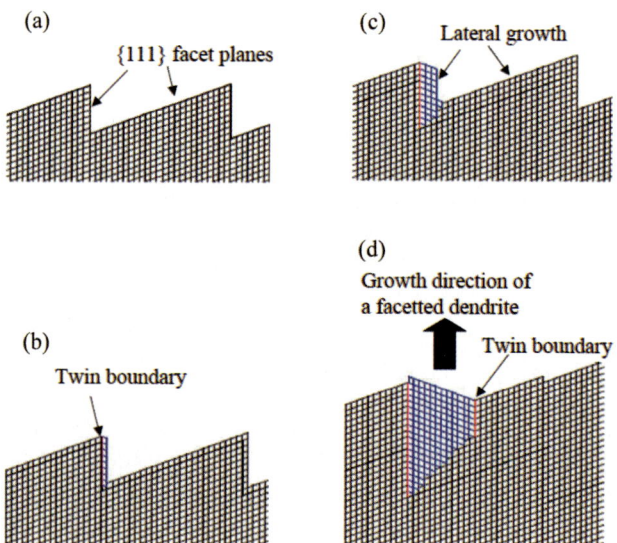

Fig. 10.10. Schematic diagrams for the generation of the parallel twins during Si melt growth. The growth interface is facetted as shown in (**a**). If a twin boundary is accidentally formed on a {111} facet plane, another twin boundary is formed parallel to the first twin after lateral growth as shown from (**b**) to (**d**)

original facet plane is formed by lateral growth along the facet plane. As a result, a twin boundary is generated in the Si crystal as shown in Fig. 10.10b. The reason that attached Si atoms freely arrange themselves on a facet plane to generate a twin is because the grain boundary energy of the Si {111} twin interface is close to zero ($\sim$30 mJ m^{-2}) [12]. When the driving force for crystal growth is large due to large supercooling, a thin layer that maintains the twin relationship to the original facet plane easily grow up by lateral growth because of the large energy gain by crystallization under a large supercooling. When such a layer continuously grows ahead by the lateral-growth mode as shown in Fig. 10.10c, it should be noted that there is another chance to generate another twin parallel to the previous twin as shown in Fig. 10.10d. In this growth mechanism, two parallel twins are certainly formed after a twin is formed on a facet plane in a growing Si crystal. When the supercooling of the Si growth melt becomes large enough, a faceted dendrite crystal starts to grow on the basis of these two parallel twins. The growing direction of the faceted dendrite crystal is parallel to the facet planes of the two parallel twins as shown in Fig. 10.10d. This growth mechanism agrees well with the in situ experimental observations of the dendrite growth as shown in Fig. 10.9. Thus our model well explains how and why the parallel twins are formed in the growing Si dendrite crystals.

10.3 Growth of High-Quality Si Multicrystals by the Dendritic Casting Method

For the conventional casting method, nucleation occurs at random sites on the bottom of the crucible, and many grains with random orientations grow from them as shown in Fig. 10.11. Grains with preferential orientations finally remain in conventional Si multicrystals. Therefore, the structure of Si multicrystals grown by the conventional casting method can not be controlled at all. Using dendrite crystals growing along the bottom of the crucible, the grain orientation and the grain size can be intentionally controlled. Figure 10.12 shows a typical dendrite crystal of Si. It has two twins with Σ3 grain boundaries and the Σ3 grain boundaries have the same {111} facets.

The orientation of dendrite crystals depends on supercooling of Si growth melt [7]. When the supercooling is smaller than 100°C, the preferential orientation of dendrite crystals is <112> or <110>. The two kinds of dendrite crystals have two twins with Σ3 grain boundaries and the Σ3 grain boundaries have the same {111} facets. So, the upper surface of the dendrite crystals is fixed to be {110} or {112} as shown in Fig. 10.13.

Figure 10.14 shows our concept to control the grain orientation and the grain size [13, 14]. At first, dendrite crystals with the <110> or <112> direction are grown along the bottom of the crucible using the proper growth conditions. Then, Si multicrystals are automatically grown on the upper

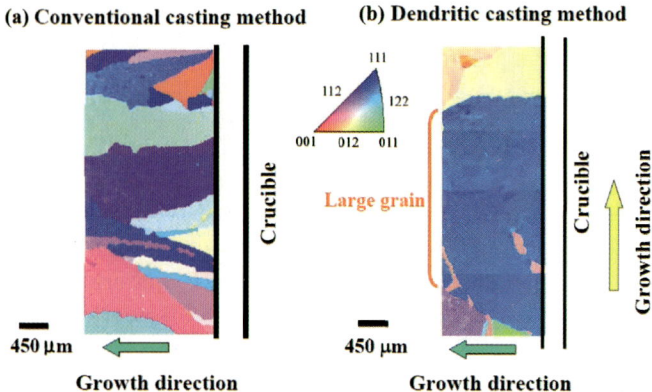

Fig. 10.11. Typical structures of Si multicrystals grown by the two types of methods. (**a**) The conventional casting method and (**b**) the dendritic casting method

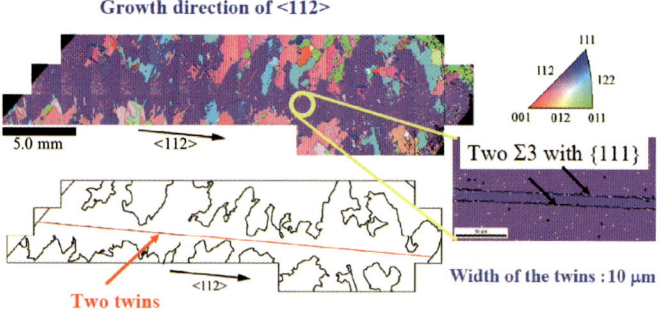

Fig. 10.12. Structure of a typical dendrite crystal of Si

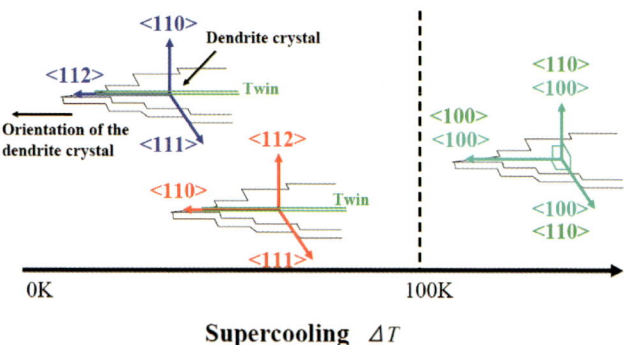

Fig. 10.13. Supercooling dependence of orientation of dendrite crystals grown in Si growth melt

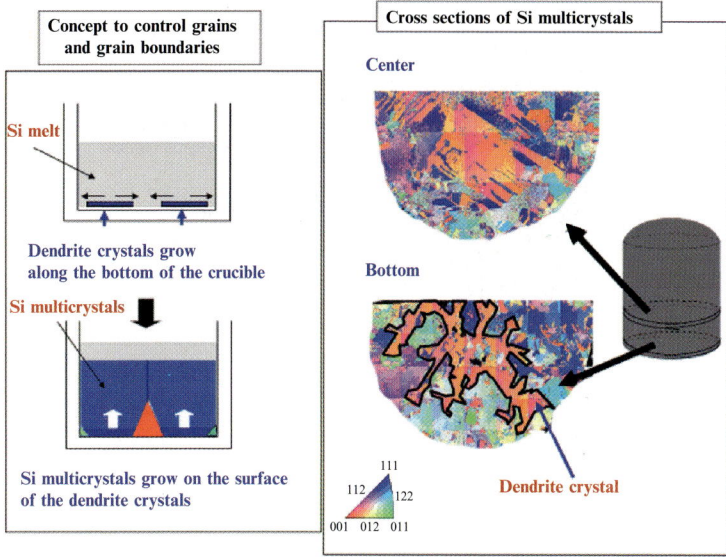

Fig. 10.14. Concept to control the grain orientation and the grain size

Fig. 10.15. Overview of an extremely high-quality textured Si multicrystal ingot with a diameter of 15 cm

surface of the dendrite crystals. Figure 10.14 shows the cross sections of the bottom and center of a Si multicrystal grown by our dendritic casting method. The orientation was analyzed by the EBSP method. A dendrite crystal with a {112} upper surface appears along the bottom of the crucible. The Si multicrystal grows on the {112} upper surface of the dendrite crystal. The grain orientation is kept constant during growth. The grain size becomes larger as the crystal grows. Figure 10.15 shows the overview of the

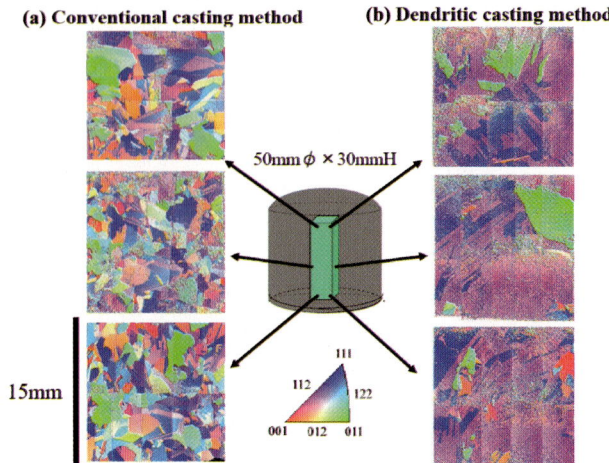

Fig. 10.16. Structures of Si multicrystal ingots grown by the two types of methods. (**a**) The conventional casting method and (**b**) the dendritic casting method

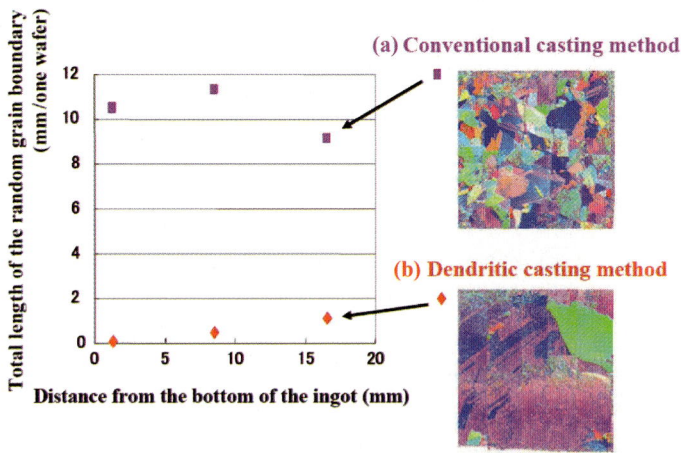

Fig. 10.17. Total length of random grain boundaries in Si multicrystal ingots grown by the two types of methods. (**a**) The conventional casting method and (**b**) the dendritic casting method

extremely high-quality textured Si multicrystal ingot with a diameter of 15 cm grown by our dendritic casting method [14]. The grain size is quite large and larger than 3 cm. Therefore, using the dendritic casting method, we can grow Si multicrystals that have larger grains with the same <112> orientation, whereas conventional Si multicrystals have smaller grain sizes with random orientations as shown in Fig. 10.16 [13].

Figure 10.17 shows the total length of the random grain boundaries in each wafer as a function of the distance from the bottom of the ingot in the

Si multicrystals with the same orientation and with different orientations. Obviously, the Si multicrystal with the same orientation has very few random grain boundaries. The structure of the bottom of the Si ingot is affected by dendrite crystals and the diffusion length is lower than 100 µm. The structure of the top of the Si ingot is not affected by dendrite crystals, and the diffusion length is larger than 150 µm.

10.4 Solar Cells Prepared by Si Multicrystals with the Same Orientation

For the preparation of solar cells using Si multicrystal wafers, the Si wafers were chemically treated in 1 HF:6 HNO$_3$ solution for 1 min at 25°C, and then etched in buffered HF solution for 2 min at 25°C. After the spin coating of a phosphorus-doped glass film [Ohka coat diffusion (OCD)], an n+ layer was formed by rapid thermal annealing at 940°C for 5 min. The samples were again etched in buffered HF solution to remove the residue of the OCD film from the surfaces. An indium-tin oxide (ITO) film of 70 nm thickness was formed as an antireflection film by sputtering. The back surface field contact was formed by printing aluminum and annealing at 840°C, and the front finger contact was formed by firing silver paste through an ITO film at 840°C. The textural structure was not formed on the surface. The conversion efficiency of the solar cells was measured using a characterization system (JASCO YQ-250BX) with an AM1.5 solar simulator (100 mW cm^{-2}) as a light source [15].

Figure 10.18 shows the conversion efficiency of solar cells prepared by Si multicrystals with the same orientation and with different orientations. Obviously, the conversion efficiency of the solar cell with the same orientation is higher than that of the solar cell with different orientations except for the

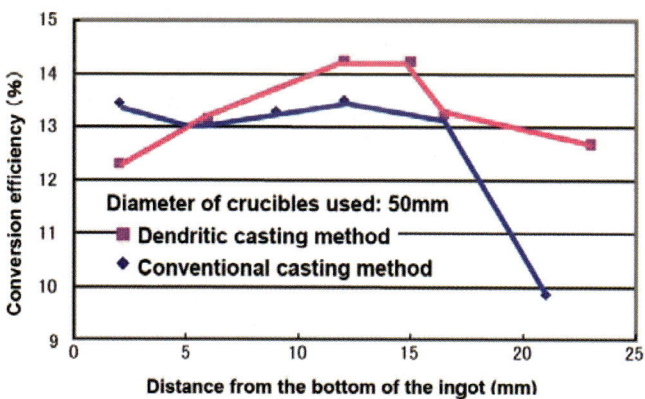

Fig. 10.18. Conversion efficiency of solar cells prepared by Si multicrystals with the same orientation and with different orientations

initially grown part. It is known that the conversion efficiency of solar cells with the top and bottom part of the crystals is small owing to the diffused and segregated impurities. This feature is reproduced for solar cells prepared by Si multicrystals with different orientations. On the other hand, the reduction in the conversion efficiency of the top part is not obvious for solar cells prepared by Si multicrystals with the same orientation. This suggests that the quality of multicrystal was remarkably improved for the finally grown part by employing the dendritic casting method. The smaller conversion efficiency at the bottom of the crucible was presumably due to inferior crystal quality of the dendrite crystals grown under relatively high supercooling. Importantly, grain boundary characters are spontaneously improved during directional solidification process when crystal growth is initiated with oriented crystals [16], which results in superior overall performance of solar cells prepared by crystals grown by the dendritic casting method.

Figure 10.19 shows the conversion efficiency of solar cells prepared by Si multicrystals with the same orientation [13]. The diameters of the crucibles used for the crystal growth were different for these solar cells. They were 50, 80, and 150 mm ϕ. Obviously, the conversion efficiency of the solar cell prepared by the 150 mm ϕ crucible is the highest. It is as high as 17% without texture structure and hydrogen passivation. This level of the conversion efficiency is very close to that of the solar cell using a Si single crystal wafer grown by Czochralski method when the same solar cell processes [17] were used for both solar cells. The conversion efficiency of the solar cell prepared by Si

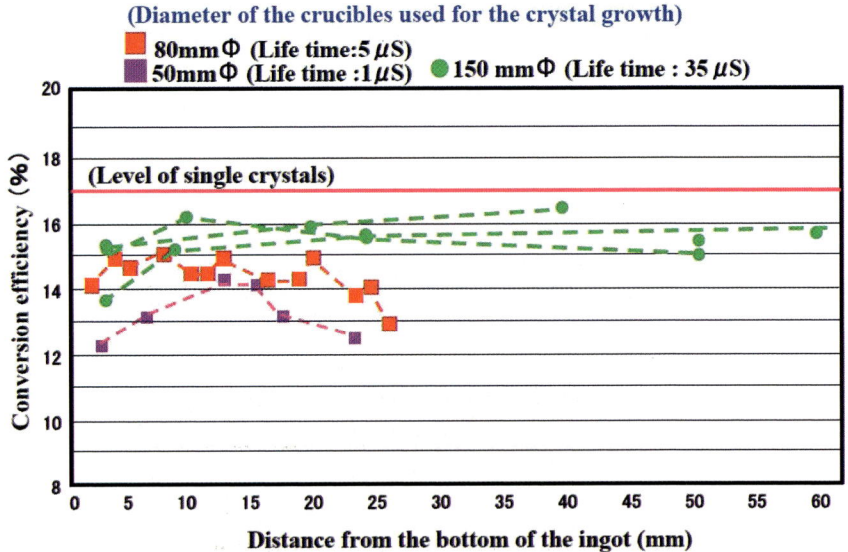

Fig. 10.19. Conversion efficiency of solar cells prepared by Si multicrystals with the same orientation using the different sizes of crucible

multicrystals with the same orientation keeps high even at the top of the Si ingot. So, these Si multicrystals have very high quality, and the yield of such high-quality Si multicrystal ingots will largely increase from 60% to 80–90%.

10.5 Quality of Si Multicrystals Grown by the Dendritic Casting Method

The upper surface of the dendrite crystals grown by the dendritic casting method can be controlled as {110} or {112} depending on the amount of supercooling, leading to the formation of Si multicrystal ingot with the same orientation. This feature permits to carry out unique fundamental research to clarify quality of Si multicrystals in terms of their microstructures and the impact on properties since spatially resolved X-ray diffraction (XRD) can be applied to plural crystal grains by choosing the scanning axis parallel to the growth direction as shown in Fig. 10.20. Spatial mapping of the peak shift and line width of rocking curves leads to visualize the orientation fluctuation and the crystal quality. In addition, sub-grain boundaries are quantified since dislocations to constitute sub-grain boundaries tend to lengthen in the growth direction and rocking curve measurements around the growth direction can detect each sub-grains as independent peaks. Figure 10.21 shows a typical cross-sectional picture of a Si multicrystal grown by the dendritic casting method, and the distribution of the peak position of the rocking curves within the dotted square around multiple twins. The sample was cut parallel to the growth direction. It is seen that fluctuation in the crystal orientation is small except the circled region around a grain boundary, where existence of sub-grain is revealed as a rocking curve with multiple peaks. Importantly, the crystals with multiple twins can be regarded as one large crystal grain, and

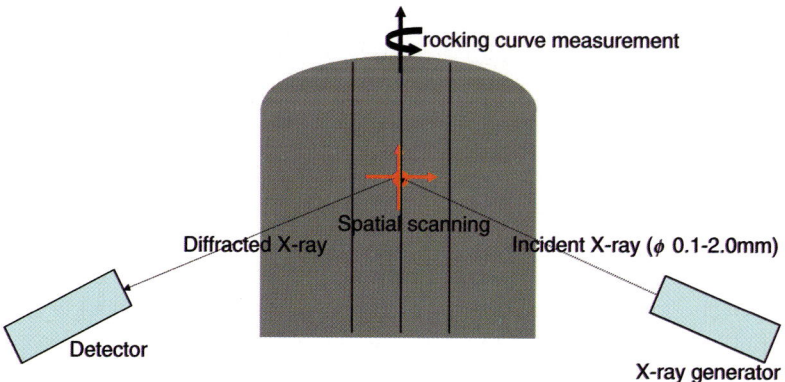

Fig. 10.20. Illustration of spatially resolved X-ray rocking curve measurement system

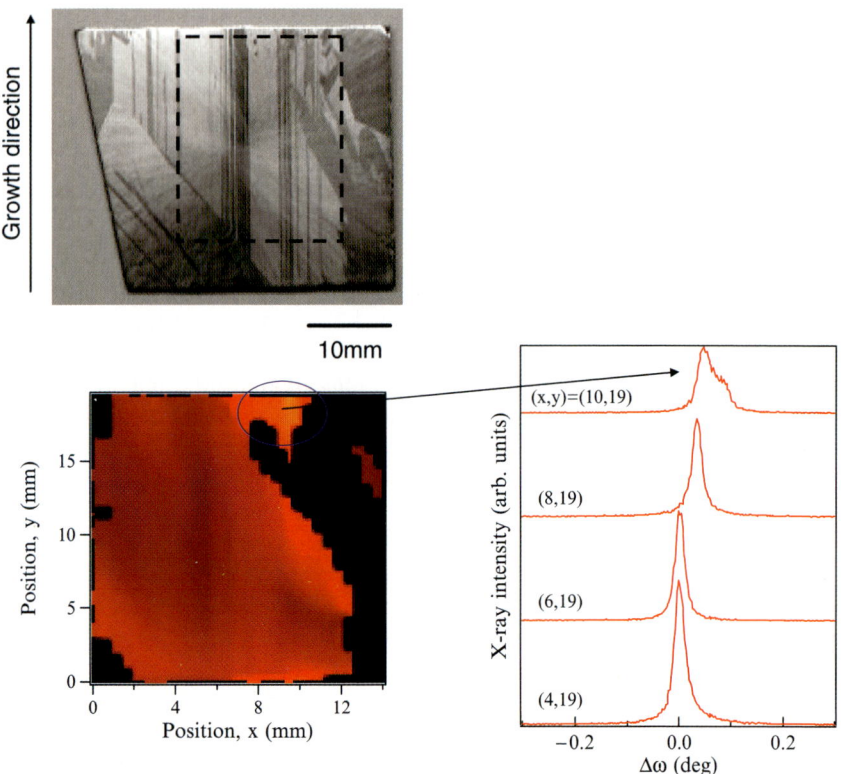

Fig. 10.21. A cross-sectional picture of a mc-Si ingot grown by DCM (*top*): The distribution of the peak position of the rocking curves in the dotted-square is displayed (*bottom/left*), indicating the fluctuation of the orientation in multiple twins. As shown in the circle and the corresponding rocking curve, existence of the subgrain boundaries is confirmed

grain boundaries inside are almost perfect $\Sigma 3$, which is known to be electrically inactive. These results confirm that the control of microstructures in Si multicrystal ingot such as grain orientations and grain boundary characters is of crucial importance for realization of high-efficiency solar cells.

The quality of Si multicrystal ingots grown by the casting method strongly depends on the crucible size. The quality of the bottom and top parts of Si multicrystal ingots is usually lower compared to that of the center part of the same ingots as shown in Fig. 10.22. It should be stressed that the diameter of our Si ingots grown by the dendritic casting method is 15 cm and the height is about 5 cm, which is still within a region of low-quality part of Si multicrystal ingots grown by the casting method in industrial scale. Importantly, high-quality crystals can be obtained even within the low-quality-part of the conventional Si multicrystal ingots using the dendritic casting method. This implies that implementation of the dendritic casting method to the industrial

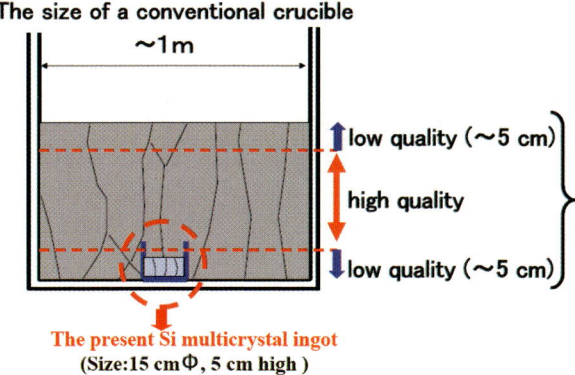

Fig. 10.22. Effect of the crucible size on the quality of Si multicrystal ingots

scale would result in realization of higher quality Si multicrystal ingots for high-efficiency solar cells.

10.6 Summary

The dendritic casting method was newly developed to obtain extremely high-quality Si multicrystal ingots and to largely increase the yield of high-quality Si multicrystal ingots. To develop such a new growth technology by understanding the crystal growth mechanism to control the grain orientation and the grain size, the in-situ observation system was newly developed to directly observe the growing interface of Si crystals at temperature higher than 1,400°C. Using the in-situ observation system, it was found that the new growth mode of Si dendrite crystals appeared along the bottom of the crucible at the initial stage of the growth. Such dendrite crystals were very effective to control the grain orientation and the grain size of Si multicrystals. Using our newly developed dendritic casting method, extremely high-quality Si multicrystal ingots were obtained, which have the same grain orientation and very large grain sizes.

The conversion efficiency of the solar cells prepared by the Si multicrystals with the same grain orientation was much higher than that of the solar cells with different grain orientations, especially for the upper parts of the Si ingots. It is worth noting that these Si multicrystals produced by the dendritic casting method had very high-quality, and the yield of such high-quality Si multicrystal ingots would be largely increased because high-quality crystals can be obtained even within the low-quality-part of the conventional Si multicrystal ingots using the dendritic casting method.

References

1. K. Fujiwara, Y. Obinata, T. Ujihara, N. Usami, G. Sazaki, K. Nakajima, J. Cryst. Growth **262**, 124 (2004)
2. E. Billig, Proc. Roy. Soc. A **229**, 346 (1955)
3. R.S. Wagner, Acta Metall. **8**, 57 (1960)
4. D.R. Hamilton, R.G. Seidensticker, J. Appl. Phys. **31**, 1165 (1960)
5. D.R. Hamilton, R.G. Seidensticker, J. Appl. Phys. **34**, 1450 (1963)
6. K.K. Leung, H.W. Kui, J. Appl. Phys. **75**, 1216 (1994)
7. K. Nagashio, K. Kuribayashi, Acta Mater. **53**, 3021 (2005)
8. K. Fujiwara, K. Maeda, N. Usami, G. Sazaki, Y. Nose, K. Nakajima, Scripta Mater. **57**, 81 (2007)
9. R.Y. Wang, W.H. Lu, L.M. Hogan, Metall. Mater. Trans. **28A**, 1233 (1997)
10. R.W. Cahn, Phil. Mag. Suppl. **3**, 363 (1954)
11. K.A. Jackson, *Growth and Perfection of Crystals* (Wiley, New York, 1958), p. 319
12. M. Kohyama, R. Yamamoto, M. Doyama, Phys. Stat. Sol. B **138**, 387 (1986)
13. K. Fujiwara, W. Pan, N. Usami, K. Sawada, M. Tokairin, Y. Nose, A. Nomura, T. Shishido, K. Nakajima, Acta Mater. **54**, 3191 (2006)
14. K. Fujiwara, W. Pan, K. Sawada, M. Tokairin, N. Usami, Y. Nose, A. Nomura, T. Shishido, K. Nakajima, J. Cryst. Growth **292**, 282 (2006)
15. K. Nakajima, K. Fujiwara, W. Pan, N. Usami, T. Shishido, J. Cryst. Growth **275**(1–2), e455 (2005)
16. N. Usami, K. Kutsukake, T. Sugawara, K. Fujiwara, W. Pan, Y. Nose, T. Shishido, K. Nakajima, Jpn. J. Appl. Phys. **45**, 1734 (2006)
17. W. Pan, K. Fujiwara, N. Usami, T. Ujihara, K. Nakajima, R. Shimokawa, J. Appl. Phys. **96**, 1238 (2004)

11

Growth of High-Quality Polycrystalline Si Ingot with Same Grain Orientation by Using Dendritic Casting Method

K. Fujiwara, W. Pan, N. Usami, M. Tokairin, Y. Nose, A. Nomura, T. Shishido, and K. Nakajima

Summary: We succeeded in the development of a new technique to grow a high-quality polycrystalline Si ingot for solar cells named as "Dendritic casting method," which utilizes the dendrite growth along the bottom of the crucible wall at the initial stage of directional growth. This method permits to obtain a textured polycrystalline Si ingot with large-size grains. Furthermore, its crystal quality was revealed to be highly maintained from the bottom to the top in the ingot, which is a great advantage from a view of the material yield. Solar cells based on this polycrystalline Si show high energy conversion efficiencies close to that of Si single crystal solar cells.

11.1 Introduction

Polycrystalline silicon (poly-Si), which has an advantage over single-crystalline silicon (sc-Si) in terms of low production cost, is widely used for solar cells, because it can be grown by casting based on directional growth. Therefore, it is demanded to improve the energy conversion efficiency of solar cells based on poly-Si without increase in the production cost. The main reason for the lower energy conversion efficiency of solar cells based on poly-Si is that the crystal quality of poly-Si is lower than that of sc-Si. The crystal structure of poly-Si is largely different from sc-Si as illustrated by formation of grain boundaries and the distribution of crystallographic orientations on a wafer surface. Generally, some grain boundaries act as recombination centers of photocarriers [1, 2]. To reduce the grain boundary density, size of grains in a poly-Si ingot should be controlled during casting. Random orientations on a wafer surface are also not convenient for solar cells. A surface textural structure for light trapping is easily formed by anisotropic chemical etching in sc-Si solar cells [3] because there is the only one crystallographic orientation on the wafer surface. On the other hand, there are many different orientations on the poly-Si wafer surface, which disturbs the formation of good textural structure due to the differences in etching rate and morphology between grains. Therefore, another special technique is used for poly-Si to form such a surface textural structure [4]. If textured poly-Si oriented in one direction were obtained, an optimum surface structure could be formed by a simple method. However, an active control of orientation distribution in a poly-Si ingot has not been carried out during casting. Moreover, research on controlling the orientation in a poly-Si ingot is limited.

In this paper, we propose a new method of growing a textured poly-Si ingot with large grains by casting. First, we investigated the melt growth behavior of poly-Si by in situ observations to obtain information on how to control the polycrystalline structure during directional growth. The results suggest that it is possible to grow a poly-Si ingot with large oriented grains by inducing dendrite growth along the crucible wall in the initial stage of directional growth. The concept of growing a textured poly-Si ingot was proposed. Second, we demonstrated to grow a textured poly-Si ingot by casting. Polycrystalline structures and solar cell properties of this ingot were investigated. It was shown that the growth concept and a textured poly-Si are promising for solar cells.

11.2 Experiments

The melt growth behavior of poly-Si was directly observed using an in situ observation system consisting of a furnace and a microscope [5]. Wafer tips of Si (p-type, $1 \sim 5 \ \Omega$ cm) were placed in a silica crucible. Then, a sample was heated in a furnace in ultrahigh-purity argon gas atmosphere. After melting, inside of the furnace was cooled at a rate between 1 and $50 \, \mathrm{K} \ \mathrm{min}^{-1}$. The images of the sample during melting and crystallization were monitored and recorded on videotape. After crystallization, orientation analysis was performed by the electron backscattering diffraction pattern (EBSP) method.

Poly-Si ingots were grown by directional growth using a Bridgman-type vertical furnace. The source materials of Si (p-type, $1 \sim 5 \ \Omega$ cm) were placed in a silica crucible of 50 or 150 mm in diameter. The sample was melted at 1,723 K in argon gas atmosphere in the furnace. Subsequently, the crucible was pulled down in a temperature gradient zone of $20 \, \mathrm{K} \ \mathrm{cm}^{-1}$. The pulling conditions were carefully chosen to control the crystal structures. Polycrystalline structures and the solar cell properties of the ingots were investigated.

11.3 Results and Discussion

11.3.1 Direct Observations of Crystal Growth Behavior of Si Melt

The crystal growth behavior of Si melt was directly observed using an in situ observation system to clarify the growth mechanism for controlling the grain orientation and the grain size.

Figure 11.1 shows the growth behavior of the sample cooled at $1 \, \mathrm{K} \ \mathrm{min}^{-1}$ [14]. The crystal growth started at the crucible wall and the growth interface

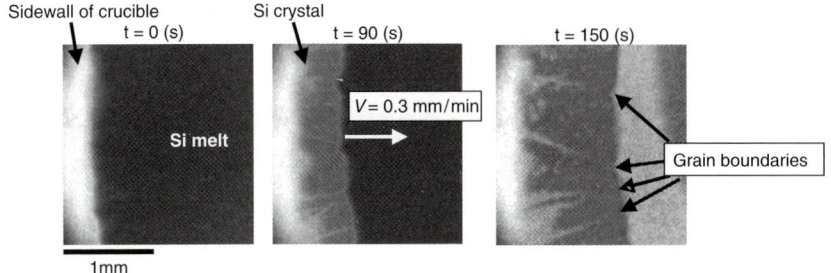

Fig. 11.1. Directional growth of poly-Si from a melt cooled at $1\,\mathrm{K}\ \mathrm{min}^{-1}$. The white lines in the growing crystal are grain boundaries

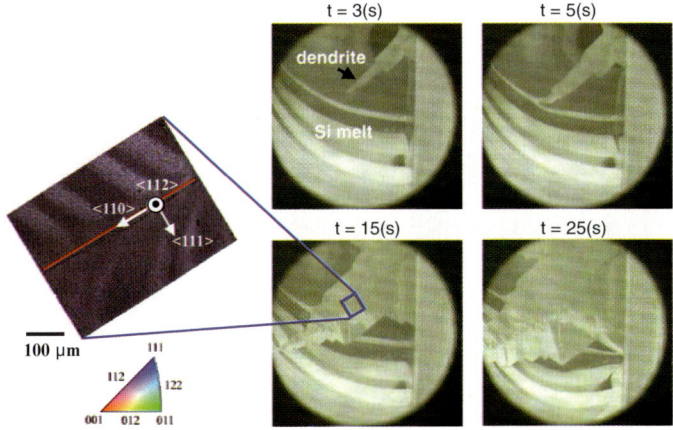

Fig. 11.2. Growth of Si-faceted dendrite from a melt cooled at $50\,\mathrm{K}\ \mathrm{min}^{-1}$ and EBSP color map. A dendrite grain rapidly grew in the initial stage of growth. The rapid growth direction of the dendrite grain was <110> and the upper plane was (112)

moved from left to right with growth. The moving velocity of the growth interface was $0.3\,\mathrm{mm}\ \mathrm{min}^{-1}$, which was similar to the growth rate of poly-Si ingot in a commercially used casting. We also observed many grain boundaries in the growing crystal. Generally, many crystal grains with different orientation are formed during casting as observed in Fig. 11.1.

Figure 11.2 shows the growth behavior of the sample cooled at $50\,\mathrm{K}\ \mathrm{min}^{-1}$ [14]. In this sample, the crystal first rapidly grew in one direction, expanded symmetrically with respect to this direction, and formed a large grain. This type of growth behavior is called faceted-dendrite growth [6]. The growth direction of this faceted dendrite was determined using EBSP. The orientation for the upper plane of the dendrite grain is shown in color in the EBSP orientation map in Fig. 11.2. It was found that the rapid growth direction was <110> and the upper plane was (112). There existed a (111) twin plane at the center of the dendrite.

There has been much research on Si or Ge faceted dendrites [6–13]. Recently, Nagashio and Kuribayashi reported four types of facet Si dendrite obtained by rapid quenching experiments, using a containerless electromagnetic levitation (EML) method [13]. They showed that when the undercooling was lower than $100\,K$, <112> or <110> dendrites were grown. The dendrite observed in Fig. 11.2 is considered to be the same as the <110> dendrite reported by Nagashio and Kuribayashi. Note that we could induce the dendrite growth at the initial stage of directional growth even in the silica crucible. This indicates that dendrite growth is possible by a casting to obtain a poly-Si ingot with large grains with specific orientations.

11.3.2 Dendritic Casting Method

In cast poly-Si, grain boundaries and various orientations are considered problems in terms of solar cell applications. To solve these problems, a poly-Si ingot should have a structure with large oriented grains. As shown in Fig. 11.2, inducing a dendrite growth in the initial stage of directional growth is effective for obtaining such a structure. If we can use this growth mechanism in the casting, we will be able to obtain a poly-Si ingot suitable for solar cells.

Figure 11.3 shows a concept of growing an ideal textured poly-Si ingot using dendrite growth during casting [14]. Dendrite growth should be induced along the bottom wall of a crucible in the initial stage of directional growth by controlling the cooling conditions. The upper orientation of a dendrite grain could be controlled by adjusting the degree of undercooling [13]. Then,

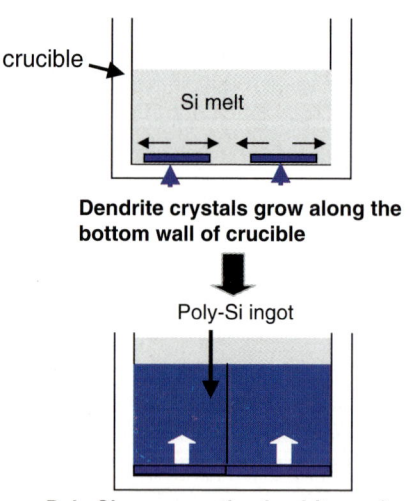

Dendrite crystals grow along the bottom wall of crucible

Poly-Si grows on the dendrite surface

Fig. 11.3. Concept of dendritic casting method to obtain textured polycrystalline Si ingots with large grains. Dendrite growth along the bottom wall of the crucible is used in the initial stage of directional growth

the upper planes of most dendrite grains should have the same orientation if the degree of undercooling were to be equal at any parts of the melt on the bottom wall of the crucible. In this way, polycrystalline substrate with dendrite grains controlled orientation is formed at the bottom of the ingot. Subsequently, crystallization should be promoted on this substrate to upper. Then, a textured poly-Si ingot with large-size grains could be obtained. This new idea for growing textured poly-Si ingot shown in Fig. 11.3 does not require the use of any seeds and can grow a large ingot by a commercial casting technique.

11.3.3 Growth of Poly-Si Ingot by Dendritic Casting Method

Poly-Si ingots were grown with a 50 mm diameter silica crucible using a small casting furnace. First, the bottom part of the crucible was cooled fast to induce dendrite growth and then the crucible was pulled down at 0.2 mm min^{-1} in the temperature gradient zone of 20 K cm^{-1}. Another ingot was grown by simply pulling the crucible down at 0.2 mm min^{-1} in the same temperature gradient zone. Figure 11.4a,b shows the orientation maps of 15 mm^2 wafers cut from the bottom, middle, and top parts of the ingots grown by (a) dendritic casting method and (b) normal casting method [14]. The (112) plane was seen to occupy almost the entire surface of the wafers and large grains

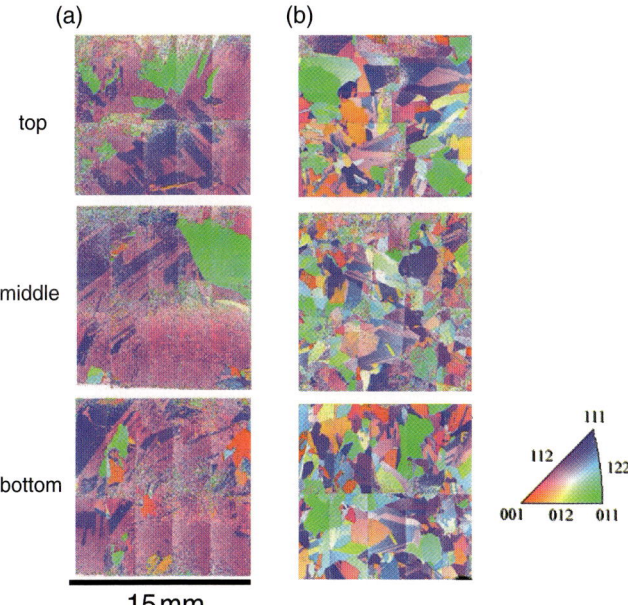

Fig. 11.4. EBSP maps of Si wafers cut from (**a**) ingot grown by dendritic casting method and (**b**) ingot grown by normal casting method

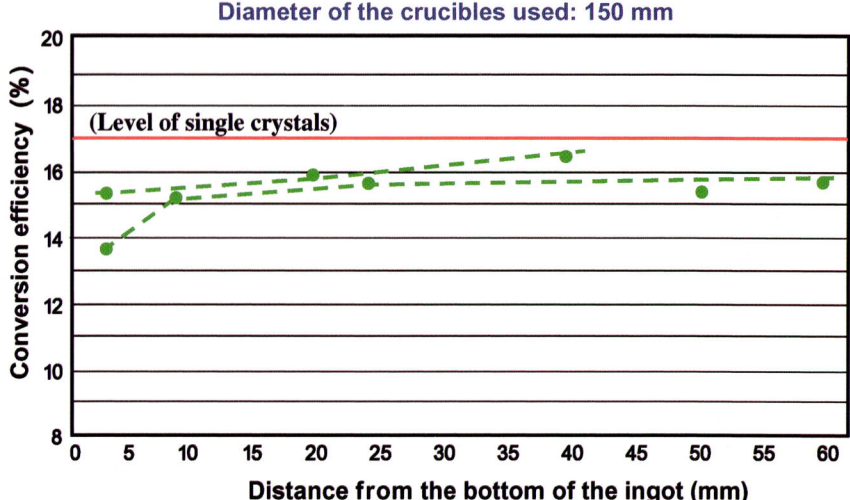

Fig. 11.5. Energy conversion efficiency of solar cells based on poly-Si wafers cut from the ingots with 150 mm diameter grown by dendritic casting method

were observed in the bottom part to top part in Fig. 11.4a, because <110> dendrites grew along the bottom wall of the crucible in the initial stage. In contrast, many equiaxed grains with random orientations were observed in the sample shown in Fig. 11.4b. It was confirmed that we could control not only the grain orientation but also grain boundary density in the poly-Si ingot by using dendritic casting method. We compared the solar cell properties in the two ingots. The higher energy conversion efficiency was obtained for the ingot grown by dendritic casting method except for initially grown part. Thus, we applied this growth method to the growth of a larger ingot of 150 mm diameter. The average grain size was very large, more than 30 mm. Figure 11.5 shows the energy conversion efficiencies of solar cells cut from the bottom part to top part of the larger ingots. The highest efficiency of the solar cells was more than 16% even without a surface textural structure or hydrogen passivation during solar cell processes. Furthermore, the efficiency was highly maintained from the bottom to the top in the ingot, which is a great advantage from a view of a material yield. For comparison, we also fabricated a solar cell based on an sc-Si wafer grown by the Czochralski method using the same solar cell process and obtained a conversion efficiency of 17%. It was confirmed that the crystal quality of our poly-Si ingot grown by dendritic casting method is as high as that of sc-Si. The results in Figs. 11.4 and 11.5 show that the dendritic casting method proposed in this study is promising for obtaining a poly-Si ingot suitable for solar cells.

11.4 Conclusion

We investigated the crystal growth behavior of poly-Si using an in situ observation system. We found that dendrite growth along the bottom wall of the crucible in the initial stage of directional growth is useful for obtaining a textured poly-Si with large grains. We have proposed the concept of poly-Si growth using dendrite growth by a casting. Based on this concept, high-quality poly-Si ingots were obtained by casting. The solar cell properties of this poly-Si were similar to those of sc-Si. The proposed concept using dendrite growth is promising for obtaining high-quality poly-Si ingots suitable for solar cells.

Acknowledgement. The authors thank Dr. G. Sazaki for helpful discussions and Dr. M. Kishimoto for technical assistance. This work was supported by the New Energy and Industrial Technology Development Organization (NEDO) and a Grant-in-Aid for Scientific Research from the Ministry of Education, Culture, Sports, Science and Technology of Japan.

References

1. A. Fedotov, B. Evtodyi, L. Fionova, Yu. Ilyashuk, E. Katz, L. Polyak, Phys. Stat. Sol. A **119**, 523 (1990)
2. Z. Wang, S. Tsurekawa, K. Ikeda, T. Sekiguchi, T. Watanabe, Interf. Sci. **7**, 197 (1990)
3. F. Restrepo, C.E. Backus, IEEE Trans. Electron. Devices **23**, 1195 (1976)
4. H.F.W. Dekkers, F. Duerinckx, J. Szlufcik, J. Nijs, in *Proceedings of 16th European Photovoltaic Solar Energy Conference*, Glasgow, 2000, p. 1532
5. K. Fujiwara, Y. Obinata, T. Ujihara, N. Usami, G. Sazaki, K. Nakajima, J. Cryst, Growth **262**, 124 (2004)
6. E. Billig, Proc. Roy. Soc. A **229**, 346 (1955)
7. D.R. Hamilton, R.G. Seidensticker, J. Appl. Phys. **31**, 1165 (1960)
8. R.S. Wagner, Acta Metall. **8**, 57 (1960)
9. G. Devaud, D. Turnbull, Acta Metall. **35**, 765 (1987)
10. C.F. Lau, H.W. Kui, Acta Mater. **41**, 1999 (1993)
11. K.K. Leung, H.W. Kui, J. Appl. Phys. **75**, 1216 (1994)
12. K. Nagashio, H. Murata, K. Kuribayashi, Acta Mater. **52**, 5295 (2004)
13. K. Nagashio, K. Kuribayashi, Acta Mater. **53**, 3021 (2005)
14. K. Fujiwara, W. Pan, N. Usami, K. Sawada, M. Tokairin, Y. Nose, A. Nomura, T. Shishido, K. Nakajima, Acta Mater. **54**, 3191 (2006)

Floating Cast Method as a New Growth Method of Silicon Bulk Multicrystals for Solar Cells

I. Takahashi, Y. Nose, N. Usami, K. Fujiwara, and K. Nakajima

Summary: We propose a new growth method named as floating cast method to realize high-quality Si bulk multicrystals for solar cells. The fundamental concept of floating cast method is to perform crystal growth from the top center of Si melt in a crucible without contacting the inner wall until the melt is entirely solidified. By using a small scale furnace, crystals were grown under the concept of floating cast method and by the conventional casting method. As a result, resistivity in the crystal grown by floating cast method was found to be higher, which implies less contamination from the coating material at the crucible wall. In addition, fewer defects such as grain boundaries and dislocations were observed. The conversion efficiency of solar cells based on the crystals grown by floating cast method was revealed to be higher than that by the conventional casting method. Therefore, floating cast method, we propose in this paper, is much promising, which could potentially solve problems in the conventional casting methods and realize high-quality Si bulk multicrystals with fewer impurities and defects.

12.1 Introduction

Si multicrystal ingots grown by directional solidification based on casting have become dominant material for solar cells, because the production cost of Si multicrystals is lower than that of Si single crystals [1]. However, the conversion efficiency of solar cells based on Si multicrystals has still been lower than that of Si single crystals for several reasons. A principal reason is the recombination of photocarriers at defects such as grain boundaries, dislocations, impurities, etc. [2–5]. These defects are introduced during the growth process, in which Si melt in the quartz crucible coated by Si_3N_4 is solidified from the bottom to the top by cooling down in a temperature gradient. In this method, grain size is limited to small size due to the generation of many nuclei at the bottom surface of the crucible. This is inevitable unless intelligent control such as "dendritic casting method" developed by Fujiwara et al. [6, 7] is not intended. Since crystals are in contact with the crucible during growth, stress is introduced originating from increase of Si volume when melt is solidified

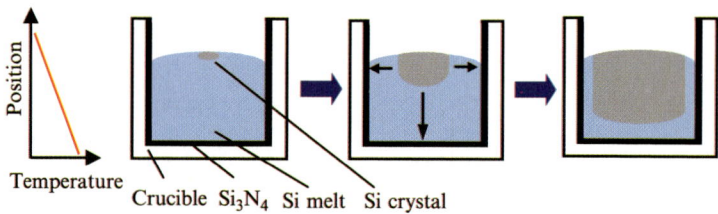

Fig. 12.1. The concept of floating cast method; First, the temperature of the melt at the top is controlled to be lower than that of the bottom. Second, nucleation is controlled to occur only at the top center of Si melt. Third, the growth process is controlled so that the crystal does not contact with the crucible until the final stage of solidification

and difference in expansion coefficient between quartz and Si. Moreover, introduction of impurities through diffusion from coating materials cannot be avoided [8].

To realize high-quality Si multicrystal ingots with less defects and impurities, we propose a new growth technique, named as floating cast method, based on solidification of Si melt in a crucible so that the advantage of the conventional casting method in terms of the production cost is maintained. Figure 12.1 shows the concept of floating cast method. First, nucleation is controlled to occur only at the top center of Si melt in a temperature gradient contrary to conventional method. In addition, since the fundamental limitations in the conventional growth method come from the fact that crystals are in contact with the crucible during growth, we control the growth process so that the crystal does not contact with the crucible until the final stage of solidification. The crystal is therefore "floating" in melt during growth.

In floating cast method, grain size is expected to increase because the crystal growth is initiated with fewer nuclei due to limitation of nucleation region at the top center of the Si melt. In addition, stress and impurity contamination will be decreased since the time when the crystal is in contact with the crucible is dramatically reduced. Hence, the conversion efficiency of solar cells is considered to be improved by decrease of the defects and impurities. In this study, we carried out crystal growth under the concept of floating cast method using a small furnace by appropriately controlling the temperature gradient. Impacts of the difference in growth methods were investigated by comparing the microstructures of Si multicrystals and solar cell properties.

12.2 Experimental Procedure

Si crystals were grown by both methods in a different temperature gradient using a small furnace, as can be seen in Fig. 12.2. Non-doped Si source was charged into the crucibles coated with a Si_3N_4 layer to prevent the crystals from sticking and cracking during the cooling process. Si source was melted

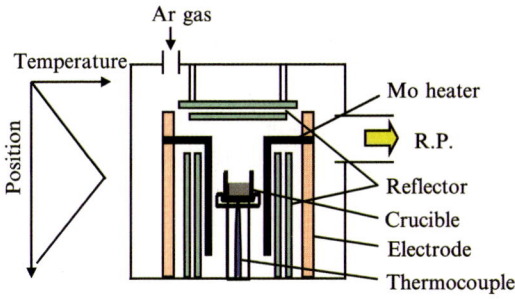

Fig. 12.2. Schematic of the apparatus; A temperature gradient can be changed by shifting the position of the crucible, since the temperature at the center of the heater is the highest

at 1,723 K in argon gas atmosphere and crystals were grown by cooling down at a rate of 0.1 K min^{-1} in the temperature gradient. The sign of the temperature gradient along the crucible could be changed by shifting the position of the crucible. For floating cast method, temperature of the top is controlled to be lower than that of the bottom in contrast to the case for the conventional method. Crystals with diameter of 23 mm and height of 18 mm were grown. After removal of their edges, they were cut perpendicular to the growth direction to 1.2 mm thickness. Grain orientation was analyzed by the electron back scattering pattern analysis (EBSP) method. Etch-pit density was measured by the method proposed by Sopori [9]. In addition, 10 mm^2 samples taken from the center part of the crystal were processed to solar cells and their properties were evaluated. Details of the processing are reported elsewhere [7]. In addition, electroluminescence (EL) imaging [10] was carried out to visualize defect distribution, which act as recombination center of carriers.

12.3 Results and Discussion

Figure 12.3 shows a comparison of resistivity along the growth direction for crystals grown by floating cast and the conventional methods. It is found that the resistivity of the crystal grown by floating cast method decreases on approaching the bottom in contrast to that by the conventional method. This could be interpreted as a consequence of difference in the growth direction, since segregated impurities will be accumulated at the bottom for floating cast method while at the top for the conventional method. We can confirm from this that we succeeded in crystal growth from the top to the bottom of ingots to satisfy the concept of floating cast method. Furthermore, resistivity of the crystal grown by floating cast method is found to be higher at each position, which implies less contamination reflecting the shorter contact time with the crucible, which could act as source of impurities.

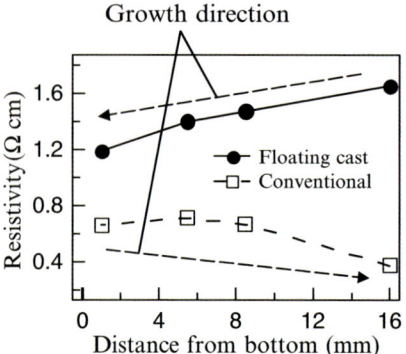

Fig. 12.3. Resistivity along the growth direction for the samples grown by floating cast and conventional methods

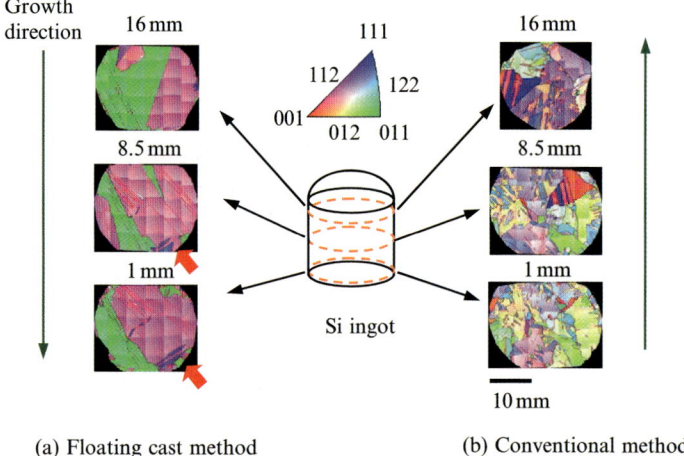

(a) Floating cast method   (b) Conventional method

Fig. 12.4. Orientation maps of the crystals grown by (a) floating cast method and (b) conventional method. Results for three samples, which were cut perpendicular to growth direction at a distance of 1, 8.5, and 16 mm from the bottom of the ingots, are displayed. Reference direction is shown in color. Σ3 grain boundaries are drawn in red and the others are drawn in black

Figure 12.4 shows orientation maps of the top, middle, and bottom parts of the samples grown by (a) floating cast method and (b) conventional method at a common cooling rate of $0.1\,K\,min^{-1}$. In these maps, grain orientation is represented in color. Note that the growth direction is opposite to each other. It is obvious that the grain size of the crystal grown by floating cast method is much larger than that by conventional method. This result is further supported by Fig. 12.5a in detail, which shows grain boundary density defined as boundary length divided by measurement points in various ingot positions. It should be also remarked that orientations of almost all the grains grown

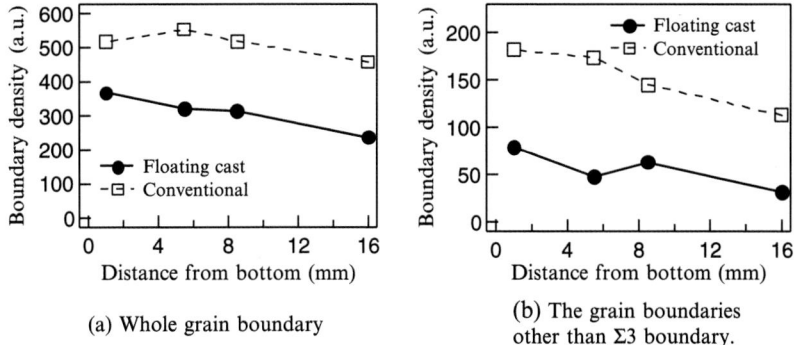

(a) Whole grain boundary

(b) The grain boundaries other than Σ3 boundary.

Fig. 12.5. Grain boundary density defined as grain boundary length divided by measurement points for (**a**) whole grain boundary and (**b**) the grain boundaries other than Σ3 boundary

by floating cast method are found to be <110> or <112>, which could be an advantage for optimization of surface texturing process to realize efficient light trapping when the crystal is processed to solar cells.

As can be seen in Fig. 12.4 and Fig. 12.5, the grain size increases as the position of the ingots rise for both samples in spite of the different growth direction. The reason behind would be different depending on the growth method. In conventional method, the reason would be competition of crystal grains with different orientation as growth proceeds, which is followed by survival and enlargement of crystals with preferential orientation. In floating cast method, unfavorable nucleation of small grains near the crucible wall as indicated by red arrow in Fig. 12.4 drastically affects the average grain, since the number of grains at the top region is so few by control of the nucleation process, which accounts for the observed phenomenon.

Moreover, grain boundary characters were found to be remarkably different between the samples grown by both methods. A ratio of Σ3 boundary length over the whole boundary length in the samples grown by floating cast method is about 85%, which is larger than that in the samples grown by conventional method of 65%. Since Σ3 boundaries are well known as inactive defects for carriers [11], we should consider the density of other grain boundaries to affect the performance of solar cells. In fact, grain boundary density other than Σ3 boundary in the samples grown by floating cast method was found to be more than twice as low as that by conventional method as shown in Fig. 12.5b. Consequently, the Si multicrystal ingot grown by floating cast method is proved to be much suitable material for solar cells.

These advantageous features in the crystals grown by floating cast method, namely the large grain size, limited number of orientation, and high fraction of Σ3 boundary, are considered to originate from the fact that less nucleation at the beginning of solidification.

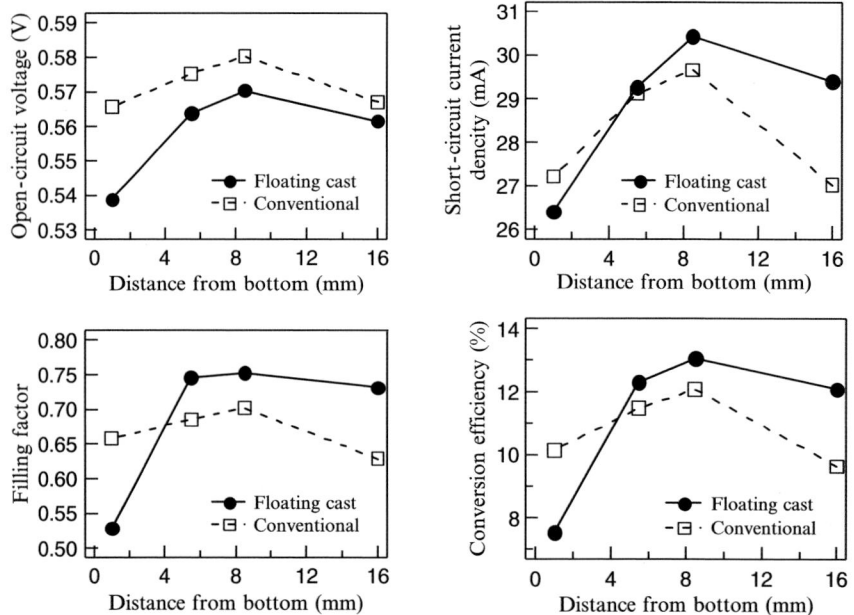

Fig. 12.6. Solar cell properties, which are open-circuit voltage, short-circuit current density, filling factor, and conversion efficiency

Etch-pit density is measured at a distance of 12 mm from the bottom for the samples grown by both methods. Etch-pit densities in the samples grown by floating cast and conventional methods are 1.2×10^4 and $3.9 \times 10^4\,\mathrm{cm}^{-2}$, respectively. The decrease of the etch-pit in the crystal grown by floating cast method could be attributed to reduction of stress originating from the contact with the crucible at the nucleation.

Solar cell properties along the ingots grown by both methods are shown in Fig. 12.6. Open-circuit voltage of the solar cells based on crystals grown by floating cast method, which is related to dopant density, is shown to be lower than that by conventional method. This result is in good agreement with resistivity measurement, where resistivity of the crystal grown by floating cast method was revealed to be higher than that by the conventional method. On the other hand, short-circuit current density of the solar cells density of the solar cells based on crystals grown by floating cast method is higher than that by conventional method, except the bottom position. This result indicates less recombination center of carriers such as the defects and impurities. Likewise, filling factor is kept high for solar cells based on crystals grown by floating cast method, except the bottom position. This result is attributed to the reduction of defects in crystals grown by floating cast method, which decreases shunt current. As a consequence, conversion efficiency of the crystals grown by floating cast method except the bottom is superior to that by conventional method.

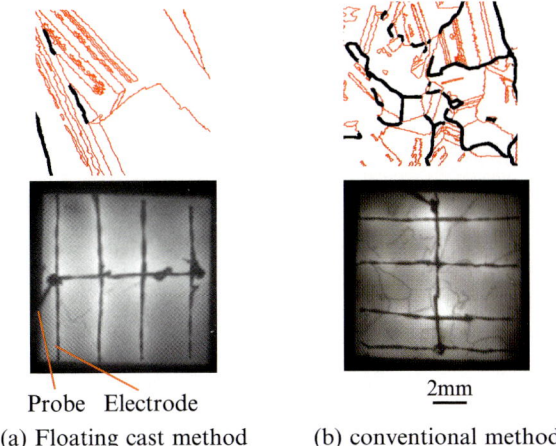

Probe Electrode $\frac{2mm}{}$
(a) Floating cast method   (b) conventional method

Fig. 12.7. EL image and grain boundary distributions of the crystal grown by (**a**) floating cast and (**b**) conventional methods; $\Sigma 3$ grain boundaries are drawn in red and the others are drawn in black

The inferred properties of the bottom part are presumably due to the stress owing to the contact with the crucible at the final stage of the growth and accumulation of impurities.

To further investigate the reason for the improved solar cell performance, EL measurements were carried out. Figure 12.7 shows comparison of EL images (bottom row) and grain boundary (top row) distributions of samples grown by both methods. Dark lines in EL images represent where carriers nonradiatively recombine. $\Sigma 3$ and the other boundaries are drawn in red and black, respectively. Although many $\Sigma 3$ grain boundaries can be found in both samples, most of them are invisible in the EL images. Most of the dark lines are confirmed to correspond with grain boundaries other than $\Sigma 3$ boundary. It should be stressed that dark lines can be seen not only at grain boundaries but also at intra-grain for the sample grown by the conventional method. This dark region suggests the existence of subgrain boundaries to consist of dislocation clusters. According to EL measurement, decrease of defects to act as recombination center of carriers is concluded to be the reason for the superior performance of solar cells based on the crystal grown by floating cast method to that by conventional method.

Finally, it should be commented that a yield of ingot, which could be utilized as solar cells, is expected to be higher for the ingot grown by floating cast method since only the bottom part is not suitable. This is in contrast with the ingot grown by the conventional method, where the top and bottom regions are not to impurities. Therefore, floating cast method is profitable not only to improve conversion efficiency but also to reduce the production cost by increasing the yield of available part. Although more advanced control of the growth process and implementation of floating cast method to larger scale furnace must be accomplished, our must be accomplished, our small-scale

experiments showed that floating cast method is promising as an alternative
practical growth technique for high-quality and high-yield Si bulk multicrystals for solar cell applications.

12.4 Summary

We proposed a new growth method named as floating cast method and demonstrated the superiority of floating cast method by growing the crystals following its growth concept. By starting with nucleation at the top center of the
melt and growing from the top to the bottom, following features of the Si bulk
multicrystal grown by floating cast method were observed:

- The crystal contains much large grains with only a few grain orientations.
- More than 85% of the grain boundaries is Σ3.
- The etch-pit density is smaller than the crystal grown by conventional
  method.

These features lead to high conversion efficiency of solar cells based on the
crystals grown by floating cast method due to lower recombination centers of
carriers as demonstrated by solar cell properties and EL images. In addition,
only the bottom part was not suitable for solar cells in contrast to the conventional method where the top and bottom parts are not available. Therefore,
we can conclude that floating cast method is considered to be much promising
for the growth of high-quality and high-yield Si bulk multicrystals for solar
cells, which could contribute to the cost reduction in the solar cell production
by increasing the solar cell performance and material yield.

References

1. P.V. News, Spring (2006)
2. J. Chen, D. Yang, Z. Xi, T. Sekiguchi, Phys. B **364**, 162 (2005)
3. J. Chen, T. Sekiguchi, R. Xie, P. Ahmet, T. Chikyo, D. Yang, S. Ito, F. Yin,
   Scripta Mater. **52**, 1211 (2005)
4. H. Sugimoto, M. Tajima, T. Eguchi, I. Yamaga, T. Saitoh, Mater. Sci.
   Semiconductor Process. **9**, 102 (2006)
5. E.R. Weber, Appl. Phys. A **30**, 1 (1983)
6. K. Fujiwara, W. Pan, K. Sawada, M. Tokairin, N. Usami, Y. Nose, A. Nomura,
   T. Shishido, K. Nakajima, J. Cryst. Growth **292**, 282 (2006)
7. K. Fujiwara, W. Pan, N. Usami, K. Sawada, M. Tokairin, Y. Nose, A. Nomura,
   T. Shishido, K. Nakajima, Acta Mater. **54**, 3191 (2006)
8. L. Liu, S. Nakano, K. Kakimoto, J. Cryst. Growth **292**, 515 (2006)
9. B.L. Sopori, J. Electrochem. Soc. **131**, 667 (1984)
10. T. Fuyuki, H. Kondo, T. Yamazaki, Y. Takahashi, Y. Uraoka, Appl. Phys. Lett.
    86, 262108 (2005)
11. W. Seifert, G. Morgenstern, M. Kittler, Semiconductor Sci. Technol. **8**,
    1687 (1993)

13

Dehydriding Reaction of Hydrides Enhanced by Microwave Irradiation

M. Matsuo, Y. Nakamori, K. Yamada, T. Tsutaoka, and S. Orimo

Summary: Effects of microwave irradiation on the dehydriding reaction of metal hydrides MH_n (LiH, MgH_2, CaH_2, TiH_2, $VH_{0.81}$, ZrH_2 and $LaH_{2.48}$) and complex hydrides MBH_4 ($LiBH_4$, $NaBH_4$ and KBH_4) were systematically investigated. Among the metal hydrides, TiH_2, $VH_{0.81}$, ZrH_2, and $LaH_{2.48}$ exhibited a rapid heating by microwave irradiation, where a small amount of hydrogen (less than 0.5 mass%) were released. On the other hand, $LiBH_4$ was heated above 380 K by microwave irradiation, where 13.7 mass% hydrogen was released. The rapid heating of metal hydrides such as TiH_2, $VH_{0.81}$, ZrH_2, and $LaH_{2.48}$ are mainly due to the conductive loss. Meanwhile the microwave heating in $LiBH_4$ is attributed to the conductive loss, which is caused by a structural transition. Furthermore, microwave was irradiated to the composites of $LiBH_4$ and TiH_2. The composites exhibited faster temperature increases than pure $LiBH_4$, which resulted in faster dehydriding reaction of $LiBH_4$. Microwave heating might be applied to hydrogen storage system, though further development of hydrides themselves and engineering techniques are required.

13.1 Introduction

Microwave irradiation techniques have attracted considerable attention in recent years as a clean and energy saving process for the preparation of inorganic and organic materials [1–3]. It has been reported that microwaves enhance the diffusion of composed elements in comparison with the external heating process by an electric furnace because microwave energy is absorbed directly by the materials [4]. Thus, reactions proceed very fast even at low temperatures, which results in a significant reduction in the processing cost and time.

Also some types of hydrides have attracted considerable attention as hydrogen storage materials for developing hydrogen energy systems. The required properties on hydrogen storage materials are low reaction temperature and fast kinetics, etc. For decreasing the reaction temperature, the stability of hydrides was controlled by alloying [5, 6], fabricating appropriate composites [7–9], and considering the electronegativity of cation [10] in metal hydrides and complex hydrides. However, the reaction kinetics generally becomes slower at lower reaction temperatures. In particular, in the complex hydrides, the fast diffusion of not only the hydrogen but also the other elements is required, which is very difficult to be achieved at ambient temperature.

Recently, we have reported that both $LiBH_4$ and TiH_2 can be heated up rapidly and dehydriding (hydrogen release) reactions proceed by multimode microwave irradiation [11]. Furthermore, it has been predicted that RH_{3-x} (R = rare earth), ZrH_2 and VH_2, can be heated up rapidly by microwave irradiation.

In this study, to prove this prediction, the effects of microwave irradiation on metal hydrides, MH_n (LiH, MgH_2, CaH_2, TiH_2, $VH_{0.81}$, ZrH_2, and $LaH_{2.48}$), were systematically investigated, where the single-mode microwaves were used to extract the effects of electric field of the irradiated microwaves. To clarify the origin of rapid heating by microwave irradiation, the measurements of the effective dielectric constant ε_r' and loss value ε_r'' were carried out for TiH_2 and $LiBH_4$, which show the rapid dehydriding reactions.

13.2 Experimental

Metal hydrides MH_n (LiH, MgH_2, CaH_2, TiH_2, $VH_{0.81}$, ZrH_2, and $LaH_{2.48}$) and complex hydrides MBH_4 ($LiBH_4$, $NaBH_4$, and KBH_4) were purchased from Aldrich Co. Ltd. Microwave irradiation for the hydrides was carried out as follows: approximately 0.3–$0.4\,cm^3$ of hydride was placed in a BN crucible, which was placed in an airtight Teflon container inside an argon glove box. The container, which was equipped with a K-type thermocouple, was inserted into a microwave cavity (IDX Corp. MS1109A-001, single-mode with 400 W, 2.45 GHz) at the point of maximum electric field, wherein microwave irradiation was carried out for 10–70 min. To investigate the dehydriding reaction by the microwave irradiation, the powder X-ray diffraction measurements (PANalytical X'PERT with Cu Kα radiation) and the hydrogen analysis (Horiba, EMGA-621W) were performed for the samples before and after microwave irradiation. The effective dielectric constant ε_r' and loss value ε_r'' were measured by the coaxial line method using a network analyzer (Advantest R3765G) in the frequency range of 0.1–3 GHz. During these experiments, the samples were handled in a glove box filled with purified argon (dew point below 183 K) by using special airtight container.

13.3 Results and Discussion

13.3.1 Microwave Irradiation on Metal Hydrides MH_n

Figure 13.1 shows the temperature changes as a function of the microwave irradiation time for LiH, MgH_2, CaH_2, TiH_2, $VH_{0.81}$, ZrH_2, and $LaH_{2.48}$.

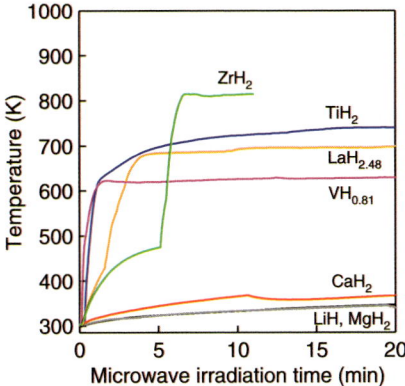

Fig. 13.1. (Color online). Temperature changes as a function of the microwave irradiation time for LiH, MgH_2, CaH_2, TiH_2, $VH_{0.81}$, ZrH_2, and $LaH_{2.48}$

The temperature changes of LiH and MgH_2 were the same as that without sample (not shown), indicating the ineffective heating by microwave irradiation. CaH_2 exhibited a slightly higher temperature than that without sample. On the other hand, rapid temperature increases were observed in TiH_2, $VH_{0.81}$, ZrH_2, and $LaH_{2.48}$ by microwave irradiation. A considerable change in the temperature increasing rate was observed at 1, 5, and 2 min for TiH_2, ZrH_2, and $LaH_{2.48}$, respectively. This change may be caused by the change of the dielectric loss values by partial dehydriding reaction. To investigate the dehydriding reaction by microwave irradiation, powder X-ray diffraction measurements were performed for TiH_2, $VH_{0.81}$, ZrH_2, and $LaH_{2.48}$, which were heated up rapidly. As shown in Fig. 13.2, the powder X-ray diffraction peaks observed after microwave irradiation became broader and smaller than those observed before irradiation. These changes are probably due to the effect of the electromagnetic field and/or the rapid cooling rate. The diffraction peaks of TiH_2 and $VH_{0.81}$ shifted to higher angles after microwave irradiation, while those of ZrH_2 and $LaH_{2.48}$ did not change by microwave irradiation. These results indicate that the progression of the partial dehydriding reactions proceeded in TiH_2 and $VH_{0.81}$ [12]; the crystal structure of TiH_2 also transforms from cubic to tetragonal by microwave irradiation [13]. The amount of released hydrogen by microwave irradiation was deduced by subtracting the amount of hydrogen after irradiation from that before irradiation. The small amount of hydrogen, 0.3 and 0.5 mass%, were released from TiH_2 and $VH_{0.81}$, while no hydrogen release was confirmed from ZrH_2 and $LaH_{2.48}$. The possible reason for the small amount of released hydrogen by microwave irradiation will be discussed later.

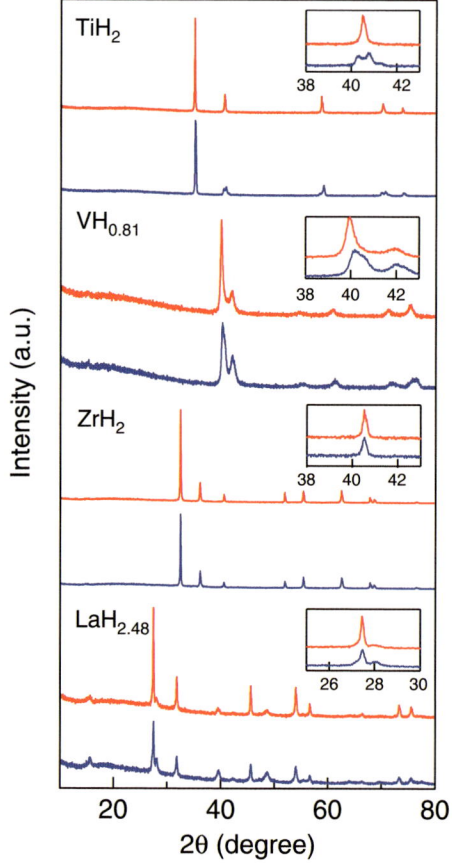

Fig. 13.2. Powder X-ray diffraction profiles of TiH_2, $VH_{0.81}$, ZrH_2, and $LaH_{2.48}$ before and after microwave irradiation. The diffraction profiles at top and bottom in each panel show before and after microwave irradiation, respectively. Insets show the enlargements of (200), (101), (002), and (111) peaks for TiH_2, $VH_{0.81}$, ZrH_2, and $LaH_{2.48}$, respectively

13.3.2 Microwave Irradiation on Complex Hydrides MBH_4

The temperature changes of MBH_4 ($LiBH_4$, $NaBH_4$, and KBH_4) during microwave irradiation are shown in Fig. 13.3. The following two tendencies were observed: ineffective heating for $NaBH_4$ and KBH_4, and a rapid heating above 380 K for $LiBH_4$. In the powder X-ray diffraction profile, small peaks of LiH were confirmed after microwave irradiation for $LiBH_4$, as shown in Fig. 13.4. Thus, the dehydriding reaction in $LiBH_4$ is expressed as follow [14]:

$$LiBH_4 \rightarrow LiH + B + 3/2H_2. \qquad (13.1)$$

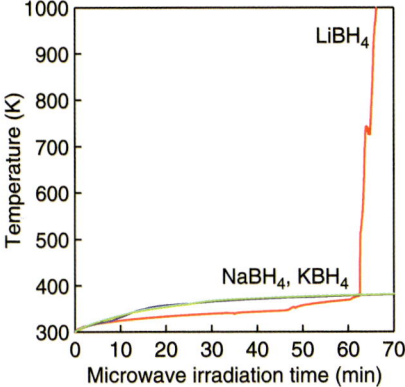

Fig. 13.3. (Color online). Temperature changes as a function of the microwave irradiation time for LiBH₄, NaBH₄, and KBH₄

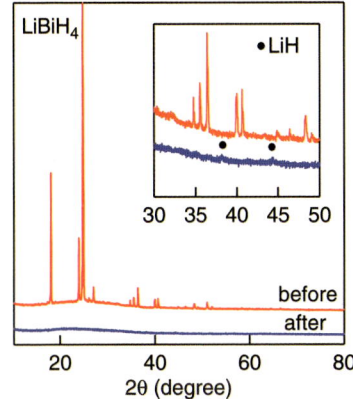

Fig. 13.4. Powder X-ray diffraction profiles of LiBH₄ before and after microwave irradiation. Insets show the enlarged X-ray pattern in the 2θ range from $30°$ to $50°$

This is the same overall dehydriding reaction as that by heating in an electric furnace, where 13.8 mass% hydrogen is released theoretically. The hydrogen analyses before and after irradiation indicated that 13.7 mass% hydrogen was released by microwave irradiation, which is in good agreement with the theoretical value obtained from (13.1). However, no diffraction peak of B was observed after microwave irradiation. Thus, the precipitated B is considered to be in an amorphous phase [15].

13.3.3 Mechanism of Rapid Heating of TiH_2 and $LiBH_4$

Metal hydrides, TiH_2, $VH_{0.81}$, ZrH_2, and $LaH_{2.48}$, were heated up rapidly by microwave irradiation, whereas small amounts of hydrogen (less than

0.5 mass%) were released by microwave irradiation. On the contrary, complex hydride, LiBH$_4$, released all amount of hydrogen at a temperature above 380 K by microwave irradiation. The possible reasons of the different behavior by microwave irradiation are briefly discussed on the basis of dielectric properties of the materials.

Generally, the generation power P by microwave irradiation is expressed as follows [4]:

$$P = 2\pi f E^2 \varepsilon''. \tag{13.2}$$

Here, f, E, and ε'' are frequency, electric field, and dielectric loss value, respectively. In the conductor, the dielectric loss ε'' can be expressed as follows:

$$\varepsilon'' = \frac{\sigma}{2\pi f} + \varepsilon''_{dipole}, \tag{13.3}$$

where σ is conductivity of the material, namely, $\frac{\sigma}{2\pi f}$ is the contribution from the conductivity loss and ε''_{dipole} is that from the dipole loss. In this study, f and E are invariable. Therefore, dielectric loss value ε'' of the material is an important factor for heating by microwave irradiation. The effective dielectric constant ε'_r and loss value ε''_r were measured for TiH$_2$ and LiBH$_4$, and the results are shown in Fig. 13.5 [11]. The ε''_r value of TiH$_2$ is inversely

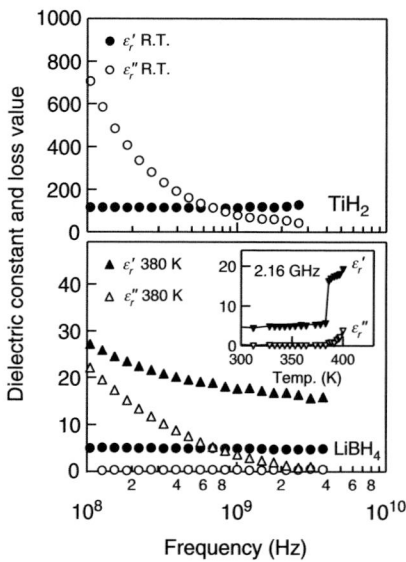

Fig. 13.5. The relative dielectric constant ε'_r (*closed symbols*) and loss value ε''_r (*opened symbols*) of TiH$_2$ (*top*) and LiBH$_4$ (*bottom*) as a function of frequency [11]. The measurements were carried out at room temperature for TiH$_2$ and at room temperature (*circles*) and 380 K (*triangles*) for LiBH$_4$. Inset shows the temperature dependence of the ε'_r and ε''_r values of LiBH$_4$ at a frequency of 2.16 GHz. Sample was pressed into toroidal shape with thickness of approximately 1 mm without sintering. Therefore, the results of ε'_r and ε''_r values are apparent values composed of the sample and argon gas

proportional to the frequency at room temperature. So, it was inferred that the microwave heating of TiH_2 is caused by the conductive loss. In fact, the electrical resistivity of $TiH_{1.86}$ is approximately $120\,\mu\,\Omega$ cm at room temperature; in other words, $TiH_{1.86}$ is a metallic conductor [16]. $VH_{0.81}$, ZrH_2, and $LaH_{2.48}$ also show metallic conduction [16]. Therefore, these metal hydrides were heated up rapidly mainly due to the conductive loss. The rest of hydrides, MH_n, LiH, MgH_2, and CaH_2, which indicated ineffective heating by microwave irradiation, have ionic bond and are insulators. These results indicate that the hydrides having a relatively high electrical conductivity can be heated up rapidly by microwave irradiation.

On the other hand, $LiBH_4$ is an insulator; its ε_r' and ε_r'' values are approximately 5 and 0, respectively, over the entire frequency range at room temperature. This is a typical frequency dependence of insulator without dielectric dipole loss. However, ε_r' and ε_r'' drastically increase at temperatures higher than 380 K, at which rapid increase in temperature was observed by microwave irradiation. The crystal structure of $LiBH_4$ transforms from orthorhombic to hexagonal at approximately 380 K, at which the $[BH_4]^-$ tetrahedron is rearranged along the c-axis [17]. Therefore, the effective heating at temperatures above 380 K is related to the change in the electrical property by this structural transition. Since the dielectric loss value ε_r'' of $LiBH_4$ is inversely proportional to the frequency at temperatures above 380 K, high electrical conductivity due to Li ion or hydrogen diffusion might be achieved at temperatures above 380 K. More detailed investigations for the conductive and dielectric losses of $LiBH_4$ are now in progress.

13.3.4 Difference of the Amounts of Released Hydrogen Between TiH_2 and $LiBH_4$

One of the reasons for the difference of the amount of released hydrogen can be related to the skin effect. From the Maxwell equation, the penetration depth of the electromagnetic wave δ is expressed as follow:

$$\delta = \left[\frac{\omega^2\varepsilon\mu}{2}\left(\sqrt{1 + \frac{1}{\omega^2\varepsilon^2\rho^2}} - 1\right)\right]^{-\frac{1}{2}}, \tag{13.4}$$

where ω is angular frequency (i.e., $2\pi f$) and μ is magnetic permeability. In the case of the metallic conductor, ρ is so small that the term $\omega^2\varepsilon^2\rho^2$ becomes small enough to unity. Hence, (13.4) is rewritten as follow:

$$\delta = \sqrt{\frac{2\rho}{\omega\mu}}. \tag{13.5}$$

The penetration depth δ of TiH_2 is deduced to be $11\,\mu m$, using $\rho = 120\,\mu\,\Omega$ cm, $\mu_r = 1$ ($\mu_r = \mu/\mu_0$; μ_0 is the magnetic permeability of vacuum), and $f = 2.45\,GHz$. On the other hand, the penetration depth of the insulator

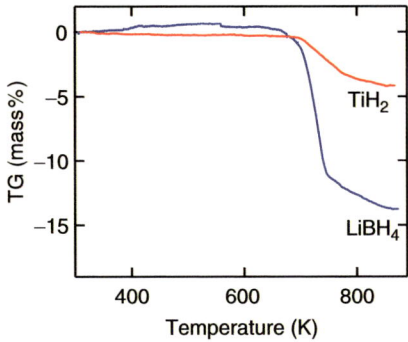

Fig. 13.6. (Color online). Thermogravimetry of TiH_2 and $LiBH_4$ upon heating in an electric furnace. (The He gas flow rate is $100\,ml\,min^{-1}$ and the heating rate is $5\,K\,min^{-1}$)

is considerably large [4]. The particle size of TiH_2 and $LiBH_4$ is approximately $45\,\mu m$ and $50–200\,\mu m$, respectively. Therefore, only a small amount of hydrogen near the surface area can be released from TiH_2, whereas almost all the hydrogen can be released from $LiBH_4$.

Another possibility is the saturation temperature by microwave irradiation. The saturation temperature of TiH_2 is $740\,K$ by microwave irradiation (Fig. 13.1), while the dehydriding reaction finishes at approximately $850\,K$ by using an electric furnace as shown in Fig. 13.6. $LiBH_4$ was heated up to above $1,000\,K$ by microwave irradiation (Fig. 13.3), at which the dehydriding reaction has proceeded completely as shown in Fig. 13.6.

13.3.5 Microwave Irradiation on the Composites of $LiBH_4$ and TiH_2

$LiBH_4$ was heated rapidly at a temperature above $380\,K$ and released $13.7\,mass\%$ hydrogen. The amount of released hydrogen is sufficient for the use of $LiBH_4$ as a hydrogen storage material [18]. However, it took more than $30\,min$ to release hydrogen because the rate of temperature increase was very slow at temperatures below $380\,K$. On the other hand, the temperature of TiH_2 increased very rapidly from the beginning of microwave irradiation, whereas the amount of released hydrogen was very small. Thus, microwaves were irradiated to the composites of $LiBH_4$ and TiH_2 to simultaneously achieve both rapid heating and large amount of hydrogen release.

The temperature changes as a function of the microwave irradiation time for $(1-x)LiBH_4 + xTiH_2$ ($x = 0, 0.5, 0.67, 0.8,$ and 1) are shown in Fig. 13.7. The composites were heated faster as the concentration of TiH_2 increased. On the other hand, with the increase in the amounts of TiH_2, the total amount of released hydrogen decreased in the composites: $5.2\,mass\%$ ($x = 0.5$), $3.3\,mass\%$ ($x = 0.67$), and $1.2\,mass\%$ ($x = 0.8$), because only a small amount

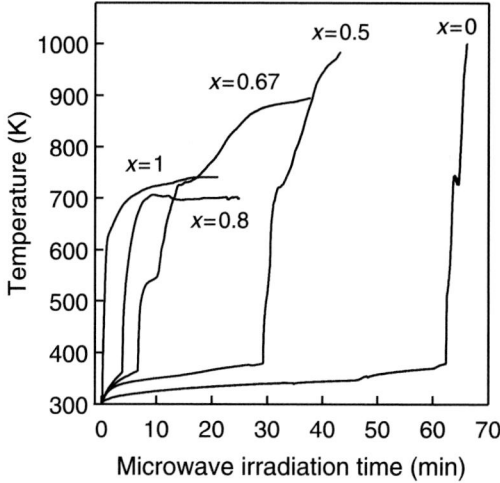

Fig. 13.7. Temperature changes as a function of the microwave irradiation time for $(1-x)\text{LiBH}_4 + x\text{TiH}_2$ ($x = 0$, 0.5, 0.67, 0.8, and 1)

of hydrogen (0.3 mass%) was released for $x = 1$ (TiH_2). This behavior is similar to a superimposition of the individual behaviors of TiH_2 and LiBH_4. Fabricating suitable composite is useful to achieve rapid heating and also to maintain the amount of released hydrogen.

13.4 Summary

The effects of the single-mode microwave irradiation on the dehydriding reactions of metal hydrides, MH_n (LiH, MgH_2, CaH_2, TiH_2, $\text{VH}_{0.81}$, ZrH_2, and $\text{LaH}_{2.48}$), and complex hydrides, MBH_4 (LiBH_4, NaBH_4, and KBH_4), have been systematically investigated. Metal hydrides TiH_2, $\text{VH}_{0.81}$, ZrH_2, and $\text{LaH}_{2.48}$ exhibited a rapid heating by microwave irradiation; a small amount of hydrogen (less than 0.5 mass%) was released. On the other hand, a rapid heating was observed in the complex hydride LiBH_4 at a high temperature above 380 K accompanied by the 13.7 mass% hydrogen release. The rapid heating of TiH_2, $\text{VH}_{0.81}$, ZrH_2, and $\text{LaH}_{2.48}$ are mainly due to the conductive loss, while the microwave heating in LiBH_4 is attributed to the conductive loss, which is caused by a structural transition. The difference in the amount of released hydrogen might be caused by the difference of the microwave penetration depth or the temperature saturation. Furthermore, microwave was irradiated to the composites of LiBH_4 and TiH_2. The composites exhibited faster temperature increases than pure LiBH_4, which resulted in faster dehydriding reaction of LiBH_4. Microwave heating might be applied to hydrogen storage system, though further development of hydrides themselves and engineering techniques are required.

Acknowledgements. This work is the Collaborative Research in Center for Interdisciplinary Research, the Inter-University Cooperative Research Program of Institute for Materials Research, Tohoku University. This study was partially supported by the Ministry of Education, Science, Sports and Culture, Grant-in-Aid for "Scientific Research on Priority Areas #18070005," "Scientific Research (A), #18206073," and by Japan Science and Technology Agency, Promotion of Seed Program.

References

1. J. Luo, C. Hunyar, L. Feher, G. Link, M. Thumm, P. Pozzo, Appl. Phys. Lett. **84**, 5076 (2004)
2. Z.J. Wang, H. Kokawa, H. Takizawa, M. Ichiki, R. Maeda, Appl. Phys. Lett. **86**, 212903 (2005)
3. R. Roy, D. Agrawal, J. Cheng, S. Gedevanshvili, Nature **399**, 668 (1999)
4. A.C. Metaxax, R.J. Meredith, Ind. Microwave Heating (Peter Peregrinus Ltd., London, 1983) 54
5. L. Schlapbach (ed.), *Topics in Applied Physics*, vol. **63** (Springer, Berlin Hiedelberg New York, 1988)
6. G. Sandrock, J. Alloys Comp. **293**, 877 (1999)
7. J.J. Vajo, S.L. Skeith, F. Mertens, J. Phys. Chem. B **109**, 3719 (2005)
8. F.E. Pinkerton, G.P. Meisner, M.S. Meyer, M.P. Balogh, M.D. Kundrat, Phys. Chem. B **109**, 6 (2005)
9. Y. Nakamori, A. Ninomiya, G. Kitahara, M. Aoki, T. Noritake, K. Miwa, Y. Kojima, S. Orimo, J. Power Sources **155**, 447 (2006)
10. M. Aoki, K. Miwa, T. Noritake, G. Kitahara, Y. Nakamori, S. Orimo, S. Towata, Appl. Phys. A **80**, 1409 (2005)
11. Y. Nakamori, S. Orimo, T. Tsutaoka, Appl. Phys. Lett. **88**, 112104 (2006)
12. H.L. Yakel Jr., Acta Crystallogr. **11**, 46 (1958)
13. R.L. Crane, S.C. Chattoraj, M.B. Strope, J. Less Common Met. **25**, 225 (1971)
14. A. Züttel, P. Wenger, S. Rentsch, P. Sucan, Ph. Mauron, Ch. Emmenegger, J. Power Sources **118**, 1 (2003)
15. S. Orimo, Y. Nakamori, G. Kitahara, K. Miwa, N. Ohba, S. Towata, A. Züttel, J. Alloy. Comp. **404**, 427 (2005)
16. K. Gesi, Y. Takagi, T. Takeuchi, J. Phys. Soc. Jpn. **18**, 306 (1963)
17. J.Ph. Soulié, G. Renaudin, R. Cerny, K. Yvon, J. Alloy. Comp. **346**, 200 (2002)
18. L. Schlapbach, A. Züttel, Nature **414**, 353 (2001)

14

Mechanically Multifunctional Properties and Microstructure of New Beta-Type Titanium Alloy, Ti-29Nb-13Ta-4.6Zr, for Biomedical Applications

M. Nakai, M. Niinomi, and T. Akahori

Summary: A new biomedical titanium alloy, Ti-29Nb-13Ta-4.6Zr, composed of nontoxic elements like Nb, Ta, and Zr, has recently been developed in order to achieve a lower Young's modulus similar to that of human hard tissues in addition to excellent mechanical properties for use as structural biomaterials. The characteristics of this material depend on the microstructures obtained by heat treatments or thermomechanical treatments. Therefore, the relationship between microstructures and basic mechanical properties, the Young's modulus, tensile and fatigue properties, etc., has been investigated with respect to Ti-29Nb-13Ta-4.6Zr. Further, the practical performance of this alloy as a metallic biomaterial has been evaluated via animal experiments. Moreover, the mechanical functionalities of this alloy, such as superelasticity, were reported recently. In the present paper, the accumulated reports on Ti-29Nb-13Ta-4.6Zr were summarized and the mechanical properties of this alloy were mainly reviewed in order to evaluate its potential for biomedical applications.

14.1 Introduction

Titanium and its alloys are currently used as structural biomaterials in biomedical devices such as artificial hip joints and dental implants, which are substitutes for failed hard tissues, on account of its excellent specific strength, corrosion resistance, and biocompatibility [1, 2]. In particular, the $\alpha + \beta$-type Ti-6Al-4V ELI [3] is widely used as a biomedical titanium alloy. This alloy was originally developed as a general structural material mainly for aerospace structures. That is, it was designed without considering its effects – harmful or otherwise – on human tissues, and therefore vanadium (V), which is a toxic element, can be found in Ti-6Al-4V ELI. As a result, new biomedical titanium alloys composed of nontoxic elements were required for use in the medical and dental fields. Ti-6Al-7Nb [4] was first developed as a biomedical titanium alloy by replacing V with niobium (Nb), which is a nontoxic and β phase stabilizing element. However, after the development of Ti-6Al-7Nb, the usage of alminium (Al) was contested as it was believed to cause

Alzheimer's disease, although this has recently been proved false. Therefore, the use of Al in titanium alloys meant for biomedical applications was temporarily discontinued, and thus certain V- and Al-free titanium alloys were developed. However, the developed titanium alloys were of the $\alpha + \beta$-type. Subsequently, mechanical biocompatibility (low Young's modulus, etc.) has come to be regarded as an important factor for metallic biomaterials. This is because there is a possibility that biomaterials with high Young's modulus cause harmful effects such as stress shielding, which delays bone healing when utilized in biomedical devices [5, 6]. Therefore, after $\alpha+\beta$-type titanium alloys were developed in the early stages of the development of biomedical titanium alloys, further research and development on β-type titanium alloys, which are advantageous to achieve a lower Young's modulus, was conducted in order to obtain a low Young's modulus similar to hard tissues. In addition to a low Young's modulus, it is important that metallic biomaterials exhibit good mechanical functionalities such as shape memory and superelasticity, particularly when used as stents, catheter orthodontic wire applications, etc. TiNi is the only shape memory alloy that is used practically. However, Ni has been shown to be a strong allergic element [7]. Therefore, Ni-free titanium alloys with not only a low Young's modulus but also good mechanical functionalities have been investigated [8] although TiNi shape memory alloys are still energetically investigated for biomedical applications [9].

We have designed an alloy – Ti-29Nb-13Ta-4.6Zr – that has both a low Young's modulus and superelasticity. In the present paper, the relationship between microstructures and mechanical properties, Young's modulus, tensile and fatigue properties, etc. in addition to mechanical functionality and biocompatibility of Ti-29Nb-13Ta-4.6Zr are reviewed.

14.2 Experimental Procedures

14.2.1 Alloying Elements and Alloy Composition

On the basis of the data on the cytotoxicity of pure metals [10] and that of the relationship between polarization resistance and biocompatibility of pure metals and typical metallic biomaterials [11], nontoxic elements were selected as suitable alloying elements for biomaterials. Also, β-phase stabilizing elements were primarily selected because β-type titanium alloys are advantageous to obtain low Young's modulus. Moreover, on determining the chemical compositions of candidate alloys, we employed the d-electron alloy design method developed by Morinaga et al. [12]. Finally, based on the results of the tensile test, Ti-29Nb-13Ta-4.6Zr was found to be the most promising alloy for biomedical applications, with a good balance between strength and ductility [13]. In fact, the cytotoxicity of Ti-29Nb-13Ta-4.6Zr in comparison with pure Ti and Ti-6Al-4V was evaluated by the NR and MTT methods [14]. The cell viability of Ti-29Nb-13Ta-4.6Zr was nearly the same as that of pure Ti and greater than that of Ti-6Al-4V in both as-extracted and filtrated extracted solutions. Namely, it was confirmed that the nontoxicity of Ti-29Nb-13Ta-4.6Zr is equivalent to that of pure Ti and is greater than that of Ti-6Al-4V.

14.2.2 Thermomechanical Treatment

The excellent mechanical performance of Ti-29Nb-13Ta-4.6Zr is achieved after subjecting it to various thermomechanical treatments [14–18]. Some typical thermomechanical processes are schematically shown in Fig. 14.1.

In Fig. 14.1a [14], the forged bars of Ti-29Nb-13Ta-4.6Zr with a diameter of 20 mm were cold rolled to plates of thickness 2.5 mm (reduction rate of 87.5%) at room temperature in air. The rolled plates were subjected to solution treatment at 1,063 K for 3.6 ks, followed by water quenching ($TNTZ_{ST}$) and subsequent aging treatments at 573, 598, or 673 K for 259.2 ks ($TNTZ_{STA573}$, $TNTZ_{STA598}$, and $TNTZ_{STA673}$). On the other hand, the solution treatment was carried out before cold rolling in Fig. 14.1b [15]. Then, the solutionized bars were cold rolled in the same manner as in Fig. 14.1a ($TNTZ_{CR}$); these rolled plates were subjected to aging treatments at 598, 673, and 723 K for 259.2 ks ($TNTZ_{CRA598}$, $TNTZ_{CRA673}$, and $TNTZ_{CRA723}$).

Further, a Ti-29Nb-13Ta-4.6Zr wire with a minimum diameter of 0.3 mm was previously fabricated. As shown in Fig. 14.1c [16], the cold rolled bar with a cross section of 10 mm × 10 mm was extruded to the wires with diameters of 1.0 mm ($TNTZ_{d1.0}$) and 0.3 mm ($TNTZ_{d0.3}$). In this case, annealing was carried out at a particular stage during wire drawing.

14.2.3 Microstructure Analysis

The constituted phases of Ti-29Nb-13Ta-4.6Zr subjected to each thermomechanical treatment were examined by X-ray diffraction (XRD) analysis and transmission electron microscopy (TEM). XRD analysis was carried out using a Cu target with an accelerating voltage of 40 kV and a current of 30 mA. TEM was carried out with an acceleration voltage of 200 kV.

14.2.4 Mechanical Tests

For measuring Young's modulus, a resonance method was employed. After proofing the apparatus using standard pure titanium, the measurement was carried out in air at room temperature. For comparison, the Young's modulus of Ti-6Al-4V ELI was also measured.

The specimens for tensile and fatigue tests were machined from the thermomechanically treated plates. The tensile axis was parallel to the rolling direction.

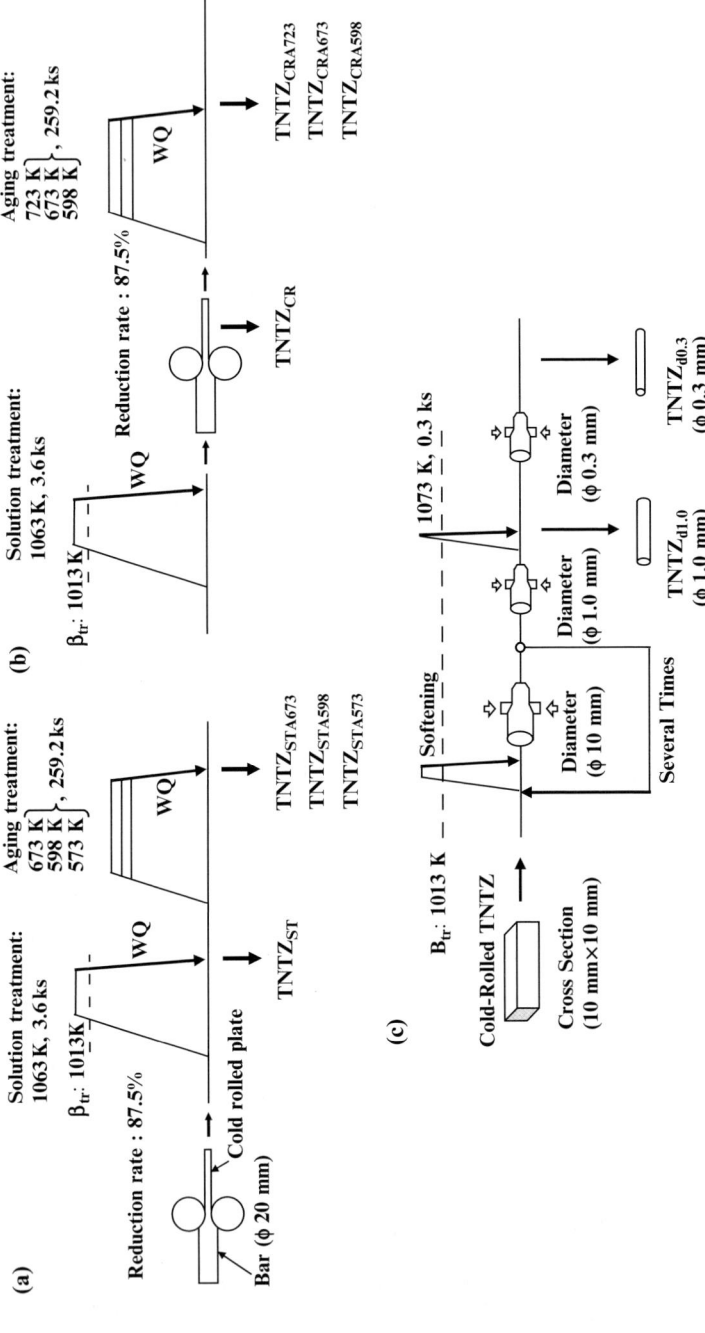

Fig. 14.1. Schematic drawings of typical thermomechanical processes for Ti-29Nb-13Ta-4.6Zr (TNTZ)

Tensile tests were carried out using an Instron-type machine with a crosshead speed of 8.33×10^{-6} m s^{-1} in air at room temperature. Also, tensile loading–unloading tests were performed in air at room temperature using the same machine that was used for the tensile tests. In the tensile loading–unloading test, loading–unloading was repeated with every 0.5% increase in the strain up to a total strain of 4.0%.

Fatigue tests were carried out using an electro-servo-hydraulic machine at a frequency of 10 Hz with a stress ratio of $R = 0.1$ in the tension–tension mode in air at room temperature and in Ringer's solution at 310 K.

14.2.5 Evaluation of Mechanical Biocompatibility

Experimental tibia fractures were made with an oscillating saw just below the tibial tuberosity of Japanese white rabbits weighing 2.5–3.0 kg, and intramedullary fixations were performed using the Ti-29Nb-13Ta-4.6Zr rod cut from the cold swaged bars as well as the Ti-6Al-4V ELI and SUS 316 L stainless steel rods. To continuously observe the state of fracture healing, X-ray pictures were taken every 2 weeks for 22 weeks.

14.3 Results and Discussion

14.3.1 Microstructure

Microstructures were observed in Ti-29Nb-13Ta-4.6Zr subjected to solution treatment at 1,063 K for 3.6 ks (TNTZ$_{ST}$) and aging treatments at 573, 598, or 673 K for 259.2 ks after solution treatment (TNTZ$_{STA573}$, TNTZ$_{STA598}$, or TNTZ$_{STA673}$). Optical microscopy revealed that these microstructures comprise equiaxed grains with an average diameter of approximately 20 μm [14]. On the other hand, when Ti-29Nb-13Ta-4.6Zr was cold rolled after undergoing solution treatment at 1,063 K for 3.6 ks (TNTZ$_{CR}$) and then subjected to aging treatments at 723 K for 259.2 ks (TNTZ$_{CRA723K}$), the average diameters of the microstructures could not be obtained by optical microscopy because of the breaking-up of the grain boundaries due to intensive cold rolling [17]. Moreover, it was difficult to observe the precipitated phases by optical microscopy because of their fineness, although α or ω phases were likely to be precipitated by aging treatments. It was also difficult to identify such precipitated phases by scanning electron microscopy.

The X-ray diffraction profiles of TNTZ$_{ST}$, TNTZ$_{STA573}$, TNTZ$_{STA598}$, and TNTZ$_{STA673}$ are shown in Fig. 14.2 [14]. For TNTZ$_{ST}$, only the peaks of the β phase are obtained. On the other hand, the ω phase is detected in addition to the β phase in every aged specimen – TNTZ$_{STA573}$, TNTZ$_{STA598}$, and TNTZ$_{STA673}$. The peak intensity of the ω phase of TNTZ$_{STA598}$ is the

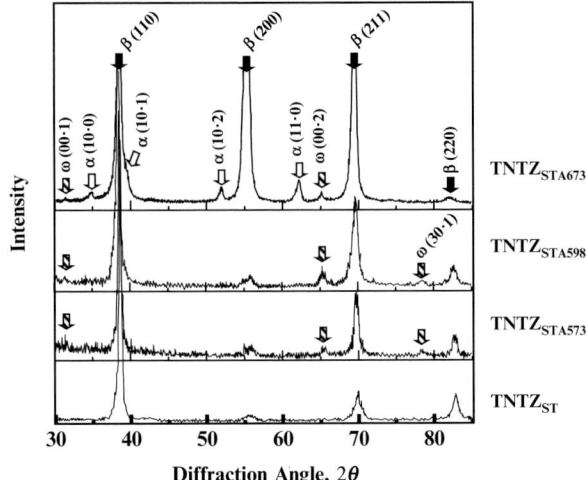

Fig. 14.2. XRD profiles obtained from Ti-29Nb-13Ta-4.6Zr subjected to solution treatment and aging treatments at 573, 598, or 673 K for 259.2 ks after solution treatment

strongest, while that of the ω phase of TNTZ$_{STA673}$ is very weak. Moreover, in the case of TNTZ$_{STA673}$, the α phase is detected in addition to the β and ω phases. Figure 14.3 shows the X-ray diffraction profiles of TNTZ$_{CR}$, TNTZ$_{CRA598}$, TNTZ$_{CRA673}$, and TNTZ$_{CRA723}$ [15, 17]. In the case of TNTZ$_{CR}$, only the peaks of the β phase are obtained, similar to the case of TNTZ$_{ST}$. However, the presence of the ω phase in TNTZ$_{CRA598}$ cannot be confirmed, while the peaks of the ω phase were clearly identified in the case of TNTZ$_{STA598}$. Moreover, the peaks of the β and α phases are obtained from TNTZ$_{CRA673}$; however, the peaks of the ω phase do not appear in the XRD profile. The β and α phases are also detected in the case of TNTZ$_{CRA723}$.

The TEM micrograph and electron beam diffraction pattern obtained from TNTZ$_{STA673}$ are shown in Fig. 14.4 [14]. Figure 14.4a,b is the observation results obtained from the same specimen, but only the diffraction condition is changed. In Fig. 14.4a,b, lath-like precipitates are observed in the β phase matrix. These precipitates were identified as the α phase. Moreover, the weak diffraction from the ω phase is detected as shown in Fig. 14.4b, although it is difficult to recognize the presence of the ω phase in the dark field image. These observation results are consistent with the results of X-ray diffraction analysis. Figure 14.5 shows the TEM micrograph and electron beam diffraction pattern obtained from TNTZ$_{CRA673}$ [15]. In spite of the fact that the same aging temperature is used in each case, a fine lath-like α phase is precipitated in the case of TNTZ$_{CRA673}$ but not in the case of TNTZ$_{STA673}$. Generally, a region in which dislocation occurs is one of the nucleation sites of the precipitate of metals and alloys. In TNTZ$_{CRA673}$, a high dislocation density must be introduced by cold rolling before the aging treatment, which

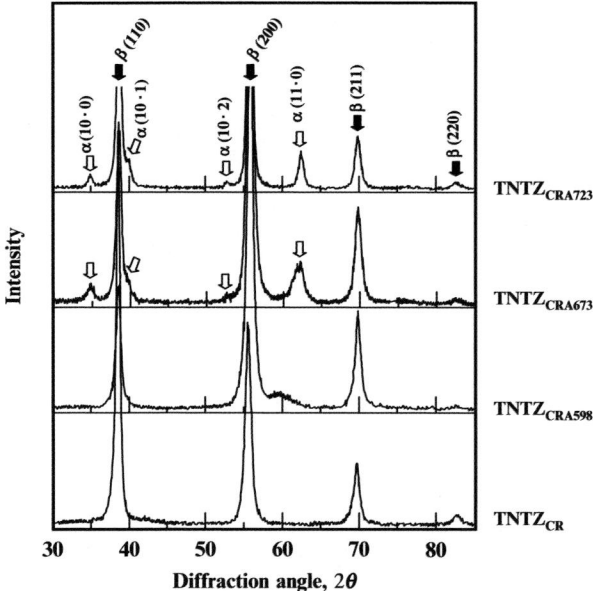

Fig. 14.3. XRD profiles obtained from Ti-29Nb-13Ta-4.6Zr cold rolled and subjected to aging treatments at 598, 673, and 723 K for 259.2 ks after cold rolling

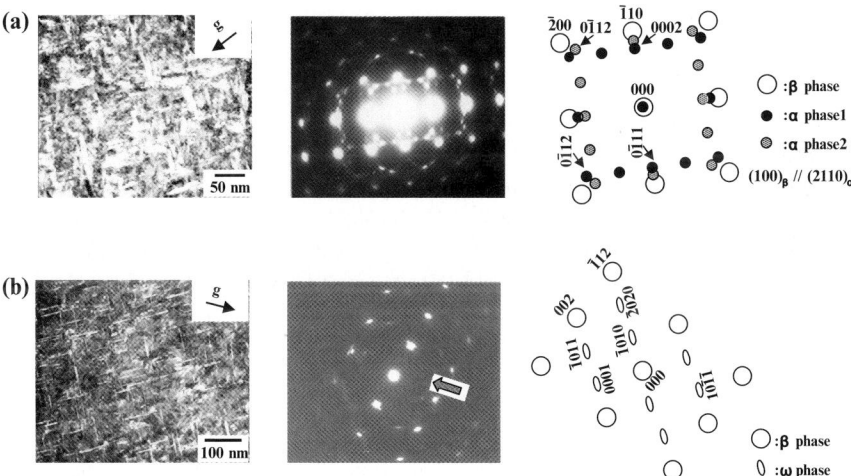

Fig. 14.4. TEM micrographs and electron beam diffraction patterns of Ti-29Nb-13Ta-4.6Zr aged at 673 K for 259.2 ks after solution treatment at 1,063 K for 3.6 ks (TNTZ$_{STA673}$): These figures, (a) and (b), are both observation results obtained from same specimen, but only diffraction condition is changed

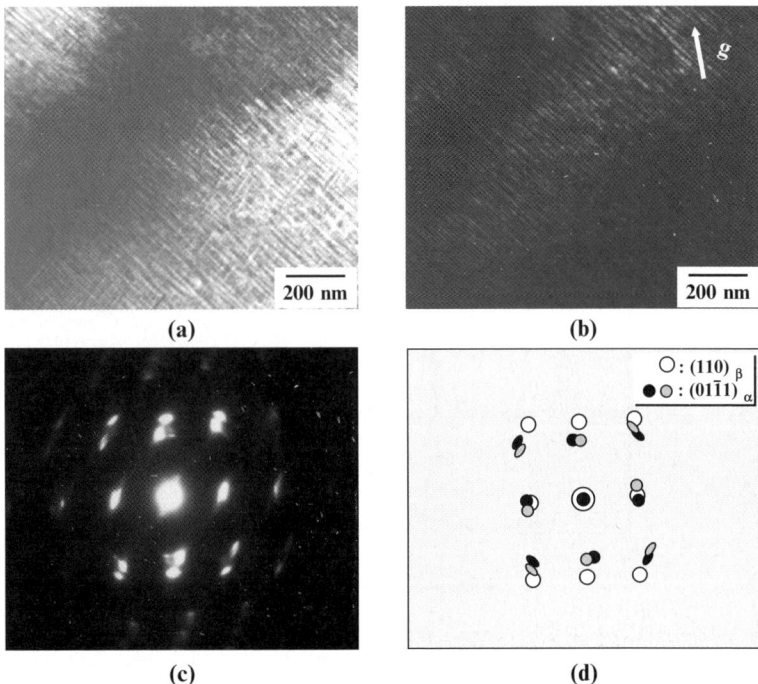

Fig. 14.5. TEM micrographs and electron beam diffraction pattern of Ti-29Nb-13Ta-4.6Zr aged at 673 K for 259.2 ks after cold rolling (TNTZ$_{CRA673}$); (**a**) bright field, (**b**) dark field, (**c**) diffraction pattern, and (**d**) key diagram

results in the increase in the number of nucleation sites. Therefore, the precipitation characteristic of TNTZ$_{CRA673}$ is different from that of TNTZ$_{STA673}$. Figure 14.6 shows the TEM micrograph and electron beam diffraction pattern obtained from TNTZ$_{CRA723}$ [15]. It is observed that only a lath-like α phase is precipitated in the case of both TNTZ$_{CRA723}$ and TNTZ$_{CRA673}$; however, the size of the precipitate of TNTZ$_{CRA723}$ is larger than that of the precipitate of TNTZ$_{CRA673}$.

14.3.2 Young's Modulus

Figure 14.7 shows the Young's moduli of Ti-29Nb-13Ta-4.6Zr obtained after solution treatment, cold rolling, and each aging treatment along with those of Ti-6Al-4V ELI subjected to solution and aging treatments [17, 18]. The Young's moduli of Ti-29Nb-13Ta-4.6Zr are lower than those of Ti-6Al-4V ELI, even under any aging condition. In particular, the lowest Young's modulus, approximately 60 GPa, was obtained in solutionized (TNTZ$_{ST}$) and cold rolled (TNTZ$_{CR}$) conditions. This value of Young's modulus is nearly half that of Ti-6Al-4V ELI. Moreover, the value of the Young's modulus

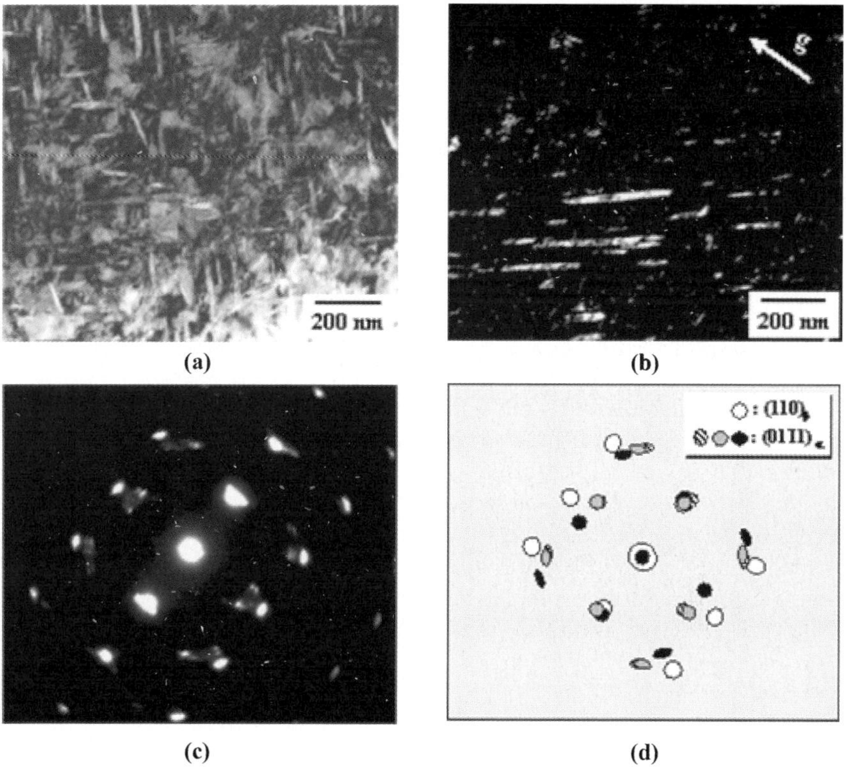

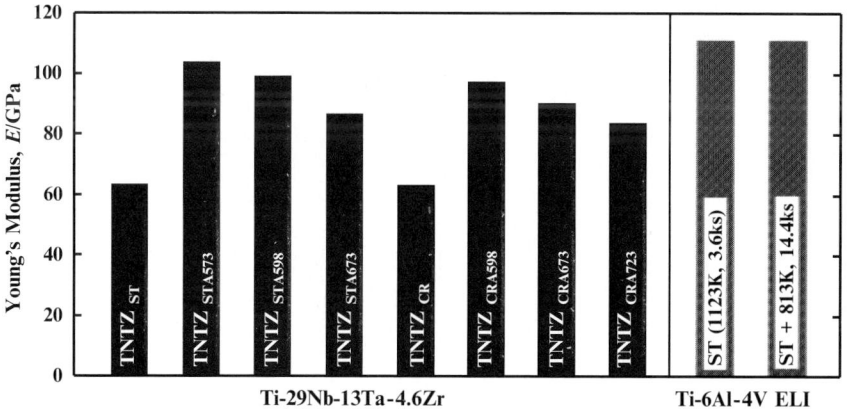

Fig. 14.6. TEM micrographs and electron beam diffraction pattern of Ti-29Nb-13Ta-4.6Zr aged at 723 K for 259.2 ks after cold rolling (TNTZ$_{CRA723}$); (**a**) bright field, (**b**) dark field, (**c**) diffraction pattern, and (**d**) key diagram

Fig. 14.7. Young's moduli of Ti-29Nb-13Ta-4.6Zr cold rolled, solutionized, and aged at various temperatures in the range of 573–723 K. This figure includes data of Ti-6Al-4V ELI in solutionized (ST) and aged conditions

of Ti-29Nb-13Ta-4.6Zr increases when subjected to aging. This result corresponds to the microstructure because α and ω phases with a higher Young's modulus were precipitated by the aging treatment as opposed to the β phase matrix. Further, the Young's moduli of aged Ti-29Nb-13Ta-4.6Zr are dependent on the aging temperature. In the range of 573–723 K, the Young's modulus tends to decrease with the increase in the aging temperature. On the basis of the result of the microstructure observation, this trend indicates that the effect of ω phase precipitation on the increase in the Young's modulus is greater than that of α phase precipitation.

14.3.3 Tensile Properties

The tensile properties of Ti-29Nb-13Ta-4.6Zr subjected to solution treatment at 1,063 K for 3.6 ks (TNTZ$_{ST}$) and aging treatment at 598 or 673 K for 259.2 ks (TNTZ$_{STA598}$, TNTZ$_{STA673}$) are shown in Fig. 14.8 [17]. The tensile strength and 0.2% proof stress of TNTZ$_{ST}$ are smaller than those of TNTZ$_{STA598}$ and TNTZ$_{STA673}$. In particular, TNTZ$_{STA598}$ has the largest tensile strength and 0.2% proof stress among these specimens. On the other hand, elongation decreases in the following order: TNTZ$_{ST}$ > TNTZ$_{STA673}$ > TNTZ$_{STA598}$. These results are derived from the type of precipitates, namely, the α and ω phases. The ω phase has a greater effect on the improvement in the tensile strength as compared to the α phase; however, the ω phase causes a significant deterioration in the ductility of Ti-29Nb-13Ta-4.6Zr.

The tensile properties of Ti-29Nb-13Ta-4.6Zr can be controlled by changing the type and amount of these precipitated phases, which in turn depend on the heat treatment or thermomechanical treatment. For example,

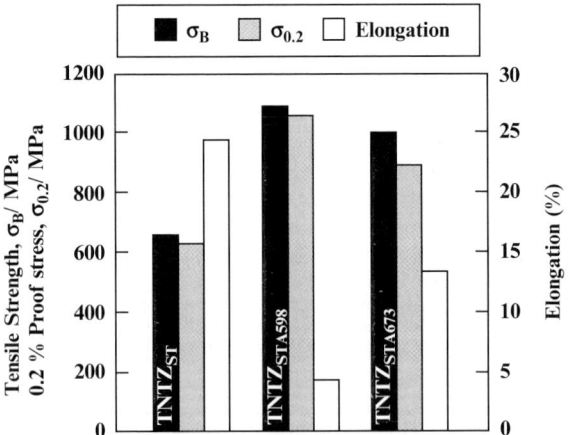

Fig. 14.8. Tensile properties of Ti-29Nb-13Ta-4.6Zr subjected to solution treatment at 1,063 K for 3.6 ks (TNTZ$_{ST}$), and aging treatment at 598 or 673 K for 259.2 ks after solution treatment (TNTZ$_{STA598}$, TNTZ$_{STA673}$)

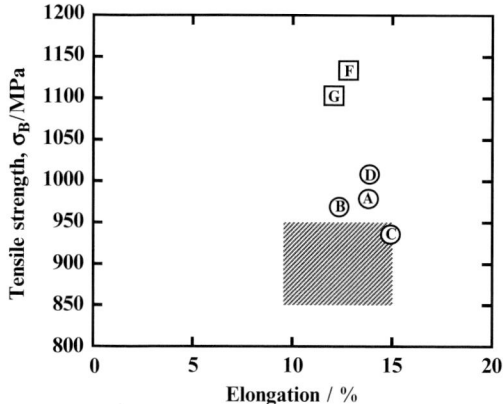

Fig. 14.9. Relationship between tensile strength and elongation of Ti-29Nb-13Ta-4.6Zr in cold rolled, solutionized, and some aged conditions: A (1,033 K, 1.8 ks + 673 K, 259.2 ks), B (1,033 K, 1.8 ks + 598 K, 100.8 ks), C (1,033 K, 1.8 ks + 723 K, 259.2 ks), D (1,063 K, 1.8 ks + 673 K, 259.2 ks), F (Cold rolling + 723 K, 100.8 ks), G (Cold rolling + 723 K, 259.8 ks). Hatched area shows range of tensile strength and elongation of Ti-6Al-4V ELI

the relationship between the tensile strength and elongation of Ti-29Nb-13Ta-4.6Zr after cold rolling, solution treatment, and each aging treatment is shown in Fig. 14.9 along with that of Ti-6Al-4V ELI [18]. When the conditions employed in the heat treatment and thermomechanical treatment are appropriate, the relationship between the tensile strength and elongation in the case of Ti-29Nb-13Ta-4.6Zr is better than that in the case of Ti-6Al-4V ELI. In particular, when Ti-29Nb-13Ta-4.6Zr is subjected to aging treatments after cold rolling (symbols F and G in Fig. 14.9), it exhibits excellent strength without deteriorating the ductility. Consequently, the relationship between the tensile properties and the cold working ratio of Ti-29Nb-13Ta-4.6Zr is shown in Fig. 14.10 [18]. The tensile strength and 0.2% proof stress increase with the cold working ratio. There is a decrease in the elongation and reduction in the area up to cold working ratio of approximately 20%, after which the values of these parameters are almost constant irrespective of the increase in the cold working ratio. Moreover, it should be noted that the Young's modulus is almost constant with an increasing cold working ratio. Therefore, the strength of Ti-29Nb-13Ta-4.6Zr can be increased by maintaining a low Young's modulus by cold working.

14.3.4 Superelasticity

Recently, the superelasticity of Ti-29Nb-13Ta-4.6Zr was reported as one of the mechanical functionalities of metallic biomaterials [16, 18]. The tensile loading–unloading curve of the wires made of Ti-29Nb-13Ta-4.6Zr with diameters of 1.0 and 0.3 mm (TNTZ$_{d1.0}$ and TNTZ$_{d0.3}$) are shown in Fig. 14.11.

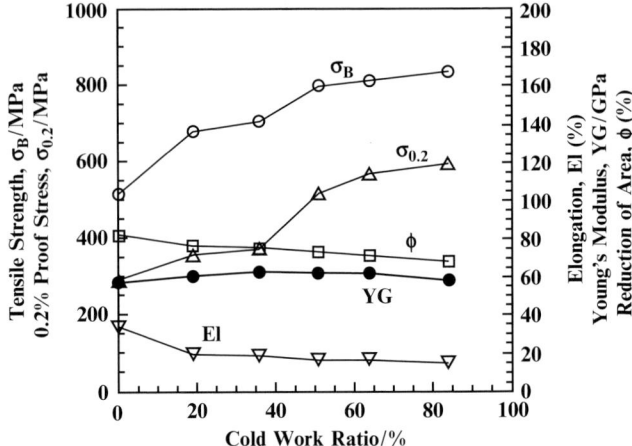

Fig. 14.10. Relationship between tensile properties and cold working ratio of Ti-29Nb-13Ta-4.6Zr

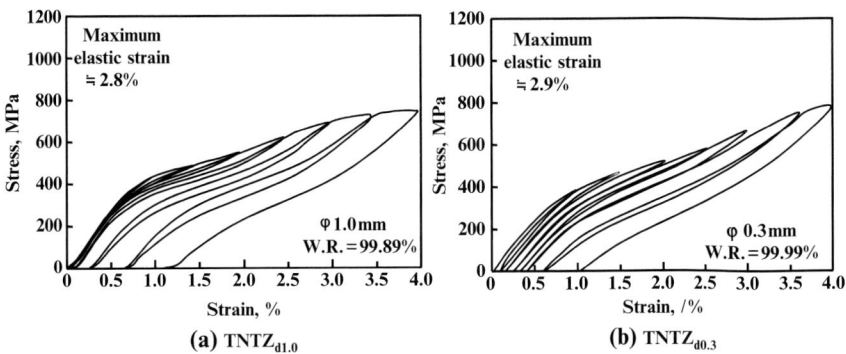

Fig. 14.11. Tensile loading–unloading curves of Ti-29Nb-13Ta-4.6Zr wire with diameters of (**a**) 1.0 mm and (**b**) 0.3 mm

The maximum elastic strain of $TNTZ_{d1.0}$ and $TNTZ_{d0.3}$ are 2.8% and 2.9%, respectively. In general, superelasticity has been attributed to deformation-induced martensitic transformation and its reverse transformation. To explain the mechanism of superelasticity of $TNTZ_{d1.0}$ and $TNTZ_{d0.3}$, differential scanning calorimetry (DSC) was carried out on these wires in the temperature range of 223–423 K. However, in this temperature range, the peak corresponding to martensitic transformation was not detected. Moreover, XRD analysis was also carried out on the deformed $TNTZ_{d1.0}$, but traces of the martensitic phase were not detected. Therefore, the mechanism of superelasticity of $TNTZ_{d1.0}$ and $TNTZ_{d0.3}$ was still unclear. The mechanism of this superelastic characteristic is interesting from the viewpoint of materials science.

14.3.5 Fatigue Properties

Stress-fatigue life (the number of cycles to failure) curves, that is, S-N curves, obtained from plain fatigue tests on Ti-29Nb-13Ta-4.6Zr subjected to solution treatment (TNTZ$_{ST}$) and aging treatments after solution treatment (TNTZ$_{STA598}$, TNTZ$_{STA673}$) and cold rolled (TNTZ$_{CR}$) and aging treatments after cold rolling (TNTZ$_{CRA673}$, TNTZ$_{CRA723}$) are shown in Fig. 14.12 [18]. The ranges of the fatigue strengths of Ti-6Al-4V ELI and Ti-6Al-7Nb for biomedical applications are shown in the same figure for the purpose of comparison. The maximum stress, i.e., the stress at the instant at which the specimen survives 10^7 cycles, is termed as fatigue limit in this paper.

The fatigue strength of Ti-29Nb-13Ta-4.6Zr in both solutionized and cold rolled condition is improved remarkably when subjected to aging treatment at temperatures in the range of 598–723 K. In particular, the fatigue strength of aged Ti-29Nb-13Ta-4.6Zr is greater in cold rolled condition than in the solutionized condition. Moreover, the fatigue strength increases proportionally with the aging temperature. The fatigue limit of TNTZ$_{CRA723}$ – approximately 780 MPa – is the highest among all specimens, and this value is approximately two times greater than that of TNTZ$_{CR}$. The improvement in the fatigue strength of TNTZ$_{CRA723}$ results from the increase in the tensile strength

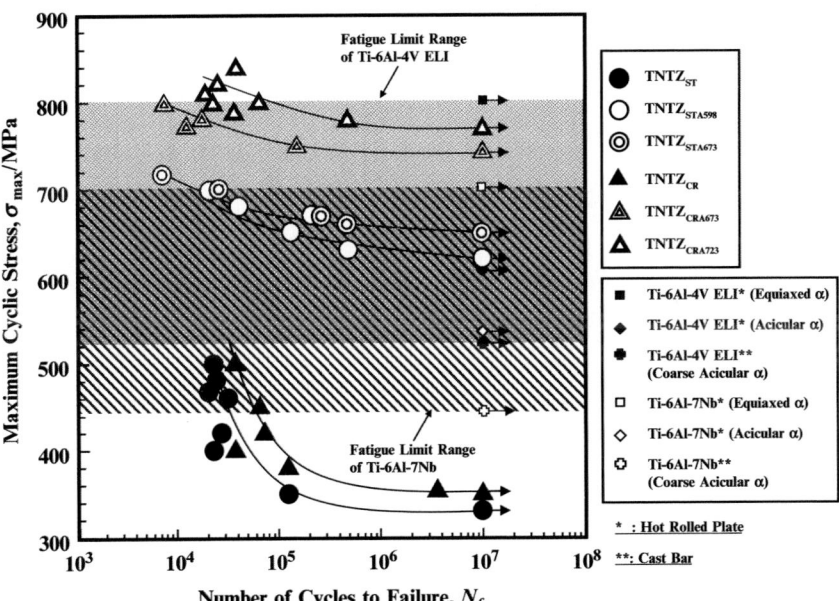

Fig. 14.12. S-N curves of Ti-29Nb-13Ta-4.6 in solutionized (TNTZ$_{ST}$), cold rolled (TNTZ$_{CR}$), and aged (TNTZ$_{STA598}$, TNTZ$_{STA673}$, TNTZ$_{CRA673}$ and TNTZ$_{CRA723}$) conditions. In this figure, data of Ti-6Al-4V ELI and Ti-6Al-7Nb are shown

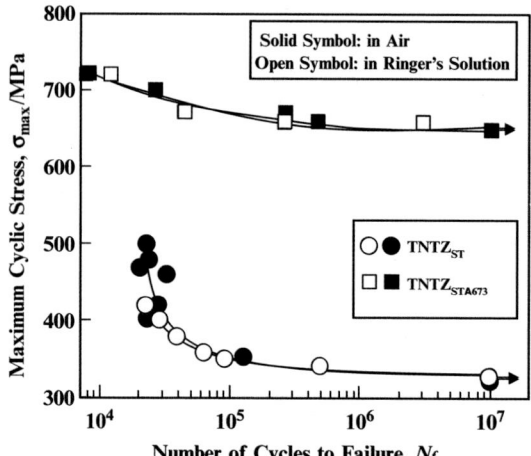

Fig. 14.13. S-N curves of Ti-29Nb-13Ta-4.6Zr solutionized ($\mathrm{TNTZ_{ST}}$) and aged at 673 K for 259.2 ks ($\mathrm{TNTZ_{STA673}}$) in air and Ringer's solution

due to a homogenously precipitated fine α phase that leads to increased crack initiation resistance and a relatively larger elongation that improves small fatigue crack propagation resistance. The fatigue limits of $\mathrm{TNTZ_{CRA673}}$ and $\mathrm{TNTZ_{CRA723}}$ are considerably greater than those of Ti-6Al-7Nb, with equiaxed α and widmanstätten α structures and are nearly equal to that of Ti-6Al-4V ELI with an equiaxed α structure, which is approximately 800 MPa. On the other hand, the fatigue limit of $\mathrm{TNTZ_{STA673}}$ is approximately two times greater than that of $\mathrm{TNTZ_{ST}}$ and is similar to that of Ti-6Al-7Nb.

The S-N curves obtained from plain fatigue tests on $\mathrm{TNTZ_{ST}}$ and $\mathrm{TNTZ_{STA673}}$ in Ringer's solution are shown in Fig. 14.13 [17, 18]. The fatigue strength in Ringer's solution is equal to that in air under both solutionized and aged conditions. Therefore, the fatigue strength of Ti-29Nb-13Ta-4.6Zr does not deteriorate in Ringer's solution.

14.3.6 Mechanical Biocompatibility

To confirm the advantage of a low Young's modulus in bone healing and remodeling, a tibial fracture was formed artificially with an oscillating saw at just below the tibial tuberosity in rabbits. Subsequently, a rod made of Ti-29Nb-13Ta-4.6Zr, Ti-6Al-4V ELI, or SUS 316L stainless steel was driven into the intramedullary canal to heal the fracture. The state of bone healing and remodeling were continuously observed with the help of X-rays every 2 weeks for a total duration of 22 weeks. The results are shown in Fig. 14.14 [18–20]. When Ti-29Nb-13Ta-4.6Zr was used, the outline of the fracture callus is very smooth in the bone remodeling stage. The amount of the callus is

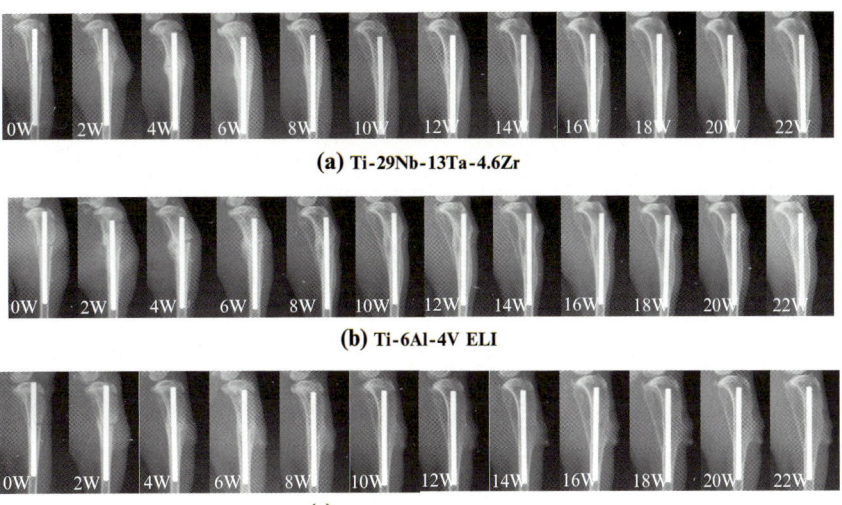

Fig. 14.14. Healing process of bone fracture from 0 to 22 weeks after surgery; (**a**) Ti-29Nb-13Ta-4.6Zr, (**b**) Ti-6Al-4V ELI, and (**c**) SUS 316L stainless steel

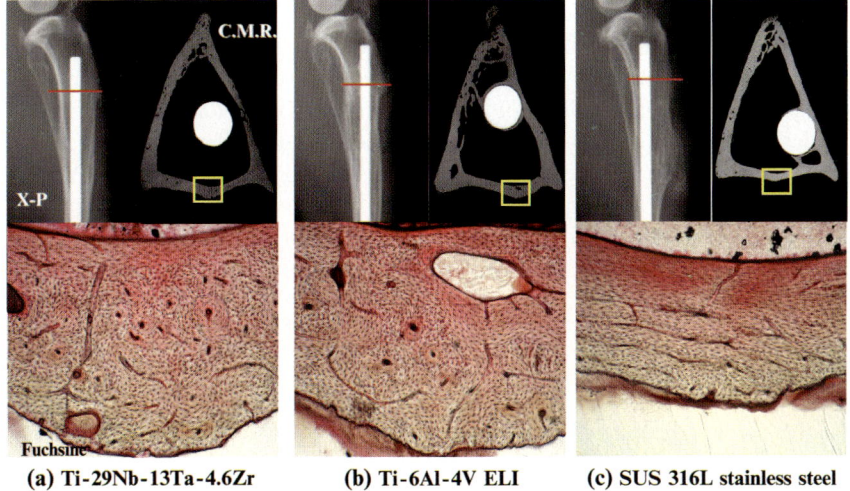

Fig. 14.15. Cross section of tibia at 24 weeks after implantation; (**a**) Ti-29Nb-13Ta-4.6Zr, (**b**) Ti-6Al-4V ELI, and (**c**) SUS 316L stainless steel

relatively small and gradually decreases after 6 weeks; then there are no traces of fracture at 10 weeks after fixation. After 10 weeks, almost no changes can be observed in bone. In Ti-6Al-4V ELI, the callus formation is almost similar to those in Ti-29Nb-13Ta-4.6Zr, but the remodeling is slightly different as compared with Ti-29Nb-13Ta-4.6Zr. In SUS 316L stainless steel, a large amount of the fracture callus is observed till the end of the follow-up period. Bone atrophy seems to occur at the posterior proximal tibial bone; this phenomenon becomes more evident every 2 weeks. The posterior tibial bone becomes very thin at 22 weeks. The cross section of the tibia implanted with each rod at 24 weeks after implantation is shown in Fig. 14.15 [18, 19]. The microstructure of the bone formed around the rod made of Ti-29Nb-13Ta-4.6Zr or Ti-6Al-4V ELI shows a number of osteons, which is the result of internal remodeling of the cortical bone. However, in the case of SUS 316L stainless steel, the bone structure shows lamination of the cortical bone with the absence of osteons. The absence of osteons implies that bone atrophy occurs when SUS 316L stainless steel is used. From these experiments on animals, the load transmission issue of the current metallic implants with a high Young's modulus may be improved by using materials with a low Young's modulus.

14.4 Conclusions

Mechanical properties and microstructures of a newly developed biomedical β-type titanium alloy comprising nontoxic elements of Nb, Ta, and Zr, Ti-29Nb-13Ta-4.6Zr were reviewed in this paper. Ti-29Nb-13Ta-4.6Zr exhibits excellent mechanical performance such as low Young's modulus, good balance of static strength and ductility (tensile properties), and high dynamic strength (fatigue properties) for metallic biomaterials. These attractive mechanical characteristics are derived from the microstructures, which are obtained through thermomechanical treatments. The advantage of a low Young's modulus for bone healing and remodeling was confirmed in this paper via animal experiments. Also, superelasticity is shown to be one of the current mechanical functionalities that are required for some biomedical applications.

References

1. M. Niinomi, T. Hattori, T. Kasuga, H. Fukui, in *Encyclopedia of Biomaterials and Biomedical Engineering*, ed. by G.L. Bowlin, G. Wnek (Marcel Dekker, New York, 2004) Online available
2. M. Long, H.J. Rack, Biomaterials **19**, 1621, (1998)
3. ASTM Designation F136–02a, Section 13 Medical Devices and Services, *Annual Book of ASTM Standards*, p. 63, 2006
4. ASTM Designation F1295–05, Section 13 Medical Devices and Services, *Annual Book of ASTM Standards*, p. 540, 2006
5. B.J. Moyen, P.J. Lahey, E.H. Weinberg, W.H. Harris, J. Bone Joint Surg. Am. **60**, 940 (1978)
6. H.K. Uhthoff, M. Finnegan, J. Bone Joint Surg. Br. **65**, 66 (1983)
7. L. Peltonen, Contact Derm. **5**, 27 (1979)

8. T. Inamura, Y. Fukui, H. Hosoda, K. Wakashima, S. Miyazaki, Mater. Trans. **45**, 1083 (2004)
9. F. El Feninat, G. Laroche, M. Fiset, D. Mantovani, Adv. Eng. Mater. **4**, 91 (2002)
10. H. Kawahara, Bull. Jpn. Inst. Met. **31**, 1033 (1992)
11. S.G. Steinemann, Periodontology 2000 **17**, 7 (1998)
12. M. Morinaga, J. Saito, M. Morishita, J. Jpn. Inst. Light Metals **42**, 614 (1992)
13. D. Kuroda, M. Niinomi, M. Morinaga, Y. Kato, T. Yashiro, Mater. Sci. Eng. A **243**, 244 (1998)
14. M. Niinomi, Biomaterials **24**, 2673 (2003)
15. T. Akahori, M. Niinomi, K. Ishimizu, H. Fukui, A. Suzuki, J. Jpn. Inst. Metals **67**, 652 (2003)
16. T. Akahori, M. Niinomi, H. Toda, K. Yamauchi, H. Fukui, M. Ogawa, J. Jpn. Inst. Metals **69**, 530 (2005)
17. T. Akahori, M. Niinomi, H. Fukui, M. Ogawa, H. Toda, Mater. Sci. Eng. C **25**, 248 (2005)
18. M. Niinimi, T. Akahori, T. Hattori, K. Morikawa, T. Kasuga, H. Fukui, A. Suzuki, K. Kyo, S. Niwa, J. Am. Soc. Testing Mater. Int. **2**, (2005). Paper ID JAI12818
19. M. Niinomi, T. Hattori, S. Niwa, in *Biomaterials in Orthopedics*, ed. by M.J. Yaszemski, D.J. Trantolo, K.U. Lewandrowski, V. Hasirci, D.E. Altobelli, D.L. Wise, (Marcel Dekker, New York, 2004), p. 41
20. T. Hattori, K. Morikawa, S. Niwa, K. Sato, M. Niinomi, A. Suzuki, J. Jpn. Soc. Clin. Biomech. **23**, 299 (2002)

Part III

Precise Control of Microscopic
and Complex Systems

Atom Probe Tomography at The University of Sydney

B. Gault, M.P. Moody, D.W. Saxey, J.M. Cairney, Z. Liu, R. Zheng, R.K.W. Marceau, P.V. Liddicoat, L.T. Stephenson, and S.P. Ringer

Summary: The Australian Microscopy & Microanalysis Research Facility (AMMRF) operates a national atom probe laboratory at The University of Sydney. This paper provides a brief review and update of the technique of atom probe tomography (APT), together with a summary of recent research applications at Sydney in the science and technology of materials. We describe recent instrumentation advances such as the use of laser pulsing to effect time-controlled field evaporation, the introduction of wide field of view detectors, where the solid angle for observation is increased by up to a factor of ~20 as well as innovations in specimen preparation. We conclude that these developments have opened APT to a range of new materials that were previously either difficult or impossible to study using this technique because of their poor conductivity or brittleness.

15.1 Introduction

In atom probe tomography (APT) single atoms are successively removed from the surface of the specimen by field evaporation [1]. The very intense electric field required for this field evaporation is achieved by preparing needle-shaped specimens, the tips of which typically have curvatures of less than a hundred nanometers. At such radii, evaporation can be generated at the apex of the tip, subjected to only a few kilovolts. To characterise ionic identity by time-of-flight mass spectrometry (tof-MS), the required electric field is generated by the superimposition of a DC voltage and either high voltage (HV) or laser pulses (Fig. 15.1). The diverging electric field at the apex gives rise to a highly magnified projection of the ions onto a position-sensitive detector (PSD) [2, 3]. Each hit on the detector can be related directly back to the specific pulse responsible for the corresponding ionization event, facilitating a highly accurate time-of-flight measurement and mass resolution. Further, the tip is cooled down to cryogenic temperatures, between 15 and 100 K, in order to prevent thermal surface diffusion of atoms prior to their evaporation, and maintained under an ultra-high vacuum (UHV) conditions. Following experiment, an inverse projection algorithm enables a three-dimensional (3D) reconstruction

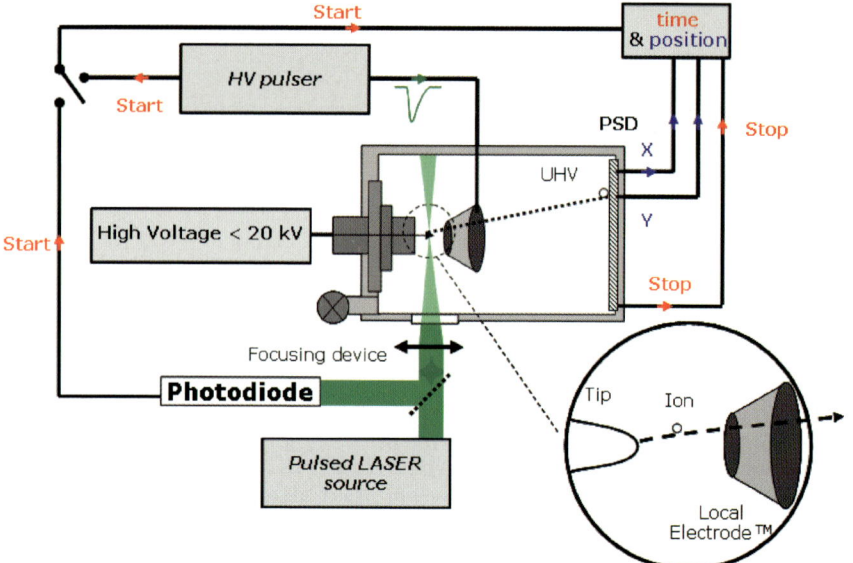

Fig. 15.1. Experimental schematic. Atom probe can be carried out using either HV or laser pulses

of the analyzed volume. Ultimately, a 3D atomic map of the material can be reproduced with sub-nanometer spatial resolution. APT, with this unique combination of very high mass and spatial resolution, has been extensively used in physical metallurgy and more generally in materials science to study structure at the nanoscale, for example in problems of interface chemistry, phase compositional analysis, segregation, diffusion and atomic clustering [4–11].

Nevertheless, APT has previously been constrained somewhat by the use of high-voltage (HV) pulses. The necessary propagation of the pulse through the material prevents its application to poorly conducting materials. Further, the HV pulses result in an energy spread, meaning ions are not evaporated and accelerated at precisely the same voltage. This spread, known as energy deficits [12], limits the mass resolution of APT. To optimise mass resolution, without the use of ion optics devices [13], the flight path is generally set to about $50\,cm$, in turn restricting the field of view to $\sim 10 \times 10\,nm^2$.

In the last 25 years, many solutions have been proposed to overcome the constraints of HV pulsing. This paper is not intended to be a complete history of the technique, as thorough overviews of the development of APT have been provided elsewhere [4, 5]. However, key advances in this area include the pulsed laser atom probe (PLAP) introduced by Kellogg and Tsong in the early 1980s [14, 15], the scanning atom probe (SAP) introduced by Nishikawa et al. in 1993 [16, 17] and the Local Electrode Atom Probe (LEAP[TM]) [18–20], the APT microscope utilised in the present research at the AMMRF. These techniques reduce or replace the necessary amplitude of the voltage pulses, thereby reducing the energy spread, significantly improving the mass resolution and enabling an improved field of view (FOV) of up to $200 \times 200\,\mathrm{nm}^2$.

The PLAP employs laser pulses instead of HV in order to increase the temperature at the tip apex of a sample subjected to a standing voltage and provoke field evaporation. In spite of many of its advantages, such as its potential application to semiconducting and other new materials, the PLAP remained almost confined to theoretical studies through the 1980s [21–26] due mainly to technological limitations. However, since 2003, with recent advances in laser technology, research programs have been undertaken dedicated to the development of practical laser-assisted atom probes utilising picosecond or femtosecond laser pulses [27–30].

APT requires the specimen to be prepared as a very sharp tip with a radius of curvature smaller than 100 nm. Typically, metallic specimens are prepared by electro-polishing. Recently, there has been extensive progress in the development of focused-ion-beam (FIB) techniques to prepare APT specimens from non-conductive materials or create site-specific samples [31–34]. It is widely recognised that APT specimen preparation is an increasingly significant issue for the the expansion of the technique within the scientific community.

In this review the basic principles of field evaporation are discussed with respect to application in modern APT techniques. Further, developments in specimen preparation utilising the FIB are described. Examples of recent research at AMMRF at the University of Sydney are presented with an emphasis on the significance of the new nanostructural insight offered by APT. Finally, the current state of APT is discussed together with some of the potential implications of the technique to an increasing wide range of scientific fields.

15.2 Field Evaporation Theory

Field evaporation is the field-induced removal of an atom from its own lattice. It corresponds to a combination of ionization and desorption of an atom from a surface subject to a strong electric field. The electric field induces a polarization of the surface atoms [35]. When the electric field is high enough to separate an electron from an atom, the so-created ion is accelerated by the surrounding field and can be removed from the surface [36].

The field evaporation phenomenon can be considered as the transition of an atom from an atomic to an ionic state (Fig. 15.2). Ions may either tunnel through the energy barrier formed by the atomic and ionic potentials energy curves, or thermal energy allow them to hop over, leading to atomic ionization. The ion tunneling mechanism is known to occur only under particular conditions, for low mass ions at very low temperature, and it is generally neglected. The probability of a thermally assisted field evaporation P can be modeled by a Maxwell–Boltzmann equation $P \propto \exp\left(-Q\left(F\right)/k_\mathrm{B}T\right)$, with $Q(F)$ the field dependant height of the barrier, k_B the Boltzmann constant

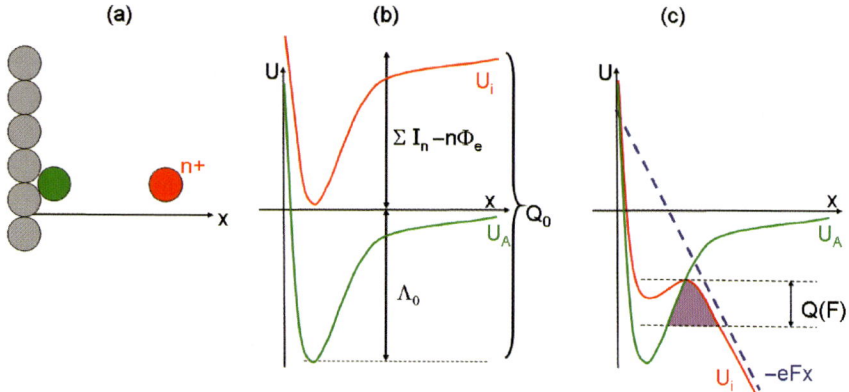

Fig. 15.2. (a) Schematic view of the field evaporation process. **(b)** U_A and U_i are, respectively, the atomic and ionic potential energy curves without electric field. **(c)** The same curves with an electric field F applied

and T the absolute temperature. Thus, the evaporation rate ν can be written as an Arrhénius law:

$$\nu = \nu_0 \exp\left(-Q\left(F\right)/k_B T\right),$$

where for ν_0, the vibration frequency of the surface atoms, a value of $\sim 10^{13}$ Hz has been experimentally measured [37]. It means that almost every $\tau_e = 10^{-13}$ s (100 fs) the atom attempts to leave the surface. In a first approximation, the field dependant height to produce a n-times charged ion can be written as

$$Q\left(F\right) = Q_0 - f\left(F\right) \text{ and } Q_0 = \Lambda_0 + \sum I_n - n\Phi_e,$$

with Λ_0 the sublimation energy, I_n the nth ionization energy, Φ_e the work function of the emitting surface and $f(F)$ a function of the electric field, which can be considered as linear around a critical electric field, named evaporation field F_e, for which the height of the potential barrier is decreased down to zero. For most metals, the evaporation field is in the range of 10–60 V nm^{-1}.

In APT, the specimen is subjected to a high voltage of a few kilovolts. To reach an electric field required to induce ionization, the specimen has to be prepared as a very sharp tip having a radius of curvature R at the apex smaller than 100 nm. The electric field is then

$$F = V/kR,$$

with V the voltage and k a constant between 2 and 8 depending on both the geometry of the tip and its electrostatic environment [38, 39]. The electric field penetration is very small in the case of metallic materials, being screened on a distance much smaller than the size of a single atom. Only the atoms at the very surface are affected by the field evaporation process that occurs almost atom-by-atom and atomic layer-after-atomic layer. It is important to note that the field evaporation process depends critically on the electric field and temperature; it is thus possible to control precisely the amount of evaporated atoms by tuning either the electric voltage or the temperature of the tip. These two different ways of controlling the field evaporation will be explored in the next section in HV pulsing and laser pulsing dedicated parts.

15.3 Atom Probe Tomography

15.3.1 Projection of the Ions: the Third Dimension

The evaporated ions are accelerated towards a detector by an electric field, in a direction normal to the surface of the specimen. Thus, the ions are projected from a highly curved surface ($R < 100$ nm) to the device used to detect them. The magnification factor, G, of the projection is

$$G = L/(m + 1) R,$$

with L the flight path of the ions from the tip to the detector, m a projection constant and R the radius of curvature of the specimen. Magnifications of a few 10^6 are easily attained, enabling the visualization of individual atoms. A companion technique to APT, known as field ion microscopy (FIM), exploits this projection more completely by using rare gas atoms to obtain an image of the specimen surface morphology and crystallography on a phosphor screen generally coupled with micro-channel plate (MCP) [4, 40, 41]. The AMMRF instrument at Sydney also provides FIM capability.

As mentioned, the ions are projected towards a detector and the development of position sensitive detectors (PSD) that are able to not only measure the ion flight time but also the position of impact on the detector represents a major development in the technique [4, 5, 42–44]. Coordinates (X_D, Y_D) PSD as designed for and applied to APT directly correspond to the (x, y) coordinates of the atom at the surface of the tip, which can be determined by an inverse projection algorithm (Fig. 15.3). As the atoms are field evaporated atomic layer-after-atomic layer, the order of arrival of the ions on the PSD can be used to determine their depth coordinate z. It is thus possible to tomographically reconstruct the third dimension and to visualise the distribution of the atoms in real space [43, 44] (Fig. 15.4). The spatial resolution is better than one inter-atomic distance in the z direction and better than 1 nm in the $x - y$ plane, allowing the observation of atomic planes (Fig. 15.5) and even the lattice using specific mathematical treatments. [45] The continued development of improved PSD detection capabilities is a significant component of ongoing instrumentation research within the APT [46–49].

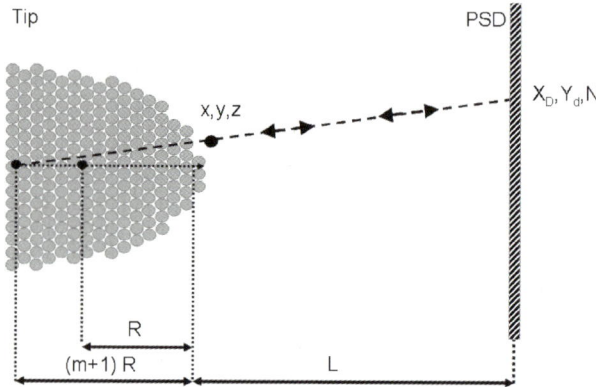

Fig. 15.3. Schematic view of the projection of the atoms from the tip surface. X_D, Y_D and N are, respectively, the coordinates of the impact of the ion on the PSD and the issue of the ion. x, y and z are its original computed positions

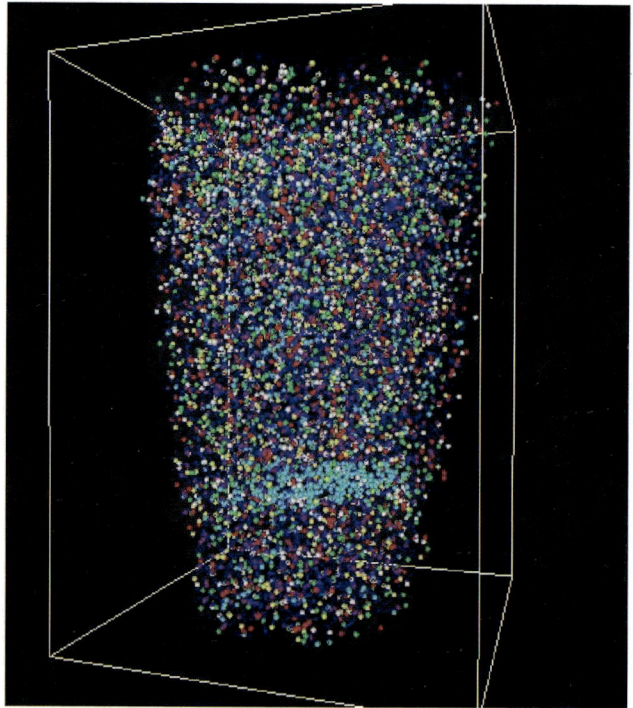

Fig. 15.4. APT reconstruction of Weldalite (Al-Cu-Li-Mg-Ag-Zr) sample, an ultra-high-strength Al-alloy used for the welded fuel tanks of space vehicles. Cu is shown as red spheres, Li as green spheres, Mg as purple spheres, Ag as yellow spheres, and Zr as blue spheres (Al atoms are not shown)

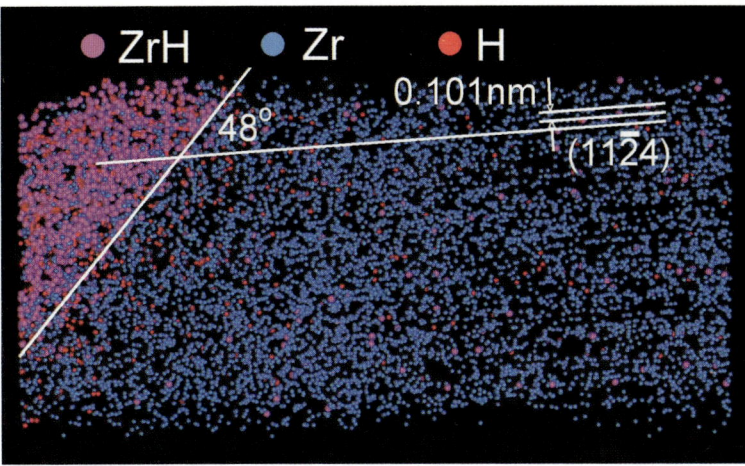

Fig. 15.5. A Zr hydride precipitate in a nuclear zirconium alloy. Identification of the crystal planes in the reconstruction allows, for example the crystallographic orientation of precipitates to be determined

15.3.2 Pulsing Modes

Two contrasting pulsing modes exist in APT: HV pulsed atom probe, often referred to as conventional atom probe, and laser assisted atom probe. In the case of voltage pulsing to achieve field evaporation, fast HV pulses of duration a few nanoseconds full-width-half-maximum (FWHM) are used. The ions are emitted at approximately the maximum of the pulse. However, the duration of the pulse is ∼1 ns and this is much greater than the characteristic field evaporation period of ∼100 fs. Thus, all atoms are not evaporated at the instant of the HV pulse maximum. Therefore, not all evaporated ions are accelerated by precisely the same voltage, and this results in an energy distribution shaped as a peak with a tail. This tail corresponds to ions evaporated with *energy deficits*, i.e. those less energetic ions evaporated at fields just before or just after the voltage maxima occurs. This phenomenon leads to a spread in the ionic times of flight, degrading the mass resolving power of the instrument.

Thermal pulsing, first proposed by Tsong [14, 15, 50], works on the principle that for a given energy barrier, increasing the thermal agitation can also provoke field evaporation. The tip is rapidly heated by absorption of the light from the laser beam, tightly focused to the apex of the specimen. The ions are accelerated only by the DC voltage, and hence there is no observable energy deficit in the time of flight distribution. The temperature of the surface decreases between pulses due to the heat dissipation along the tip shank. The pulse of heat is generated at the apex so as to enable the time of the evaporation to be controlled.

The peak temperature has been experimentally estimated [14, 22] and theoretically calculated using thermal diffusion models [51, 52]. The work of

Vurpillot et al. [53] and Gault et al. [54] has demonstrated that the time for the tip to cool down to a cryogenic temperature is between a few hundred nanoseconds and a few tens of microseconds, depending on the illumination conditions. Thermal pulsing has produced some very significant results, enabling the analysis of pure silicon [14], III–V semiconductors [55, 56] and other poorly conductive materials [4, 57, 58]. However, this very simple principle will only be efficient with laser pulses of duration longer than the electron–lattice coupling time.

During the very first few femtoseconds of the laser pulse, the lattice may be considered as frozen, whereas the electrons are free to absorb photons and to acquire energy from the pulse. Therefore, the electrons and the lattice may be considered as two different systems. The electron cloud is warmed by light absorption, whereas the temperature of the lattice remains unchanged. Progressively, the electron cloud becomes coupled with the lattice, with increasing time towards some characteristic time scale smaller than a picosecond for most metals [59, 60]. For laser pulses longer than a picosecond, the electron cloud and the lattice become coupled during the laser pulse, leading to a direct increase of the tip temperature.

However, in the case of shorter laser pulses, some physical effects that concern only the electrons can occur. Thereby Vella et al. [61, 62] demonstrated that a second-order non-linear optical effect can provoke an ultrafast ion emission. This phenomenon, termed optical rectification, corresponds to the generation of a DC polarization during the laser pulse, equivalent to an ultrafast HV pulse. As its duration is comparable with the period of vibration of the atoms at the surface, it is possible to activate field evaporation of surface atoms. We can thus refer to optical field evaporation corresponding to ultrafast HV pulsed field evaporation. Once the laser pulse is finished, the energy acquired by the electron cloud is transferred to the lattice, the temperature of which increases. A thermal pulse is then generated with femtosecond laser pulses, and it can also provoke the field evaporation [63].

Bunton et al. [28] have shown that in the case of picosecond laser pulses, the atom probe performance in terms of spectral resolution is highly dependant on the specimen nature and geometry and on the illumination conditions. The spatial resolution may also be affected, but no results have been reported on this point up to now. Even if a consensus now exists among the community that field evaporation process depends on the illumination conditions such as pulse duration, wavelength, polarization, etc, there is still much experimental and theoretical exploration to be undertaken to understand and optimise these effects.

15.4 Specimen Preparation

Specimen preparation is one of the most significant issues in characterisation techniques that strive to provide nanomolecular resolution and this is also true for APT. Metallic specimens are commonly prepared by electrochemical polishing of either wire of diameter $\sim$300 µm or of square-sectioned blanks with similar edge dimensions and lengths >5 mm. [4] A neck is progressively formed, decreasing the width of the blank until appropriate dimensions are developed. The size and the radius of curvature of tips prepared in this way meet the requirements of an APT specimen, i.e. a smoothly tapered needle with a circular cross-section and an end radius of $\sim$50–150 nm have been widely and successfully used. Other methods must be used to fabricate APT tips from materials that exhibit low electrical conductivity. Direct chemical methods have been demonstrated in certain semiconductor materials [64, 65] although reproducibility was never really achieved. Another issue that requires consideration in the selection of strategy for specimen fabrication is that of the site-specificity of the phenomenon under study. Site specific APT of, for example sparsely distributed interfaces, second phases etc. is difficult with chemical or electrochemical methods and requires an iterative inspection of the specimen using transmission electron microscopy (TEM) to assess whether the feature of interest is within the likely field of 3D view of a subsequent APT experiment. A useful guide to assess whether the features of interest require a site-specific specimen preparation strategy is to consider the likelihood of intersecting these features when a 100 nm circle is placed at random on a typical TEM micrograph of the microstructure.

The use of dual-beam scanning electron microscope (SEM)/focused ion beam (FIB) for specimen preparation by annular milling has offered new opportunities to APT research. This technique enables sample preparation for almost any kind of material. Additionally, it allows isolation and analysis of precise regions of microstructure, which can be aligned very close to the apex of the tip and hence into the FoV of the instrument. Two of the main methods used in Sydney [33, 34] are presented in the two next sections.

15.4.1 'Cut-Out' Method

The local electrode design of the LEAP allows specimens to have a 'micro-tip' geometry whereby the apex of the needle may be as close as a few tens of microns from the bulk of the specimen without compromising the uniformity or intensity of the electric field. This enables a significant part of the specimen preparation to be undertaken with use of a FIB, as only relatively small amounts of material need to be removed (milling using FIB removes material at a relatively slow rate). A 'cut-out' specimen preparation method was developed to avoid the long milling times required using many other FIB-based techniques [66].

Specimens are first prepared from bulk material using standard TEM preparation techniques to create a thin knife-edge with a wedge angle of a few degrees and a thickness of <10 µm at the leading edge (Fig. 15.6a). It is important that the thickness and wedge angle are not made too large as the speed of this technique depends on cutting through this wedge, using line cuts to remove whole sections, rather than milling away all of the excess material

196   B. Gault et al.

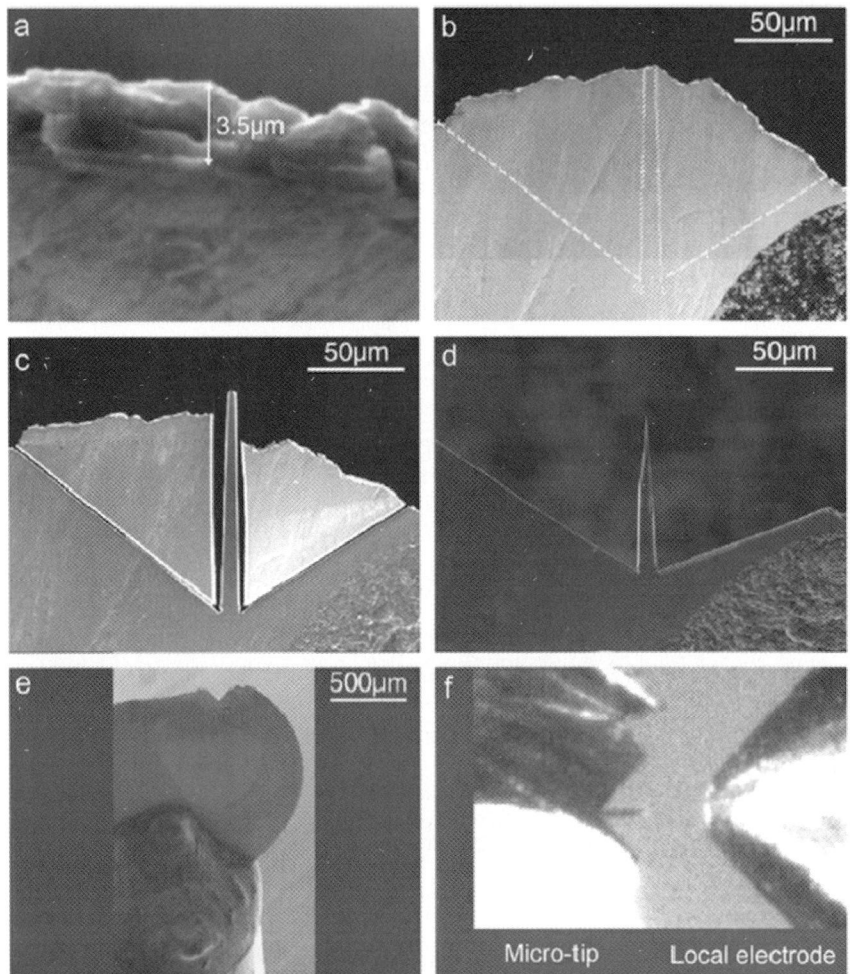

Fig. 15.6. (a) The thin end of the wedge following tripod polishing. The labelled dimension is the cross-section thickness. (b) A suitable region is selected for cut-out of a micro-tip post. The post is tapered to give greater mechanical strength near the base. (c) The triangular sections are cut free from the wedge using the ion beam. (d) The micro-tip after annular milling. (e) The wedge specimen is mounted on an atom probe specimen stub with silver epoxy. (f) The micro-tip specimen is aligned to the local electrode in the LEAP

(Fig. 15.6b–c). Once the triangular sections have been removed (less than 15 min on a suitable specimen), the central post is sharpened to a point using an annular milling pattern (Fig. 15.6d). Using this method a single micro-tip can be prepared from a sharp wedge in around 40–60 min. If the wedge is sufficiently thin over a distance of ~1 mm or more, then several micro-tips can be fabricated from a single specimen. This also leads to improved efficiency when analysing the tips in the atom probe, since several specimens can be examined without replacing the specimen mount.

This technique provides a simple means of fabricating multi-tip specimens from bulk material using methods that are similar to those used in standard TEM specimen preparation. It is particularly useful for materials that cannot be easily electropolished. In addition, this method may be used to prepare site-specific atom probe specimens in cases where the region of interest can be resolved in the FIB/SEM. These may include features such as grain boundaries or other interfaces, device structures, particles and precipitates.

15.4.2 'Lift-Out' Method

Another technique for fabricating micro-tip specimens from bulk material has been developed involving a lift-out procedure. However, unlike most FIB-based lift-out techniques, this method does not require the use of pre-fabricated micro-tip arrays in order to produce micro-tip specimens for the LEAP.

In this method, a bulk sample is prepared by polishing its top surface to a roughness of <1 µm (if required). FIB line cuts are used to undercut a long post from the surface of the material, as shown in Fig. 15.7. One end of

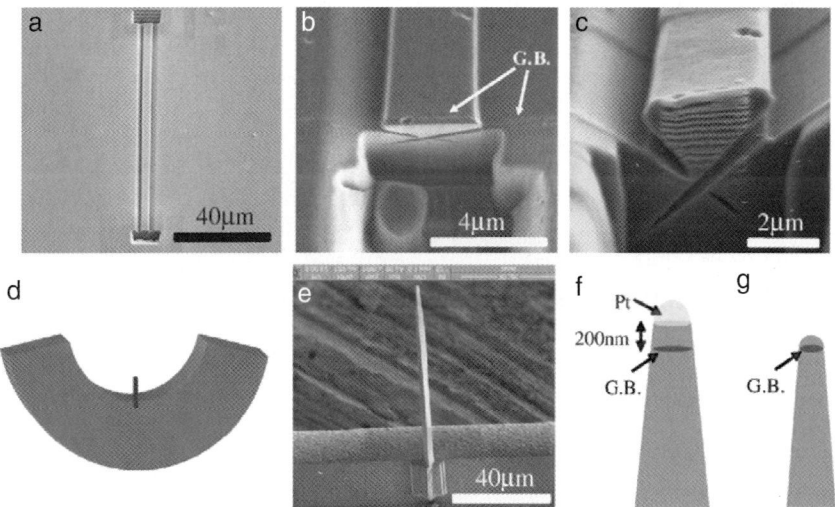

Fig. 15.7. Preparation of a specimen from a grain boundary. (**a**) A triangular post is cut loose from the bulk, (**b**) careful cleaning mills are carried out on the end of the post, so that the grain boundary is located 200 nm from the end of the post, (**c**) a small amount of platinum is deposited onto the end of the post for use as a marker, (**d**) and (**e**) the post is lifted from the bulk, placed onto a Cu grid and welded on with platinum in the FIB, (**f**) and (**g**) careful annular mills are used to slowly remove material, creating a tip with grain boundary <100 nm from the apex

the post is freed from the bulk by milling a cross section. The other end is freed similarly after a manipulator needle has been attached to the post using Pt deposition. Once the post is completely free, it is placed on a sectioned copper slot grid using the manipulator needle and attached by deposition of Pt. Transfering the needle from the bulk specimen to the sectioned slot gird may also be perfomed using 'ex-situ' lift-out method with micro-manipulators equipped with a glass needle. The manipulator is then cut away from the post using the FIB and the tip is sharpened using annular milling patterns.

For both this technique and the cut-out method, the tip is tapered to provide increased strength, making it less likely that the specimen will fracture during an acquisition. Compared to the cut-out method, the lift-out method requires less preparation of the sample material, but requires more skill and has a lower success rate.

In most cases, the lift-out method is more suitable for site-specific specimen preparation than the cut-out method, since the region of interest has to be only near the surface of the original bulk sample. The cut-out method requires the region of interest to be positioned close to the knife-edge formed by mechanical grinding. The lift-out method also provides more flexibility in the orientation of the region of interest with respect to the specimen axis.

Both techniques provide relatively straight-forward methods for creating micro-tip specimens suitable for the LEAP, without the need for pre-fabricated micro-tip arrays or other tools not common to TEM specimen preparation. Multiple tip specimens can provide economies of scale in both their manufacturing stage and their analysis in the local electrode atom probe.

15.5 Experimental

The AMMRF's atom probe laboratory at the University of Sydney operates two new generation Local Electrode Atom Probes, the LEAP^{TM} 3000X and 3000XSi (Fig. 15.8), manufactured by Imago Scientific Instruments Corporation. The instruments are high-performance atom probe microscopes, with the latest Delay Line Detector (DLD) technology, providing an improved FoV. This detector has a large surface area enabling a FoV of more than 200 nm for an Al specimen. The flight path is adjustable between 80 and 160 mm. Improvements to ultra high vacuum (UHV) instrumentation mean that pressures of ~2.0×10^{-11} Torr are routinely achieved. During experiments, the temperature of the tip can be maintained as low as 17 K, by means of a cryogenerator, with no significant vibration of the specimen.

The voltage pulser is operable at very high repetition rates (up to 200 kHz) enabling truly massive data-sets to be acquired, of the order 10^8 atoms per run, in just a few hours. Further, the mass resolution has been greatly improved by the use of new electronics, and 900 FWHM is observed routinely, with noise levels generally below 10 counts per million signals per nanosecond

Fig. 15.8. The two LEAP 3000X installed in the AMMRF national atom probe laboratory at the University of Sydney

of time-of-flight. The combination of such high mass resolution and low background noise enables extremely reliable measurements of composition.

Significantly, one of the instruments (the 3000X Si) has been upgraded to incorporate the option of ultra-fast laser pulsing with very high repetition rates of up to 500 kHz. The laser pulses have a duration of 10 ps and deliver an energy pulse of between 1 and 20 nJ at a wavelength of 532 nm. Data acquisitions rates of 5 million ions per minute are achievable with this instrument. The mass resolution is significantly increased when using laser pulsing, with an upper-limit towards 1,500 FWHM for tungsten. This is demonstrated in Fig. 15.9, where the signal-to-noise ratio is higher for laser pulsing. Note the comparison of the W^{180} peak in the 3+ charge state (peak at 60 amu): it is barely detectable in voltage pulsing, but is clearly observed in laser mode.

What is particularly significant is that the use of laser pulsing opens up the APT technique up to a much wider field of scientific and technological problems because materials with low electrical conductivity such as semiconductors, ceramics and other insulating materials may be studied.

The LEAP 3000X Si (Fig. 15.10) has also been designed to incorporate noble gases into the experimental chamber, providing a FIM capability. An image is formed by collecting several million of gas ions per second on the DLD. It is therefore possible to obtain high quality digital FIM images and an example from pure W sample is provided in Fig. 15.11. All parameters of the FIM experiment can be computer controlled.

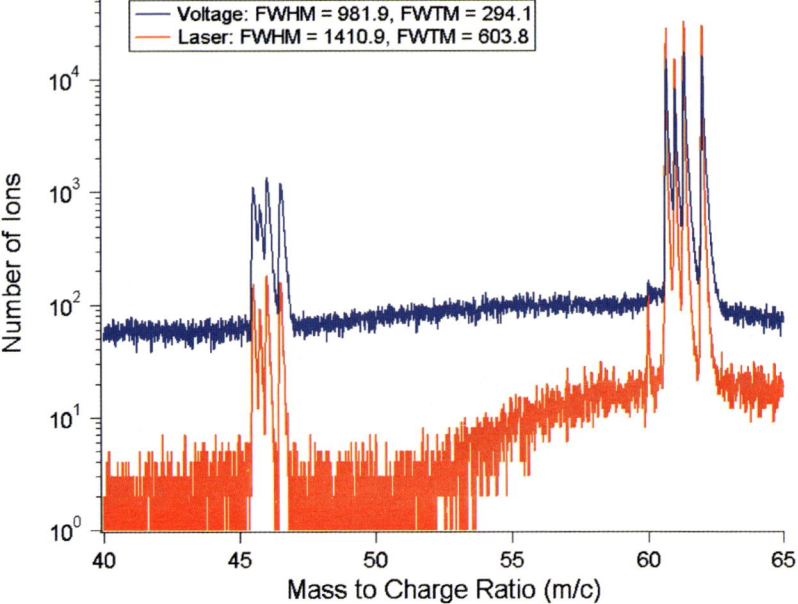

Fig. 15.9. Voltage vs. laser pulsing mode mass resolutions for a tungsten specimen. Note that the signal-to-noise ratio is higher for the laser system. Note the comparison of the W^{180} peak in the 3+ charge state (peak at $60\,\mathrm{m}\,\mathrm{c}^{-1}$), it is barely detectable in voltage, but very prominent in laser mode. The natural abundance of this isotope in the literature is 0.12%

Fig. 15.10. The LEAP 3000X Si, with laser pulsing capabilities

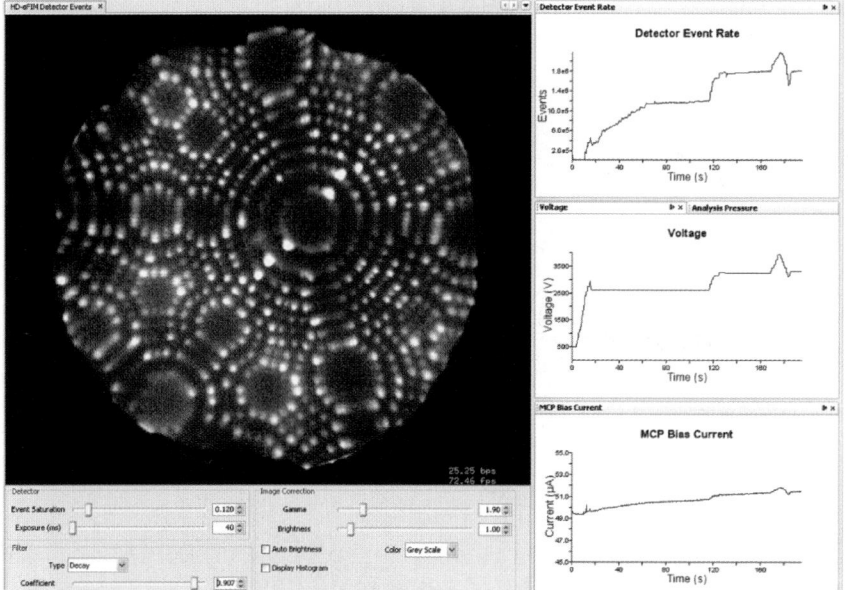

Fig. 15.11. Digital field ion images (FIM) is of pure tungsten atoms observed with He in a {011} orientation, obtained with the LEAP 3000X Si currently installed at the University of Sydney

15.6 The Dual Beam SEM/FIB

Electrochemical polishing methods for the preparation of specimen for APT are routinely employed at the University of Sydney laboratory. However, as previously discussed, poorly conductive material or site-specific investigations require alternative approaches, namely FIB techniques. A new generation, multifunctional FEI QUANTA 200 3D integrated equipment platform was also installed recently in our laboratories (Fig. 15.12). This dual SEM/FIB workstation is capable of operating in 'environmental' mode and interfacing with a custom-designed Renishaw Raman spectrometer. This integrated equipment platform consists of a dual electron/ion column, and so it may be used as both a SEM and FIB system. The ion column brings an extra dimension to microscopy, allowing precise microfabrication of the sample at the resolution at which it is being imaged. The electrons may be accelerated by a voltage of up to 30 kV, allowing to reach a resolution of only 3.5 nm. A Kleindiek MM3A-EM micro-manipulator with a high resolution of 0.25 nm and an extensive working range of 100 cm^3 can be used in combination with a Pt source to prepare APT specimens from almost any material or device using the lift-out technique described above.

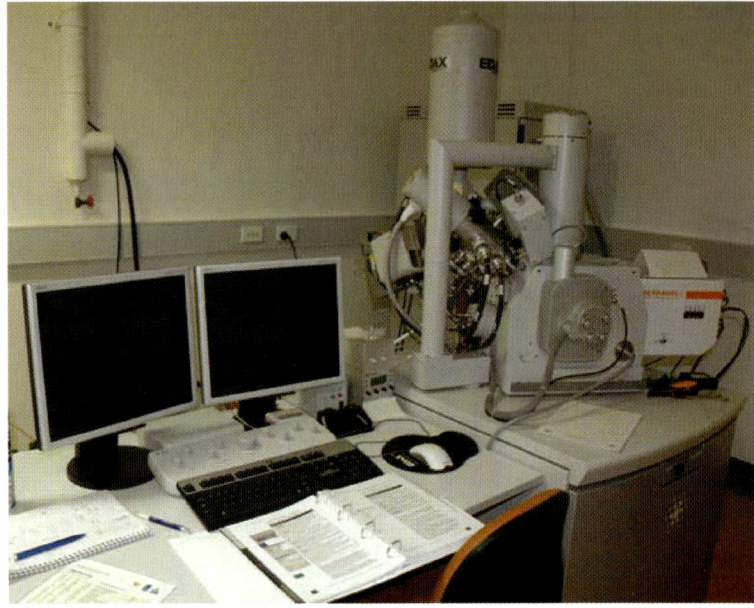

Fig. 15.12. FEI QUANTA 200 3D integrated equipment platform. This dual SEM/FIB workstation is currently installed at the University of Sydney

15.7 Applications: Nanostructural Analysis of Materials

The unprecedented 3D atomic level insight offered by APT is stimulating increasing interest in the role of nanostructure such as atomic clustering, in the evolution of microstructure and new ideas for structure–property relationships. However, the sheer volume of data produced by APT combined with the subtle fine-scaled nature of these phenomena means that considerable information remains unresolved by mere visual inspection of the atom maps. The continued development of sophisticated data mining tools specific to multi-gigabyte 3D atom probe data sets is essential to this research. [5, 9]

Moody et al. recently investigated frequency distribution techniques like contingency table analysis [5, 67, 68] (CTA) to identify the presence of nanostructure, and track how it quantitatively changes in a system subject to thermal treatments, for example. These analyses see the APT data set divided into a grid of blocks with an equal amount of atoms. In CTA the number of A and B type atoms simultaneously occurring in each block is counted. An A–B Contingency Table is thus a 2D histogram tabulating the total number of blocks simultaneously containing X number of A atoms and Y number of B atoms. Compared to a generated theoretical expected table, correlative effects in the distribution of these atomic species, co-/anti-segregation, can be identified and quantified. Research has focused on how to optimise CTA for APT, and in particular on the direct comparison of analyses between multiple data sets and the interpretation of results, including the validity of significance testing and the applicability of random comparators.

Grid based methods provide an effective method for assessing the amount of clustering present in a system; however, more sophisticated computatively intensive algorithms are needed to characterise the nature of individual clusters in terms of size, chemistry, morphology, etc. The most commonly applied clustering algorithm in APT is the maximum separation method [69], based on the principle that clustered solute atoms are closer than matrix solute atoms, represented by some characteristic separation d_{max}. Stephenson et al. have investigated the application of density based cluster identification algorithms. The density based scanning algorithm [70] (DBSCAN) is based on the principle that clustered solute atoms are denser than matrix solute atoms. This research culminated in the development of the *core-linkage algorithm* [71] (Fig. 15.13), in which an initial 'core' step that identifies high density regions of solute atoms is followed by the 'link' step, which links together, and effectively clusters, any solute atom within d_{link} of a core atom. This novel algorithm has several key advantages over maximum separation, notably its applicability to systems containing both fine scaled solute clusters and much larger precipitates or microstructure, and a significantly decreased sensitivity to fluctuations in the reconstructed spatial coordinates of the atomistic data.

Data mining techniques are sensitive to user-controlled parameters (like d_{max}) and considerable research has gone into developing objective analytical methodologies for their optimised and objective selection. Further, data at the surface of the reconstruction or within obvious crystallographic artifacts can detrimentally bias the results of the described data mining analyses and is identified and removed using techniques developed around high-order nearest neighbour distance distributions [71]. This data preparation is critical to obtain quantitative results applying analytical techniques, particularly with the large size and complexity of data sets obtained with wide FOV, high-throughput atom probe tomography.

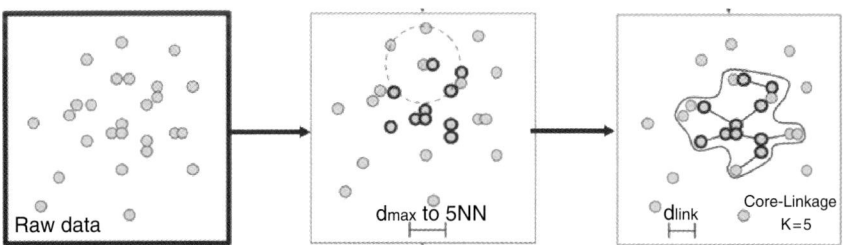

Fig. 15.13. Schematic of core-linkage algorithm. Initial 'core' step identifies high density regions of solute atoms within the raw data, i.e. solute atoms within d_{max} of another solute atom. The following step identifies clusters by linking any solute atom within d_{link} of a core atom, where d_{link} is chosen either using heuristical approaches or analyses based on complete spatial randomness

These statistical algorithms have been successfully applied to APT results generated from a range of different material systems, including the examples discussed in the proceeding sections.

15.7.1 Al-Cu-Mg Alloys

Al-Cu-Mg alloys that lie in the $\alpha + S$ phase field and have a low Cu:Mg ratio exhibit a remarkably rapid hardness increase during ageing at elevated temperatures [72]. This rapid hardening phenomenon (RHP) occurs in Al-1.1Cu-xMg alloys that contain at least 0.5 Mg (at.%) and may provide >50% of the total hardness increment within the first 100 s of ageing. Hardness measurements have been made as part of the study at the University of Sydney to highlight the rapid hardening, and these hardness vs. time plots are shown in Fig. 15.14. The RHP does not occur in binary Al-Cu and Al-Mg alloys or in general within ternary Al-Cu-Mg alloys with phase compositions in the $\alpha + \theta$ field (high Cu:Mg ratio), even though all are age hardenable. This intriguing phenomenon has been studied in detail only relatively recently and proposals to explain this include the concept of solute cluster strengthening [8] and, more recently, the suggestion that hardening arises from precipitation of very fine scale GPB zones [73].

APT of Al-Cu-Mg alloys at the University of Sydney has been conducted with the aim to characterise in detail any atomic clustering during the early stages of ageing where the supersaturated solid solution is first decomposing. Particular importance has been assigned to the task of differentiating

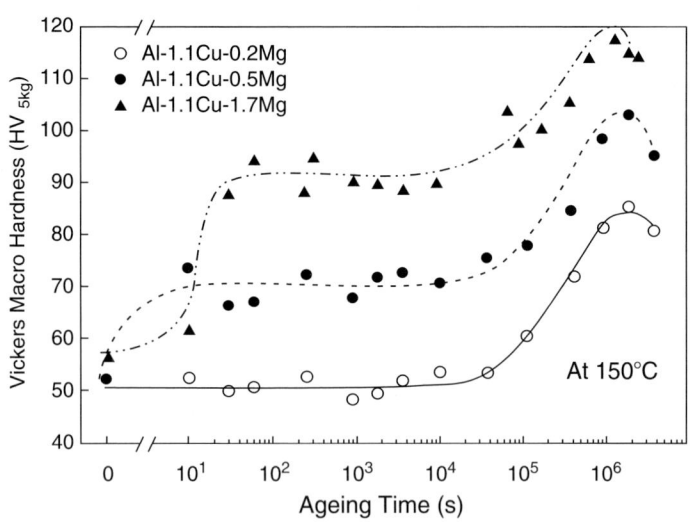

Fig. 15.14. Age hardening curves for Al-1.1Cu-xMg (x = 0.2, 0.5, 1.7 at. %) following solution treatment for 1 h at 525°C, quenching and subsequent ageing at 150°C

experimental observations of clustering from those that could be expected from random compositional fluctuations in the solid solution, and parameters such as cluster chemistry, size distribution and relative number density have been considered from this viewpoint. Hence, a data mining approach using statistical techniques and cluster-finding algorithms has been implemented.

An initial investigation by Marceau et al. [11] implanting contingency table analysis (CTA) for Cu-Mg co-clustering was performed on a systematic series of AP data sets following quenching and ageing for 60 s and for 1 h at 150°C. In this analysis, like-atom species clustering (Cu-Cu and Mg-Mg) was also investigated by comparing the experimentally observed frequency distributions to a binomial distribution that models random configuration. At the 99.9% level of significance there is a preferred interaction between Cu and Mg from the earliest possible stage of phase-decomposition (i.e.) in the as-quenched state. This preferred interaction was also observed in data from samples aged 60 s and 1 h at 150°C. Moreover, a preferred Cu-Cu and Mg-Mg interaction was observed for all three conditions. Figure 15.15 summarises the statistical analysis of the APT data for the Al-1.1Cu-0.5Mg alloy and is a plot of the chi-squared coefficient that can be used to assess quantitative trends from one CTA to another [67, 68] A relative increase in clustering is represented by corresponding increase in the value of the coefficient. As ageing proceeds, the level of co-clustering of Cu-Mg remains similar overall, as does that for the Mg-Mg clusters. However, the Cu-Cu clusters possess a stronger interaction after 60 s at 150°C. Figure 15.15b is an atom map reconstruction from the Al-1.1Cu-0.5Mg alloy aged 60 s at 150°C and reveals a region of local clustering between Cu and Mg. The clusters are seen to be pre-precipitate aggregations of solute, possessing a diffuse shape and form. Rod-shaped GPB zones were not detected in the APT experiments.

The results indicate that Cu-Cu, Mg-Mg and Cu-Mg clusters are formed in the ternary alloy already in the as-quenched state and during ageing at 150°C.

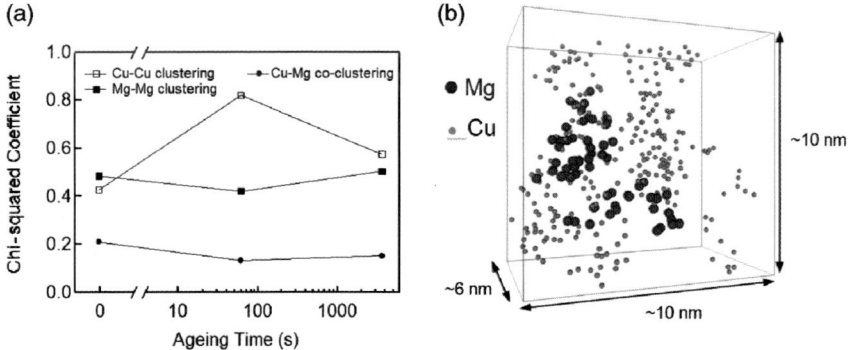

Fig. 15.15. (a) Chi-squared coefficient calculated from APT data for Cu-Mg, Mg-Mg and Cu-Cu clustering for samples in the as-quenched condition and after ageing for 60 s and 1 h at 150°C. (b) A typical example of Cu-Mg co-clusters in the material aged 60 s at 150°C

The use of the chi-squared-coefficient suggests that a key change between the clusters present in the as-quenched samples and those in samples aged for 60 s at 150°C is related to the degree of Cu-Cu clustering interactions. Given that zone-like precipitate structures were not detected and statistically significant dispersions of clusters during ageing were found, Marceau et al. [11] attribute the change in hardness to these clustering reactions. Amongst several key questions that remain is the question of the relative significance of cluster chemistry (like-atoms vs. mixed) and their size and relative number density and these are currently being explored.

15.7.2 Fe-Base Alloys: Fenimnal Spinodal Alloy

Phase separation and decomposition behaviour in novel Fe-Ni-Mn-Al spinodal alloys have been investigated using the LEAP. These alloys exhibit a spinodal phase separation over a range of compositions and can exhibit extraordinary yield strengths of up to 2.3 GPa. Four alloys were prepared by arc melting and casting and their compositions in atomic % are listed in Fig. 15.16. This

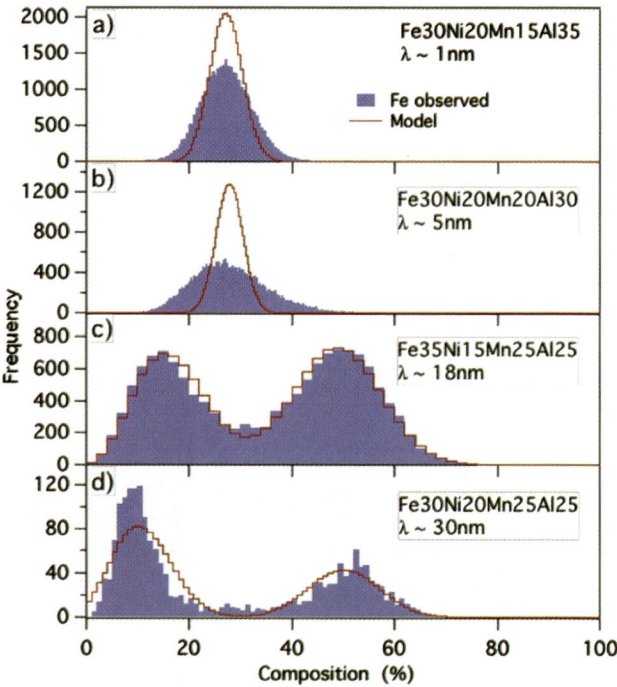

Fig. 15.16. Frequency histograms for the concentration of Fe over the analysed volume for four alloy compositions in the as-cast condition. The approximate wavelength of the spinodal is also provided. For (**a**) and (**b**), the binomial distribution for a random arrangement of Fe atoms is also provided. For (**c**) and (**d**), a fit to the data using the LBM model for spinodal decomposition is provided

research is part of a broader study into the evolution of microstructure in this system subject to ageing treatments at elevated temperature, and the optimization of bulk material properties. The ultimate goal of this research is to develop improved microstructure-property relationships for these types of alloys. Following the initial work of Cahn [74], most models of the functional relationship between the critical resolved shear stress (CRSS) σ_c and microstructure include a strong dependence on λ, the wavelength or periodicity of the compositional fluctuations. The present results are focussed on the initial starting microstructures in the 'as-cast' condition.

Figure 15.16a–d compares frequency distributions of the Fe concentration across the analysed volumes for each of the four alloy compositions. Figure 15.16a,b also include the expected binomial distribution, i.e. that expected for a random arrangement of Fe atoms. In Fig. 15.16c,d, where the Mn:Al ratio is higher, the profiles also include the Fe frequency distribution predicted by the Langer/Bar-on/Miller (LBM) model for spinodal decomposition [74]. It is clear that the substitution of Mn for Al tends to enhance the spinodal decomposition (Fig. 15.17). Figure 15.18 is a typical 3D atomic reconstruction of the as-cast microstructure. The wide FoV of the LEAP allows several wavelengths of the spinodal decomposition to be viewed in a single analysis, as is evident from the isoconcentration surface in Fig. 15.19. The

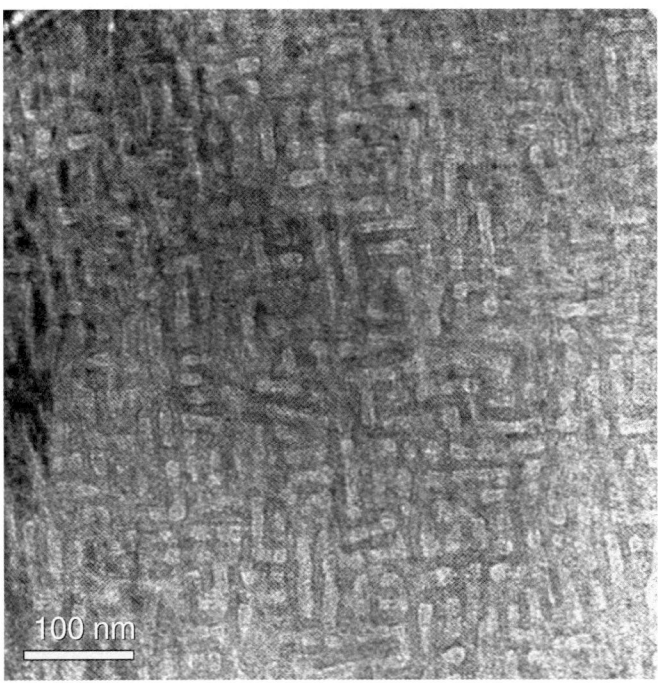

Fig. 15.17. TEM image of a Fe35Ni15Mn25Al25 alloy. The spinodal decomposition is clearly visible

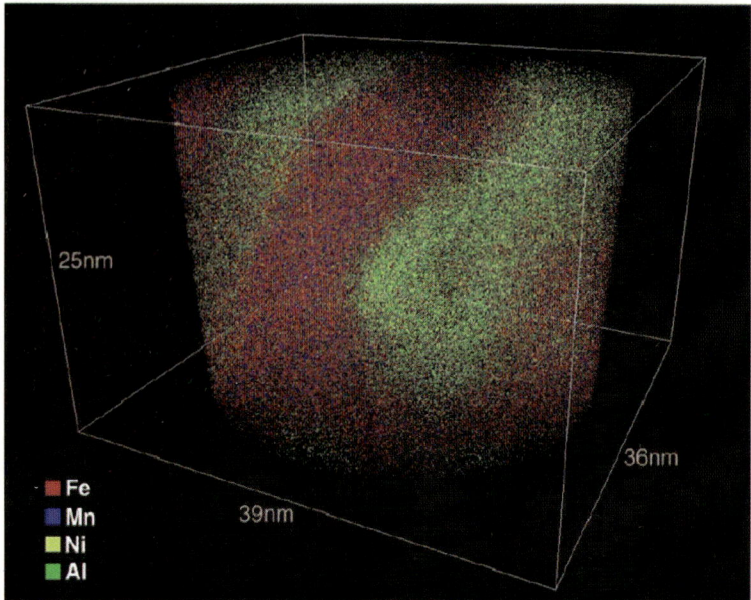

Fig. 15.18. Atom map of the as-cast Fe35Ni15Mn25Al25 alloy showing the spinodal decomposition into Fe-rich and Ni-rich phases

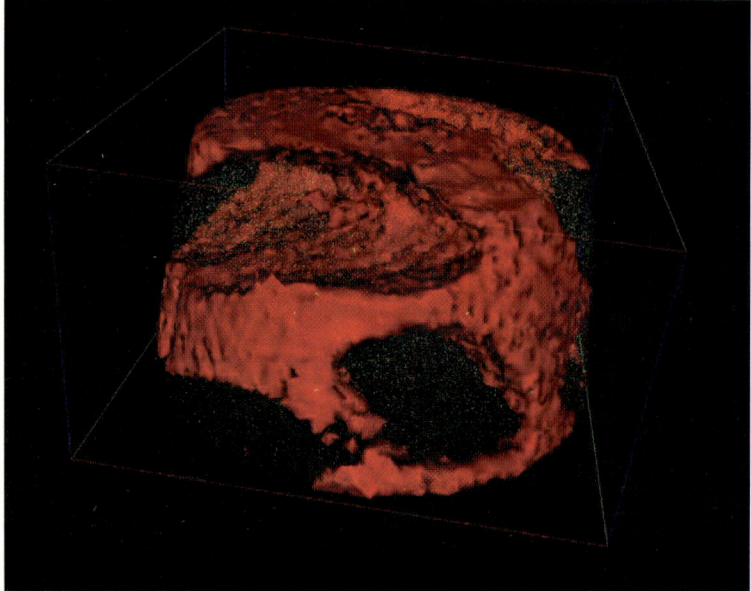

Fig. 15.19. An isoconcentration surface corresponding to a concentration of Fe-20% in the same alloy as Fig. 15.11. Five percent of the detected Ni and Al atoms are also shown (*yellow dots*). The box dimensions are $30 \times 30 \times 40 \, \mathrm{nm}^3$

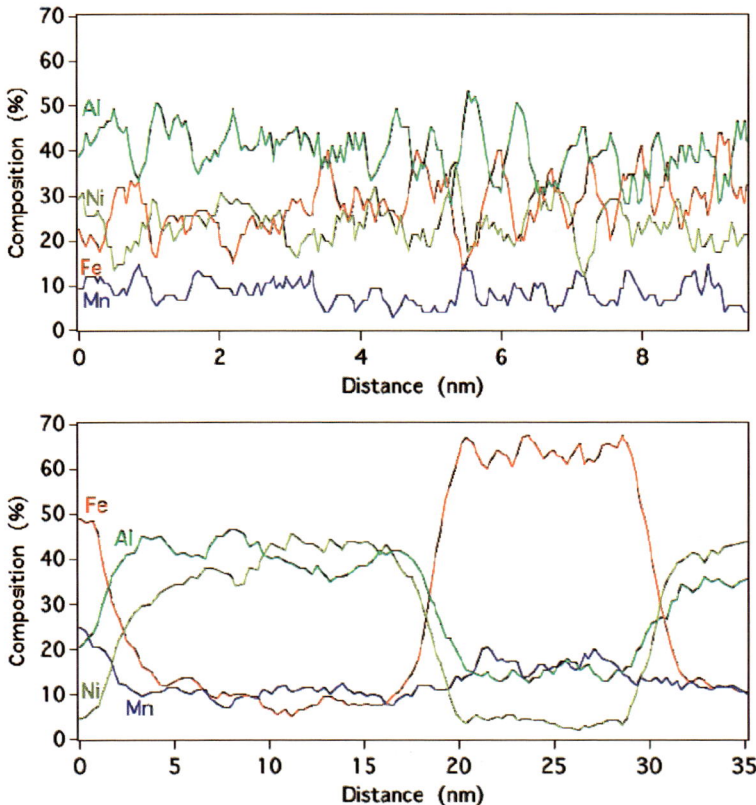

Fig. 15.20. One-dimensional composition profiles from as-cast Fe30Ni20Mn15Al35 (*up*) and as-cast Fe30Ni20Mn25Al25 (*down*), providing indicative wavelength of decomposition

large number of atoms collected also provides good statistics for quantitative analyses of the wavelength and amplitude of the compositional modulations, Fig. 15.20.

15.7.3 Al-Zn-Mg-Cu Alloys

The addition of Cu provides an essential enhancement of the stress-corrosion cracking resistance to Al-Zn-Mg alloys [75] and also increases significantly their age hardening response [76]. Recent work [77] on the mechanism of this enhanced age hardening response focussed on whether or not Cu modifies the existing ageing processes in base ternary Al-Zn-Mg alloys and/or introduced additional precipitation processes that normally occur in the Al-Cu-Mg system [8]. On the basis of the observation of an early rapid hardening reaction, which is also a characteristic of artificially aged Al-Cu-Mg alloys [72], and the observation of both GPB zones and S phase precipitation albeit after

(a)   (b)

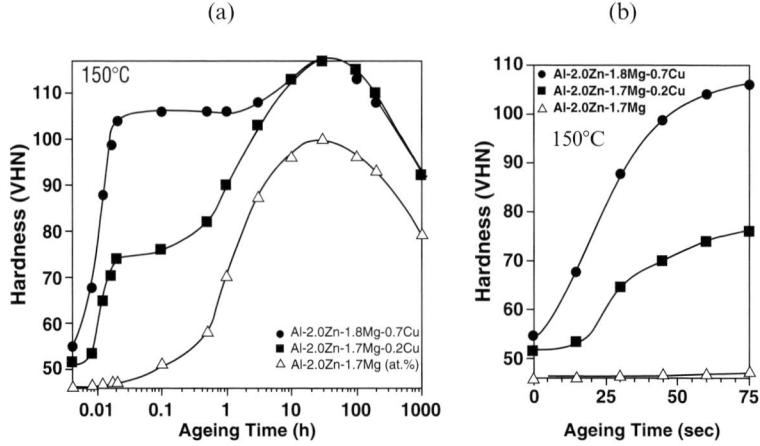

Fig. 15.21. T6 hardness curves at 150°C of selected Al-Zn-Mg(-Cu) alloys: (**a**) 200 h ageing (**b**) 75 s ageing

substantial ageing times (24 h) at elevated temperature ageing (190°C), it was concluded that the Cu introduces clustering and precipitation processes from the Al-Cu-Mg alloys.

The age hardening curves in Fig. 15.21a reveal the extent of the enhancement of the age hardening response associated with the addition of Cu. Of particular note is the initial rapid hardening in the Cu-bearing alloys, which seems to be primarily responsible for the overall augmentation in hardening, compared to the ternary alloy. Figure 15.21b is an enlarged section of the early stages of hardening and provides details of the extent of the rapidity of the initial hardening response. We observe that the addition of 0.2 Cu results in ∼65% of the overall hardening achieved in the initial rapid hardening reaction. The addition of 0.7 Cu provided ∼90% of the overall hardening in the initial rapid hardening reaction. These measurements have been used to select samples for detailed characterisation, presented following.

Figure 15.22a is a <110>$_\alpha$ bright field (BF) TEM image recorded from a specimen of the ternary Al-1.9Zn-1.7Mg alloy in the as-quenched (AQ) condition. This image is typical and affirms that there is no quenched-in precipitate structure and that the solution treatment was effective in dissolving second phases, as might be expected. Further to this, there are no streaks or diffuse intensity in the inset selected area electron diffraction (SAED) patterns that suggest precipitation of zones or other phases. Figure 15.22b provides APT data in the form of atom maps for the Zn and Mg species. This data is also typical for AQ alloys and reveals what appears to be a uniform solute distribution. As with the TEM data, there are no zones or other precipitates to be observed. Figure 15.22c summarises the results of our compositional frequency histogram for Zn-Zn like-atom clustering. At the 0.001 level of significance, it was found that there are, in fact, preferred solute–solute distribution. This

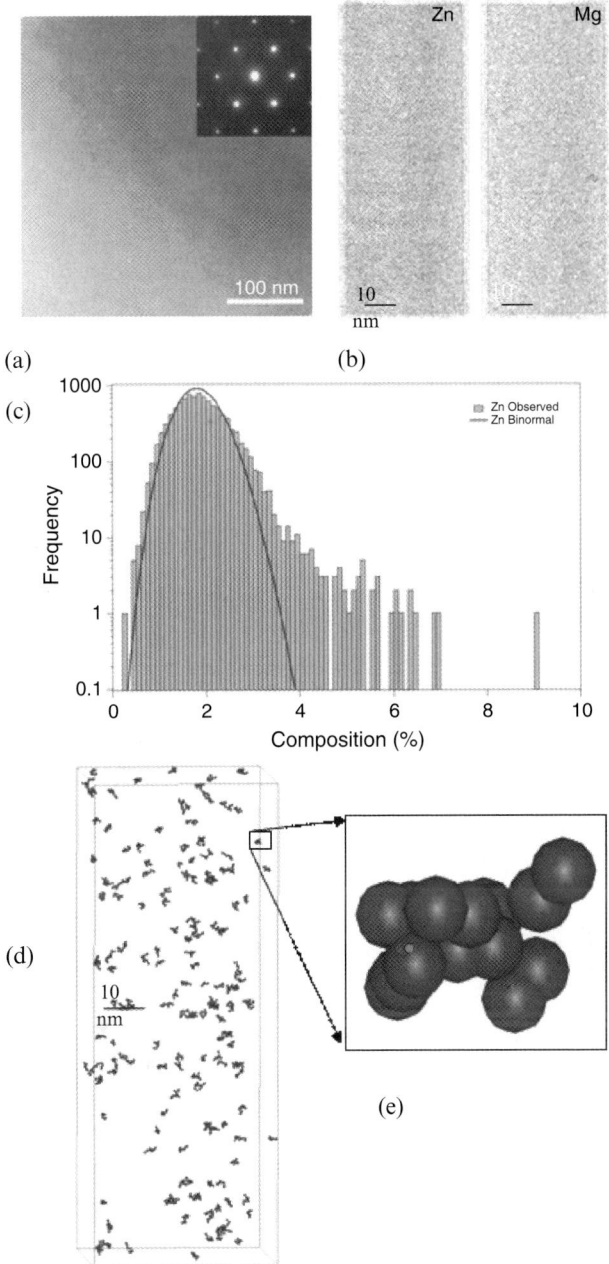

(a)

(b)

(c)

(d)

(e)

Fig. 15.22. The figure is extensively described in the text

suggests that the solution treatment at 460°C for 1 h and cold-water quench, which is a highly typical heat treatment procedure, may actually possess a discrete nanostructure in the form of preferred solute–solute interactions and atomic co-clusters. This seems significant because this quenched-in nanostructure will likely play a major role in the subsequent decomposition of the solid solution, including the subsequent vacancy migrations and vacancy supersaturation, solute clustering and precipitation processes. Figure. 15.22d was generated by using a d_{max} threshold of 0.65 nm so as to discriminate clusters as when solute atoms are located within 0.65 nm of each other and when there is at least nine atoms per group. The dispersion reveals that a high number-density of these co-clusters exist and Fig. 15.22e is a high resolution APT image of a typical individual cluster.

15.8 Summary

The high spatial resolution of the 3D atomistic reconstructions make APT a uniquely powerful tool for the analysis of nanostructure. APT routinely enables the identification of fine scale phenomena beneath the resolution of conventional electron microscopy and further has the potential to accurately characterise the distribution of such features on an atomic scale in terms of dispersion, number, chemistry, size and morphology. The significance of this capacity for quantitative and qualitative analysis of nanostructure is highlighted in the research examples above. In particular, opportunities arising from increased understanding of the role of atomic clustering during the early stages of phase decomposition and the roles of these clusters in both influencing transformation pathways in subsequent precipitation reactions as well as the direct influence of solute clusters on macroscopic material properties present significant design opportunities. To predict and control the development of such phenomena will enable design of new materials via new compositions and/or novel thermo-mechanical treatments, the benefits of which are pervasive, with scientific, industrial, economic and environmental benefits.

Most recently, the combination of advances in instrumentation together with developments in FIB specimen preparation techniques have presented a host of potential new research opportunities. Laser assisted atom probe is opening up APT to a range of new materials, with semiconductors especially an emerging area where there is much current international research interest. Moreover, relatively little effort has been devoted thus far to ceramic materials or minerals. Combined with increasingly wide angle detectors, ultrafast laser pulse repetition is rapidly increasing acquisition rates with data sets routinely of the order of 10^8 atoms. The significance of maximizing the volume of analysed microstructure is apparent from the preceding section on spinodal decomposition. APT semiconductor analysis has recently produced some striking visualizations, and the technology has advanced to such an extent that the reconstruction of an entire device is within reach. Further, the impact of FIB specimen preparation techniques can not be overestimated. These methods facilitate an approach to fabrication of APT specimens from poorly conducting and other materials unsuited to electropolishing. Equally significant is the capability they provide to conduct site specific investigations. The ability to identify then isolate specific microstructure, such as grain boundaries, and prepare a sample around this feature can significantly increase the effectiveness of APT analyses.

While the current and potential impact of APT to an increasingly wide range of research applications is clear, it remains an emerging technique. Work is actively being undertaken to further enhance the results of APT, such as optimising the experimental conditions, for example specimen tip shape, laser pulse size and shape. Improving instrumentation must remain a priority, with emphasis on pulsing technology and detector technology. There is also a need for improved, accurate and efficient 3D reconstruction and the physics of reconstruction is widely regarded as an area of considerable research potential. Finally, there remains much to do in the development of data mining techniques of the very large 3D data-sets [71, 78]. Given our capacity to acquire and position tens and hundreds of millions of atoms and our goal to develop quantitative nanostructure-activity relationships for materials, it is proposed that APT will play a key role in the emerging era of what we regard as materials informatics as a tool for the design of new materials. Further, the role of computational simulations is becoming increasingly important in APT research. The complementary nature of 3D atomistic APT digital data sets and computational atomistic simulations is starting to provide new insights into the ways that materials work. Researchers will continue to push the boundaries with the materials to which APT is applied. Initial work is already proceeding on organic materials, with the study of biological systems via atom probe, a longer term goal.

Acknowledgements. This research was partly funded by the Australian Research Council (ARC). ZL and RZ are grateful for ARC support towards a QEII and ARC Fellowship, respectively. The AMMRF is supported by the Department of Science, Training & Industry NCRIS fund. The authors are grateful for the support from The University of Sydney and particularly for the technical and scientific inputs of Messrs. Phillip Gowlett and Alex La Fontaine (Atom Probe Engineers). We are also grateful to Prof. Ian Baker and colleagues from Dartmouth College for collaboration on the Fe-Ni-Mn-Al Alloys.

References

1. E.W. Müller, Phys. Rev. **102**, 618 (1956)
2. A. Cerezo, T.J. Godfrey, G.D.W. Smith, Rev. Sci. Instrum. **59**, 862 (1988)
3. D. Blavette, A. Bostel, J.M. Sarrau, B. Deconihout, A. Menand, Nature **363**, 432 (1993)

4. M. Miller, A. Cerezo, M. Hetherington, G. Smith, *Atom Probe Field Ion Microscopy* (Oxford Science Publications, UK, 1996)
5. M. Miller, *Atom Probe Tomography: Analysis at the Atomic Level* (Kluwer, New York, 2000)
6. D. Blavette, E. Cadel, A. Fraczkiewicz, A. Menand, Science **286**, 5448 (1999)
7. E.V. Pereloma, A. Shekhter, M.K. Miller, S.P. Ringer, Acta Mater. **52**, 5589 (2004)
8. S.P. Ringer, T. Sakurai, I.J. Polmear, Acta Mater. **45**, 3731 (1997)
9. S.P. Ringer, Mater. Sci. Forum **519–521**, 25 (2006)
10. P.V. Liddicoat, T. Honma, L.T. Stephenson, S.P. Ringer, Mater. Sci. Forum **519–521**, 555 (2006)
11. R.K.W. Marceau, R. Ferragut, A. Dupasquier, M.M. Iglesias, S.P. Ringer, Mater. Sci. Forum **519–521**, 197 (2006)
12. E.W. Müller, S.V. Krishnaswamy, Rev. Sci. Instrum. **45**, 1053 (1974)
13. S.J. Sijbrandij, A. Cerezo, T.J. Godfrey, G.D.W. Smith, Appl. Surf. Sci. **94–95**, 428 (1996)
14. G.L. Kellogg, T.T. Tsong, J. Appl. Phys. **51**, 1184 (1980)
15. T.T. Tsong, S.B. McLane, T. Kinkus, Rev. Sci. Instrum. **53**, 1442 (1982)
16. O. Nishikawa, M. Kimoto, Appl. Surf. Sci. **76–77**, 424 (1994)
17. M. Huang, A. Cerezo, P.H. Clifton, G. Smith, Ultramicroscopy **89**, 163 (2001)
18. T.F. Kelly, P.P. Camus, D.J. Larson, L.M. Holzman, S.S. Bajikar, Ultramicroscopy **62**, 29 (1996)
19. T.F. Kelly, D.J. Larson, Mater. Charact. **44**, 59 (2000)
20. T.F. Kelly, T.T. Gribb, J.D. Olson, R.L. Martens, J.D. Shepard, S.A. Wiener, T.C. Kunicki, R.M. Ulfig, D.R. Lenz, E.M. Strennen, E. Oltman, J.H. Bunton, D.R. Strait, Microsc. Microanal. **10**, 373 (2004)
21. G.J. Kellogg, Chem. Phys. **74**, 1479 (1981)
22. G.J. Kellogg, Appl. Phys. **52**, 5320 (1981)
23. G.J. Kellogg, Phys. Rev. B **28**, 1957 (1983)
24. T.T. Tsong, T.J. Kinkus, Phys. Rev. B, **29**, 529 (1984)
25. T.T. Tsong, T.J. Kinkus, S.B. McLane, J. Chem. Phys. **78**, 7497 (1983)
26. T. Hashizume, Y. Hasegawa, A. Kobayashi, T. Sakurai, Rev. Sci. Instrum. **57**, 1378 (1986)
27. B. Gault, F. Vurpillot, A. Vella, M. Gilbert, A. Menand, D. Blavette, B. Deconihout, Rev. Sci. Instrum. **77**, 043705 (2006)
28. J.H. Bunton, J.D. Olson, D.R. Lenz, T.F. Kelly, Microsc. Microanal. **13**, 418 (2007)
29. A. Cerezo, P.H. Clifton, A. Gomberg, G.D.W. Smith, Ultramicroscopy **107**, 720 (2007)
30. P. Stender, C. Oberdorfer, M. Artmeier, P. Pelka, F. Spaleck, G. Schmitz, Ultramicroscopy **107**, 726 (2007)
31. D.J. Larson, D.T. Foord, A.K. Petford-Long, H. Liew, M.G. Blamire, A. Cerezo, G.D.W. Smith, Ultramicroscopy **79**, 287 (1999)
32. M.K. Miller, K.F. Russell, Ultramicroscopy **107**, 761 (2007)
33. J.M. Cairney, D. Saxey, D. McGrouther, S.P. Ringer, Phys. B **394**, 267 (2007)
34. D.W. Saxey, J.M. Cairney, D. McGrouther, T. Honma, S.P. Ringer, Ultramicroscopy **107**, 756 (2007)
35. T.T. Tsong, E. Müller, Phys. Rev. **181**, 530 (1969)
36. T.T. Tsong, J. Chem. Phys. **54**, 4205 (1971)

37. G.L. Kellogg, Phys. Rev. B **29**, 4304 (1984)
38. T. Sakurai, E.W. Müller, Phys. Rev. Lett. **30**, 532 (1973)
39. T. Sakurai, E.W. Müller, J. Appl. Phys. **48**, 2618 (1977)
40. E.W. Müller, K. Bahadur, Phys. Rev. **102**, 624 (1956)
41. T.T. Tsong, *Atom Probe Field Ion Microscopy* (Cambridge University Press, UK, 1990)
42. A. Cerezo, T.J. Godfrey, G.D.W. Smith, Rev. Sci. Instrum. **59**, 862 (1988)
43. A. Cerezo, J.M. Hyde, M.K. Miller, G. Beverini, R.P. Setna, P.J. Warren, G.D.W. Smith, Surf. Sci. **266**, 481 (1992)
44. D. Blavette, B. Deconihout, A. Bostel, J.M. Sarrau, M. Bouet, A. Menand, Rev. Sci. Instrum. **64**, 2911 (1993)
45. F. Vurpillot, L. Renaud, D. Blavette, Ultramicroscopy **95**, 223 (2003)
46. A. Cerezo, T.J. Godfrey, J.M. Hyde, S.J. Sijbrandlj, G.D.W. Smith, Appl. Surf. Sci. **76–77**, 374 (1994)
47. L. Renaud, G. da Costa, M. Bouet, B. Deconihout, Nucl. Instrum. Meth. Phys. Res. A **477**, 150 (2002)
48. O. Jagutzki, A. Cerezo, A. Czasch, R. Dörner, M. Hattab, M. Huang, V. Mergel, U. Spillmann, K. Ullmann-Pfleger, T. Weber, H.S. Böcking G.D.W. Smith, IEEE Trans. Nucl. Sci. **49**, 2477 (2002)
49. G. da Costa, F. Vurpillot, A. Bostel, M. Bouet, B. Deconihout, Rev. Sci. Instrum. **76**, 013304 (2004)
50. T.T. Tsong, J.H. Block, M. Nagasaka, B. Viswanathan, J. Chem. Phys. **65**, 2469 (1976)
51. H.F. Liu, T.T. Tsong, J. Appl. Phys. **59**, 1334 (1984)
52. H.F. Liu, H.M. Liu, T.T. Tsong, Rev. Sci. Instrum. **55**, 1779 (1984)
53. F. Vurpillot, B. Gault, A. Vella, M. Bouet, B. Deconihout, Appl. Phys. Lett. **88**, 094105 (2006)
54. B. Gault, A. Vella, F. Vurpillot, A. Menand, D. Blavette, B. Deconihout, Ultramicroscopy **107**, 713 (2007)
55. A. Cerezo, C.R.M. Grovenor, G.D.W. Smith, Appl. Phys. Lett. **46**, 567 (1985)
56. A. Cerezo, C.R.M. Grovenor, G.D.W. Smith, J. Microsc. **141**, 155 (1986)
57. M.K. Miller, P. Angelini, A. Cerezo, K.L. More, Colloque de Physique **50**, 459 (1989)
58. J. Liu, T.T. Tsong, Phys. Rev. B **38**, 8490 (1988)
59. J.G. Fujimoto, J.M. Liu, E.P. Ippen, N. Bloembergen, Phys. Rev. Lett. **53**, 1837 (1984)
60. J. Hohlfeld, S.S. Wellershoff, J. Güdde, U. Conrad, V. Jähnke, E. Matthias, Chem. Phys. **251**, 237 (2000)
61. A. Vella, F. Vurpillot, B. Gault, A. Menand, B. Deconihout, Phys. Rev. B **73**, 165416 (2005)
62. A. Vella, M. Gilbert, A. Hideur, F. Vurpillot, B. Deconihout, Appl. Phys. Lett. **89**, 251903 (2006)
63. M. Gilbert, F. Vurpillot, A. Vella, H. Bernas, B. Deconihout, Ultramicroscopy **107**, 767 (2007)
64. A.J. Melmed, R.J. Stein, Surf. Sci. **49**, 645 (1975)
65. G. Kellogg, Phys. Rev. B **28**, 1957 (1983)
66. M.K. Miller, K.F. Russel, G.B. Thompson, Ultramicroscopy **102**, 287 (2005)
67. B.S. Everitt, *The Analysis of Contingency Tables*, 2nd edn. (Chapman and Hall, London, 1992)

68. M.P. Moody, L.T. Stephenson, P.V. Liddicoat S.P. Ringer, Microsc. Res. Technol. **70**, 258 (2007)
69. M.K. Miller, E.A. Kenik, Microsc. Microanal. **10**, 336 (2004)
70. M. Ester, H.P. Kreigel, J. Sander, X. Xu, in *Proceedings of the 2nd International Conference on Knowledge, Discovery and Data Mining*, Portland, Oregon, 1996
71. L.T. Stephenson, M.P. Moody, P.V. Liddicoat, S.P. Ringer, Microsc. Microanal. **13**, 448 (2007)
72. J.T. Vietz, I.J. Polmear, J. Inst. Met. **94**, 410 (1966)
73. C.Y. Zahra, M. Dumont, Phil. Mag. **85**, 3735 (2005)
74. J.W. Cahn, Acta Metall. **11**, 1275 (1963)
75. E. H. Dix Jr., Trans. Am. Soc. Met. **52**, 1057 (1950)
76. I.J. Polmear, J. Inst. Met. **89**, 51 (1960–1961)
77. S.K. Maloney, I.J. Polmear S.P. Ringer, Mater. Sci. Forum **331–337**, 1055 (2000)
78. R.A. Karnesky, D. Isheim, D.N. Seidman, Appl. Phys. Lett. **91**, 013111 (2007)

16

A Study on Age Hardening in Cu-Ag Alloys by Transmission Electron Microscopy

E. Shizuya and T.J. Konno

Summary: We have prepared Cu-1, 2, 4 at% Ag alloys by arc-melting and investigated their age-hardening and precipitation behaviors by using hardness tests, conductivity measurements, and TEM. The evolution of hardness during annealing exhibited strong composition dependence: the 4 at% Ag alloy aged for 20 min at 450°C already showed the maximum hardness, while the 2 at% Ag alloy began to harden after 9 h ageing. The maximum hardness was followed by gradual decrease, while electrical conductivity steadily increased to nearly 90% IACS. The observed precipitates in the 4 at% Ag alloy aged for 20 min consist of rod-shaped fcc Ag precipitates, which grow in the $[110]_{Cu}$ direction, and whose cross-sections are elliptic with an ellipticity greater than two. Widely interspaced precipitates then appear abruptly in the 4 at% Ag alloy aged for 27 h. These observations have been discussed from the view points of the morphology and crystallography.

16.1 Introduction

Cu-Ag alloys have been developed as high strength and conductivity materials with an optimum composition in the range of Cu-7 $\sim$ 45 wt% Ag. For example, a Cu-24 wt% Ag alloy of ultimate tensile strength of 1.5 GPa and conductivity of 65% international annealed copper standard (IACS) has been reported [1, 2]. Recently, Cu-Ag alloys of lower Ag contents were found to exhibit comparable mechanical and electrical properties (e.g., the Cu-2 wt% Ag alloy of ultimate tensile strength of 1.2 GPa and conductivity of 81.7% IACS) [3]. The age-hardening of Cu-Ag alloys has also long been studied by a number of researchers [4–13]. This can be summarized as follows: a metastable phase, often designated as α', is reported to appear during ageing treatments below 375°C [6, 14]. It is a nonequilibrium solid solution of Ag and Cu, and its unit cell parameter approximately corresponds to that of a cubic Ag-16 at% Cu alloy. Except for this case, fcc Ag phase is known to precipitate in the fcc Cu matrix with the orientation relationships $[100]_{Ag} // [100]_{Cu}$, $(010)_{Ag} // (010)_{Cu}$, or its twin-variant [9, 10]. The precipitation mechanism has largely been categorized into continuous and discontinuous cases. The latter has a shorter incubation period but lower peak hardness than the former [6]. Thus, in fine grain alloys, the mechanical properties are governed by discontinuous precipitation, and the subsequent grain boundary reaction leads to softening [6]. The discontinuous precipitates consist of Ag precipitates and surrounding Cu matrix where Ag is depleted. Räty and Miekk-oja reported the growing mechanism of these planar precipitates, which form mainly on $\{111\}_{Cu}$ and $\{100\}_{Cu}$ planes in single crystals, by using transmission electron microscopy (TEM) [5]. They proposed that vacancies provided by prismatic unit dislocations and dislocations with Burgers vector a $\langle 100 \rangle$ play a significant role. On the other hand, Monzen et al. ascribed the growth of rod-shaped precipitates to the relaxation of the misfit strain between the precipitates and matrix in the rod direction based on TEM observations using single crystals [12, 13, 15]. Both the two research groups suggested that the fcc Ag phase precipitates discontinuously and grows in $\langle 110 \rangle_{Cu}$. However, the precipitation behaviors in polycrystalline materials are not the same at those observed in a single crystal. Recently, three-dimensional (3D) tomography has been introduced in the field of materials science, which can provide 3D information of microstructure, such as size, shape, and distribution of precipitates. For example, it has been applied to Al-based alloys, and revealed the evolution of precipitates in detail [16, 17]. The purpose of the present study is therefore to investigate the morphology and crystallography of discontinuous precipitates in polycrystalline Cu-1, 2, 4 at% Ag alloys by TEM and 3D tomography, and shed light to the nature of phase evolution.

16.2 Experimental Procedure

Alloy ingots (Cu-1, 2, 4 at% Ag) were prepared by arc-melting and solution-treated by 6 h annealing at 780°C followed by a water quench. Ageing treatments for several durations at 450°C were carried out and terminated by a water quench. The specimens then underwent hardness tests, conductivity measurements, and TEM observations. The hardness tests were repeated 10 times and mean value and standard deviation were calculated. The conductivity measurements were carried out using four-terminal method. Specimens for TEM observations were prepared as follows: 3 mm discs were cut out of sheets ground to thickness of about 100 μm. The discs were thinned by a twin-jet method using a solution of 20% nitric acid and 80% methanol. They were polished with an ion mil to remove the residual contamination. Thin areas of the specimens were observed by a JEOL JEM-2000FX, JEM-3010 TEMs, and FEI Tecnai 30F TEM. Three-dimensional tomographic features were reconstructed using a series of scanning transmission electron microscopy high-angle annular detector dark field (STEM-HAADF) images.

16.3 Results

Figure 16.1 shows changes of Vickers hardness of the Cu-Ag alloys annealed at 450°C as a function of ageing time. As seen, the changes exhibited strong composition dependence: the Cu-4 at% Ag alloy showed the maximum hardness already at the 1.2×10^3 s (20 min) annealing, suggesting that the aging started almost immediately after the anneal, whereas the Cu-2 at% Ag alloy began to harden only after 3.24×10^4 s (9 h) ageing. The Cu-1 at% Ag alloy did not even harden during 9.72×10^4 s (27 h) of ageing. These observations indicate that the incubation period depends strongly on the Ag content, or the chemical potential of the solute Ag.

Figure 16.2 shows the conductivity change of the Cu-4 at% Ag alloy as a function of ageing time. The conductivity of as-quenched alloy was 74% IACS, which can be ascribed to quenched-in defects, such as excess vacancies and dislocations, as well as to supersaturated Ag atoms. With annealing, it monotonously increased and reached nearly 90% IACS, which can be compared to 94% IACS, a theoretical value if 0.8 at% of Ag atoms exist as solute.

Figure 16.3a,b shows a bright field (BF) image and the corresponding $[11\bar{2}]_{Cu}$ selected area diffraction (SAD) pattern observed for as-quenched Cu-4 at% Ag alloy. (In the latter, the arrowed diffraction spots originates from Cu_2O, which demonstrate that the oxide grows epitaxially on the matrix.) As seen in Fig. 16.3a, there are a number of dark contrasts arising from the strain field, but no precipitates were observed, which can be verified from the absence of diffraction spots expected from fcc-Ag.

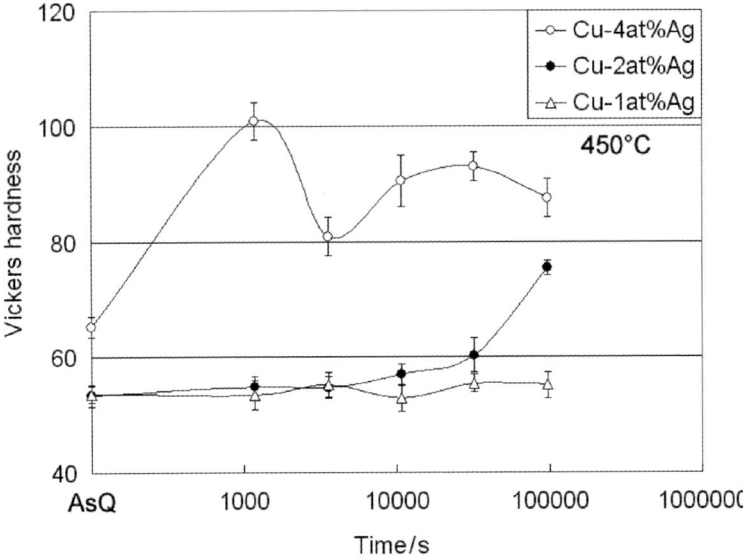

Fig. 16.1. Vickers hardness changes of Cu-1, 2, 4 at% Ag alloys as a function of ageing time at 450°C

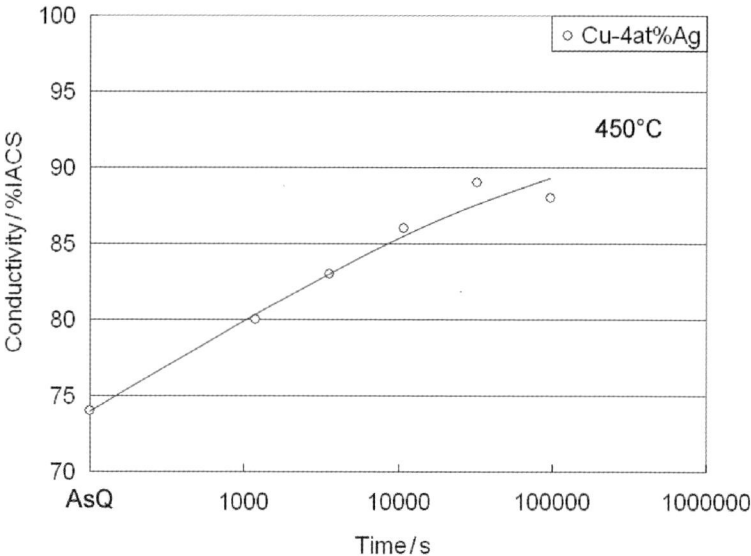

Fig. 16.2. Conductivity change of Cu-4 at% Ag alloy as a function of ageing time at 450°C

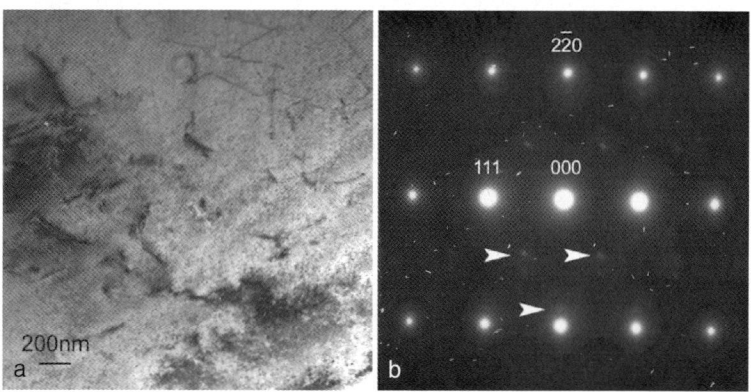

Fig. 16.3. (a) BF image and (b) corresponding SAD pattern of as-quenched Cu-4 at% Ag alloy. The incident beam direction is $[11\bar{2}]_{Cu}$. Note the presence of contrasts arising from dislocations. The arrows in (b) indicate spots are due to Cu_2O

Figure 16.4a,b shows a BF and dark field (DF) images observed in Cu-4 at% Ag alloy aged for 20 min at 450°C; while Figs. 16.4c and d are, respectively, the corresponding $[11\bar{2}]_{Cu}$ SAD pattern and an SAD pattern taken with the specimen tilted slightly along its $[\bar{2}20]_{Cu}$ axis. (We carried out the tilting experiment in order to confirm that most of the spots flanking the $\bar{2}20_{Cu}$ arise from double diffraction, as well as to exclude the effect of the oxide.) In the BF image (Fig. 16.4a), the dark contrasts with inter-lamellar spacing

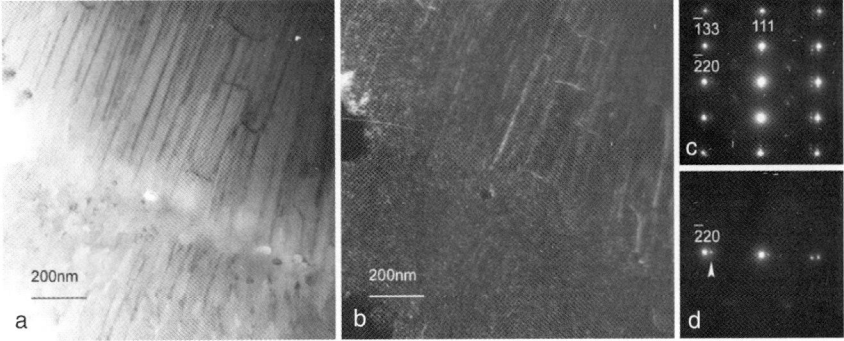

Fig. 16.4. (a) BF and (b) DF images of Cu-4 at% Ag alloy aged at 450°C for 20 min. The incident beam direction is $[11\bar{2}]_{Cu}$; (c) and (d) corresponding SAD patterns, where the latter was taken by tilting the specimen. The arrow in (d) indicates the selected spot for the DF image (b). The lamellar contrasts originate from Ag precipitates and nearly parallel with $(\bar{1}31)_{Cu}$

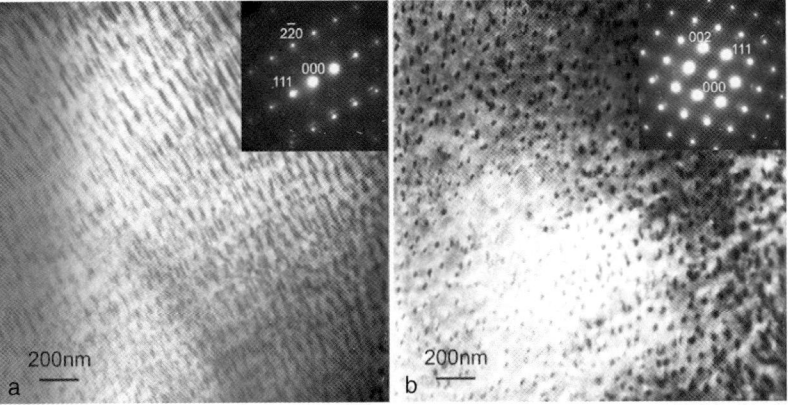

Fig. 16.5. BF image and inset SAD pattern of Cu-4 at% Ag alloy aged at 450°C for 20 min. The incident beam directions are (a) $[11\bar{2}]_{Cu}$ and (b) $[\bar{1}10]_{Cu}$

of several 10 nm are seen. The DF image (Fig. 16.4b), which was taken with the arrowed $\bar{2}20_{Ag}$ spot (Fig. 16.4d), also showed bright lamellar contrasts; thus suggesting that these precipitates are due to the precipitation of fcc-Ag. The SAD pattern shown in Fig. 16.4c can thus be viewed as the superimposition of $11\bar{2}_{Cu}$ and $11\bar{2}_{Ag}$ patterns (as well as that of the oxide), which in return suggests that the precipitates possess an orientation relationship (OR) of $[11\bar{2}]_{Cu} // [11\bar{2}]_{Ag}$, $(\bar{2}20)_{Cu} // (\bar{2}20)_{Ag}$. This is simply the OR frequently referred to as "cube-on-cube" OR. It can also be realized in the figure that the interfaces between the precipitates and matrix are nearly parallel to $(\bar{1}33)_{Cu}$.

Figures 16.5a,b are other sets of a BF image and an inset SAD pattern, observed for the same specimen. Figure 16.5a, which was taken with the incident beam along the $[11\bar{2}]_{Cu}$ axis, once again exhibited the same OR of $[11\bar{2}]_{Cu}//[11\bar{2}]_{Ag}$, $(\bar{2}20)_{Cu}//(\bar{2}20)_{Ag}$, but the interface of the precipitates found in this case are nearly parallel with $(111)_{Cu}$. Together with the aforementioned interface (Fig. 16.4a), it can then be concluded that the Ag precipitates are not in a lamellar shape but possess a rod-like morphology. It can further be deduced, considering the zone-axis of $\{311\}_{Cu}$ and $\{111\}_{Cu}$, that these rod-like precipitates grow in $\langle 110 \rangle_{Cu}$ directions. To confirm this observation, we took a BF image along the $[110]_{Cu}$ axis, which is seen in Fig. 16.5b. As shown here, the precipitates indeed exhibit elliptic dark contrasts, suggesting that the fcc Ag having a cube-on-cube OR grows in the $[110]_{Cu}$ direction. This observation is consistent with previous results, which have been carried out for single crystals [5, 12].

In addition, Fig. 16.6a,b is STEM-HAADF images of the same specimen, where bright contrasts corresponds to the Ag precipitates. These images show that the Ag precipitates, which grew in the $[110]_{Cu}$ direction, are in fact more than 1 μm in length.

Figure 16.7 is a snapshot of reconstructed 3D-tomographic STEM-HAADF images. The reconstruction was carried out using more than 70 individual images. As can be seen in this snapshot, the precipitates in fact possess a "flat" rod shape, i.e., the cross-section of the rods are not circular but elliptic. We found that the aspect ratio or the ellipticity is more than two, the origin of which is still under question, and will be briefly discussed later.

Figure 16.8 is a BF image and inset is SAD pattern observed in Cu-4 at% Ag alloy aged for 27 h at 450°C, with the incident beam direction of $[\bar{1}10]_{Cu}$. It can be realized that the basic features resemble those taken for the specimen annealed for a shorter time: the "apparent" interface between the precipitates and the matrix is (111), and the OR expected from the SAD pattern is essentially the same as those previously described, that is, cube-on-cube OR.

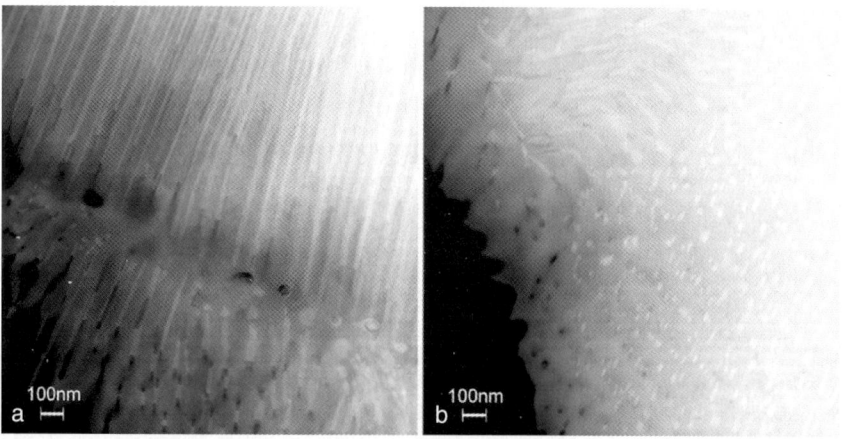

Fig. 16.6. STEM-HAADF images of Cu-4 at% Ag alloy aged at 450°C for 20 min

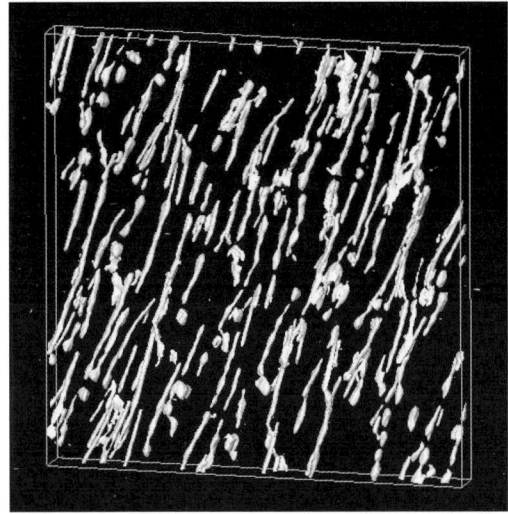

Fig. 16.7. A snapshot of 3D tomographic image, reconstructed from a series of STEM-HAADF images of Cu-4 at% Ag alloy aged at 450°C for 20 min. The frame is about 600 nm in width

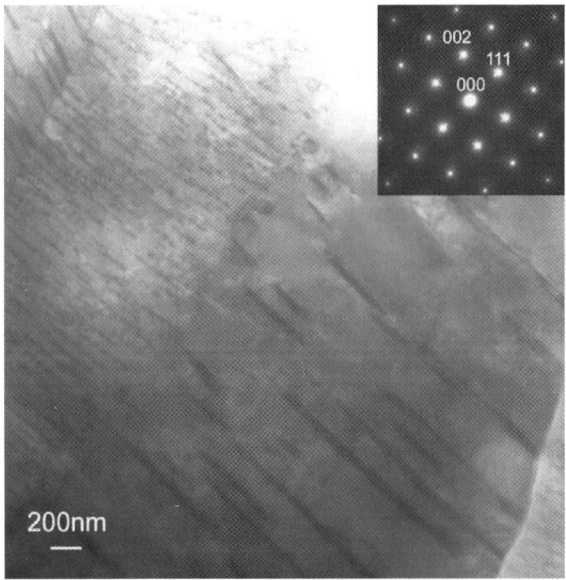

Fig. 16.8. BF image of Cu-4 at% Ag alloy aged at 450°C for 27 h, showing an abnormal growth of precipitation and discontinuous boundary between fine and coarse precipitate regions

However, the precipitates shown in the center of the image possess over 100 nm of interspacing between them, which is in contrast with the upper-left image in the same picture, showing rod-like precipitates in the same orientation but with inter-rod spacing of only several 10 nm. This observation suggests that the rod-like precipitates grow discontinuously.

16.4 Discussion

In the present study, we reported that age-hardening and an accompanying increase in electrical conductivity in Cu-Ag alloys are due to the precipitation of rod-shaped fcc-Ag phase within the Cu matrix. The precipitation behaviors as monitored by the changes in Vickers hardness exhibited strong composition dependence: there exist a long incubation period for the 1 at% Ag alloy, whereas the precipitation occurred almost immediately within the 4 at% Ag alloy. This observation can be understood by considering a change in driving force of precipitation within a supersaturated solid solution. Figure 16.9 shows schematically a hypothetical free energy diagram, where an α phase is precipitated within a β matrix, and the equilibrium composition is designated by x_0. At a slightly supersaturated composition of x_1, the driving force for the precipitation increase markedly with the increase in the chemical potential of the solvent metal, which can be expressed as $\mu'_\alpha - \mu_\alpha$, as shown on the right-hand side ordinate in the diagram. Thus, driving force for the precipitation,

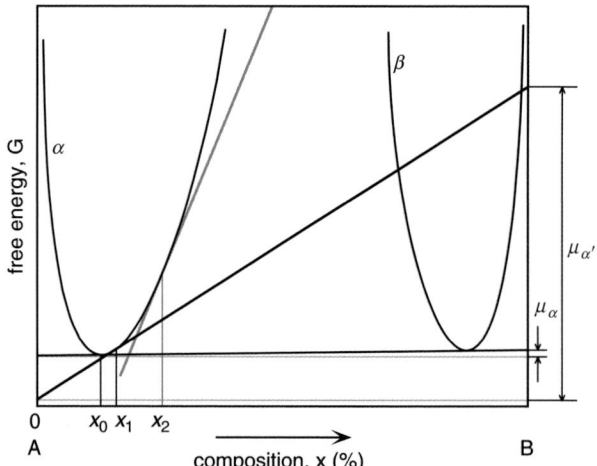

Fig. 16.9. Hypothetical free energy vs. composition diagram to describe the chemical potential of an α phase, which precipitates in a β matrix. x_0 is assumed to be the equilibrium composition. The change in chemical potential of the solvent element A upon precipitation from a supersaturated composition x_1 is given by the corresponding change in tangents: $\mu'_\alpha - \mu_\alpha$

and therefore the incubation period, depends sensitively on the shape of the free energy curve.

In Cu-4 at% Ag alloy aged for 20 min, observable thin area was covered with rod-shaped precipitates. According to previous reports, they can be thought as discontinuous precipitates that give rise to the peak hardness (designated as nodule (I) in [6]). By analogy, discontinuous growth shown in Fig. 16.8 seems to suggest that subsequent grain boundary reaction and the precipitation of widely interspaced particles are discontinuous one, which does not contribute to hardening (nodule (II) [6]). Also it should be mentioned that the metastabele α' phase and continuous precipitates, which have been reported previously [5], were not observed in present study.

The growing mechanisms of discontinuous precipitations suggested by both Räty and Miekk-oja, and Monzen et al. are concerned with planar colonial structure. For example, the former group reported that planar colonial structures in $\{100\}_{Cu}$ and in $\{111\}_{Cu}$ in polycrystalline materials are replaced by growing coarse 3D colonies. However, it has been difficult to elucidate the distribution, morphology, and structure of these coarse 3D colonies. In the present study, TEM images of rod-shaped Ag precipitates and 3D-tomography demonstrated clearly that the cross-sectional shape of Ag precipitates in the 3D colonies is in fact elliptic.

As for the origin of the ellipticity, it is difficult to give a conclusive comment, but a close examination of Fig. 16.5b reveals that the minor axes of a number of cross-sectional image of precipitates are in fact in the $\langle 111 \rangle$ or $\langle 110 \rangle$ direction, that is, the precipitates can elongate along $\{111\}_{Cu}$ and $\{110\}_{Cu}$. Considering that the surface energy of these planes is low, we may suggest that the Ag precipitates grow in a stable manner along $\{111\}_{Cu}$ than the other directions, resulting in an elliptic cross-section of the rod structures.

16.5 Conclusions

Cu-1, 2, 4 at% Ag alloys were investigated from the view points of their age-hardening and precipitation behaviors during an annealing at 450°C by using hardness tests, conductivity measurements, and TEM. The results can be summarized as follows:

1. The hardness changes exhibited strong composition dependence: the 4 at% Ag alloy showed the maximum hardness at the 20 min annealing, the 2 at% Ag alloy began to harden after 9 h ageing, and the 1 at% Ag alloy did not harden during 27 h of ageing.

2. The 4 at% Ag alloy aged for 20 min exhibited the maximum hardness followed by gradual decrease, while electrical conductivity increased steadily to reach nearly 90% IACS after 9 h of annealing.

3. The observed precipitates in the 4 at% Ag alloy aged for 20 min consist of rod-shaped fcc Ag precipitates, which possess the cube-on-cube OR,

grow in the $[110]_{Cu}$ direction, and whose cross-sections are elliptic with an ellipticity greater than two.

4. These precipitates did not show a gradual growth, but;idely interspaced precipitates appeared abruptly after ageing for 27 h.

Acknowledgement. We thank Mr K. Inoke of FEI Company Japan for 3D tomography.

References

1. Y. Sakai, K. Inoke, T. Asano, H. Maeda, J. Jpn. Inst. Metals **55**, 1382 (1991)
2. Y. Sakai, H.J. Schneider-muntau, Acta Mater. **45**, 1017 (1997)
3. *National Institute for Materials Research Press Release* No. 132 (2005)
4. W. Leo, Z. Metallk. **58**, 456 (1967)
5. R. Räty, H.M. Miekkoja, Phil. Mag. **18**, 1105 (1968)
6. N. Kainuma, R. Watanabe, J. Jpn. Inst. Metals **33**, 198 (1969)
7. R. Wirth, H. Gleiter, Acta Metall. **29**, 1825 (1981)
8. W. Gust, J. Beuers, J. Steffen, S. Stiltz, B. Predel, Acta Metall. **34**, 1671 (1986)
9. G. Rao, D. Zhang, P. Wynblatt, Scripta Metall. Mater. **28**, 459 (1993)
10. G. Rao, J.M. Howe, P. Wynblatt, Scripta Metall. Mater. **30**, 731 (1994)
11. S. Spaic, M. Pristavec, Z. Metallk. **88**, 925 (1997)
12. R. Monzen, K. Murase, H. Nagayoshi, C. Watanabe, Phil. Mag. Lett. **84**, 349 (2004)
13. R. Monzen, H. Nagayoshi, C. Watanabe, S. Onaka, Phil. Mag. Lett. **85**, 13 (2005)
14. S. Nagakura, S. Toyama, S. Oketani, Acta Metall. **14**, 73 (1966)
15. C. Watanabe, R. Monzen, H. Nagayoshi, S. Onaka, Phil. Mag. Lett. **86**, 65 (2006)
16. K. Inoke, K. Kaneko, M. Weyland, P.A. Midgley, K. Higashida, Z. Horita, Acta Mater. **54**, 2957 (2006)
17. K. Kaneko, R. Nagayama, K. Inoke, E. Noguchi, Z. Horita, Adv. Sci. Tech. **7**, 726 (2006)

17

Rubber-Like Entropy Elasticity of a Glassy Alloy [1]

M. Fukuhara, A. Inoue, and N. Nishiyama

Summary: The thermoelastic effect is monopolistically characteristic of rubbers in solids. Here we report observations of thermal-induced entropy elasticity for a glassy alloy, $Pd_{40}Cu_{30}Ni_{10}P_{20}$ in terms of acoustoelasticity and the Gough–Joule effect. Seven kinds of elastic parameters of the glassy alloy have been simultaneously measured as a function of temperature ranging from 298 to 673 K. The decreases in elastic moduli, Poisson's ratio, and the increase in tension below the second-order like-phase transition temperature suggest rubber-like thermal dynamic micro-Brownian stretching, described as $F(Pa) = 0.282\,T + 562$, which may be associated with the rotational and vibrational motions of polyhedron clusters. The glassy alloy also showed the Gough–Joule effect at room temperature. In short, the glassy alloy has some rubbery characteristics that we have never before observed for ordinary crystalline alloys and inorganic materials.

17.1 Introduction

Glassy alloys are peculiar metallic alloys that lack the long-range cyclic order of normal, crystalline alloys. Much attention has been devoted to the glass forming-ability of atoms of different types in glassy alloys [2], that is, their cluster structures [3], structural relaxation during annealing [4], and potential applications [5]. Indeed, their exceptional magnetic and mechanical properties have been used commercially. However, in addition to these characteristics, it is also important to observe their acoustic properties, especially the rubber-like elasticity of glassy alloys, in order to understand structural relaxation of disorder phase during heating.

Our interest lies in simultaneously determining seven kinds of elastic parameters of a chilled glassy alloy $Pd_{40}Cu_{30}Ni_{10}P_{20}$ (with elongation of ~1,400% [6]) and experimental detection of Gough–Joule effect in terms of thermoelasticity. As far as we know, however, no research work has been carried out previously on the simultaneous measurement of all the elastic parameters and detection of Gough–Joule effect for glassy alloys. Rubbers are characterized by five characteristics: (1) low Young's and shear moduli, (2) high elongation ($\sim$350%), (3) thermal-induced elasticity, (4) a large Poisson's ratio (0.40$\sim$0.45), and (5) generation or absorption of heat under adiabatic dilation or compression, respectively (the Gough–Joule effect).

The second potential energy between atoms directly dominates elastic moduli, and hence the elastic moduli are important parameters for thermodynamic evaluation of glassy alloys. Since glassy alloys without crystal structure can be treated as an isotropic elastic medium (a material that does not have a preferred orientation), Young (E), shear (G), and bulk (K) moduli, Lamè parameter (λ), compressibility (κ), Poisson's ratio (ν), and Debye temperature (Θ_D) of the glassy alloy can be precisely calculated by use of both longitudinal and transverse velocities with the same frequency [7, 8].

17.2 Experimental

The specimen (density of $9.285\,\text{mg m}^{-3}$, $T_g = 560\,\text{K}$) produced by an injection casting method [9] was in the form of a long rod (10 mm in diameter and 8 mm in length) fastened to a stainless steel waveguide with threads of pitch 1.5 mm, using a domed cap nut made of stainless steel. Naphtenic hydrogen oil with viscosity of 400 Pa s [10] was used as a couplant medium between the specimen and the waveguide to improve nonequilibrium of heat distribution for sample in previous paper [11]. We used a longitudinal wave generation PZT transducer with 7 MHz frequency as an optimum frequency. The specimen was measured from 298 K up to 673 K at a low heating rate of $0.03\,\text{K s}^{-1}$ in an argon atmosphere at an ambient pressure. The difference between the top and the bottom of the specimen was $\pm 0.5\,\text{K}$ during measurement. The experimental procedure is described in previous papers [7, 12, 13]. For calculation of the Debye temperature, we used 4 as number of mass point per quasi-crystalline unit cell and $5.06 \times 10^{-29}\,\text{m}^3$ as the average atomic volume.

The Gough–Joule effect was investigated by use of a glassy alloy bolt. The specimen was in the hollow, straight form of a long rod (3.8 mm in diameter and 8 mm in length) unified by a screw bolt with a hexagonal head, long rod (5 mm in diameter and 13 in. length), and a threads of pitch of 0.8 mm (Fig. 17.4, insert), to apply homogeneously as much tension as possible [14]. A load cell was inserted between the bolt head and a spherical brass holder. The specimen was horizontally torqued by a stainless steel nut, which was tightened using a spanner. The heat transfer change was measured by a spherical resistance thermocouple (with accuracy of 1 mK) of 0.1 mm in diameter, which contacts with the surface of the narrow rod by varnish drop.

17.3 Results and Discussions

17.3.1 Temperature Dependence of the Elastic Parameters

Temperature dependence of four elastic moduli – E, G, K, and λ – is shown in Fig. 17.1a–d, respectively. The uniaxial volume-preserving E and G moduli show a negligible increase up to around 470 K and then a subtle decrease. By striking contrast, the three-dimensional volume-nonpreserving K and λ moduli decrease noticeably with increasing temperature, then jump up to around 500 K, and decrease again more gently. The order of magnitude in the four elastic moduli is K, λ, E, and G, showing that the glassy alloy readily undergoes dilational and shear deformation, but resists volumetric (bulk) deformation, and so G (or E) $\ll K$. We have observed the same order for elastic moduli of amorphous polymers and rubbers [8, 15].

From comparison of the large increase in bulk moduli with the small one in shear moduli at 500 K, the peak appears to correspond to the second-order like-phase transition, associated with the three-dimensional volume-nonpreserving volumetric distortion [16, 17] rather than the volume-preserving one, because the thermodynamics of the second-order phase transition yields an anomaly in K at the transition temperature [18]. However, the potential energy minimum among atoms in the multicomponent glassy alloy is not as rigid as that of the crystal one so that the transition point is strictly not the

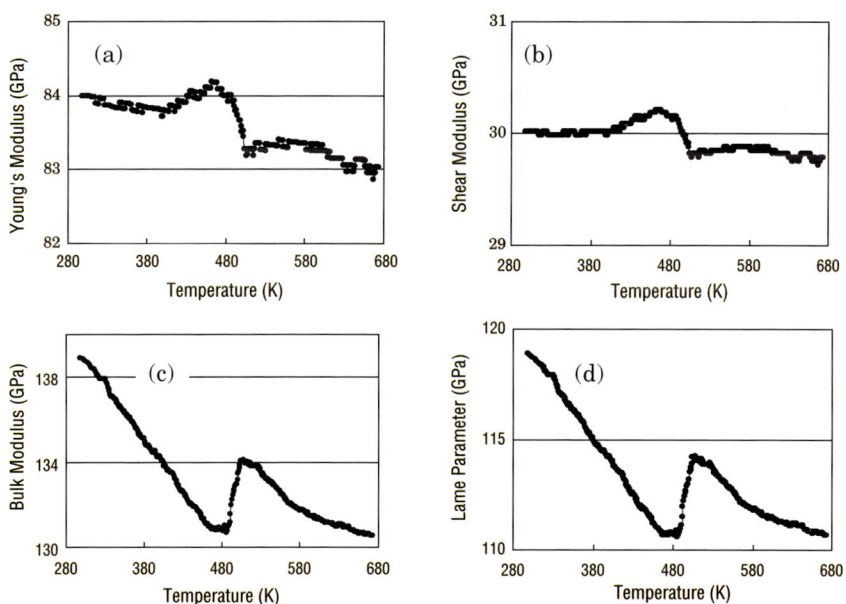

Fig. 17.1. Temperature dependence of Young (**a**), shear (**b**), and bulk moduli (**c**) and Lamè parameter (**d**) for glassy alloy $Pd_{40}Cu_{30}Ni_{10}P_{20}$

second order. Therefore, the transition may be the β-softening glass transition corresponding to relaxation motion, which is often observed in amorphous polymers [8]. The transition temperature of 500 K is 60 K lower than the calculated DSC value (560 K) in extrapolated-heating rate of 0.033 K s^{-1} without ultrasound field [19]. The 500 K transition is not due to ultrasound-induced transition, because of tiny pressure effect far below 0.2 MPa in ultrasonic transducer. Furthermore, the temperature of 500 K would be corresponded to starting temperature for thawing of frozen free volume [20]. In high polymers containing rubbers, furthermore, the decrease in elastic moduli has been observed at lower temperature below their glass temperature (Polyimide (PI), $0.84T_g$ [8]; Polycarbonate (PC), $0.61T_g$ [21]), because rubbery elasticity does not occur until frozen molecular segments begin to thaw. The specimen used in this study was observed at $0.84T_g$, which is the same as that of PI.

Since the elasticity of materials cannot be evaluated by the elastic moduli themselves alone, the ratio G/K can be conveniently taken as a measure of elasticity [22]. The ratio of the glassy alloy at room temperature was compared with those of other representative materials [7, 8, 13]: glassy alloy, 0.22; carbon steel, 0.49; Ti-6Al-4V, 0.35; Inconel 718, 0.45; α-alumina, 0.64; zirconia, 0.42; β'-sialon, 0.51; quartz, 0.86; polymethyl methacrylate, 0.41; PI, 0.34; urethane rubber, 0.35. The lowest value, which bears a close resemblance to high elasticity in volume-nonpreserving distortion, indicates rubbery characteristics for the glassy alloy [23].

Next we show the temperature-dependence elastic Debye temperatures Θ_D in Fig. 17.2. The Debye temperature shows almost no change within a temperature variation of ±1 K over the whole determined temperature range, indicating negligibly tiny change in the maximum frequency allowed, that is,

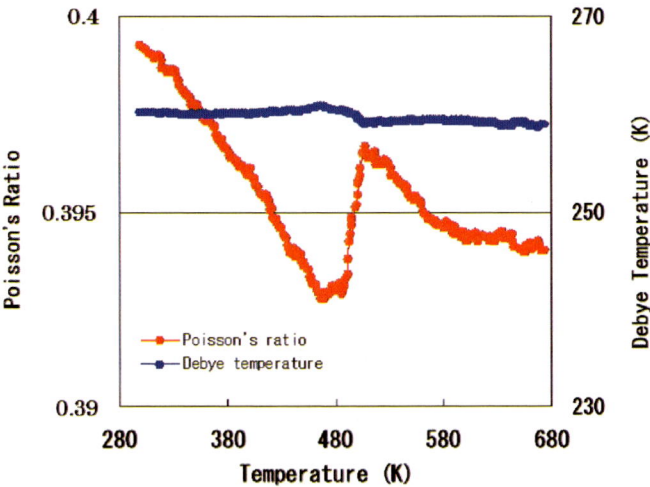

Fig. 17.2. Temperature dependence of Poisson's ratio and Debye temperature for glass alloy $Pd_{40}Cu_{30}Ni_{10}P_{20}$

the effective atomic distance for the transition [13, 15, 17]. This means no structural change in the temperature region of interest.

From the three-dimensional volume-nonpreserving elastic deformability of view, we then evaluate Poisson's ratio. Temperature-dependence of Poisson's ratio (Fig. 17.2) resembles that of K and λ as shown in Fig. 17.1c,d. The decrease up to 480 K demonstrates enhancement of toughness in the volume-nonpreserving distortion. Such a decrease is an unusual behavior for metals and alloys, which are generally softened on heating. The nature of the toughness in a qualitative way can be interpreted as analogous to cases of amorphous rubber stretching [24], as well as amorphous polymers [8, 15]. Here it is well known that entropy elasticity in rubber with chain polymers is attributed to micro-Brownian rotation based on the thermoelastic effect [25]. Hence, by analogy, we infer that the entropy elasticity in glassy alloy is related to micro-Brownian stretching, corresponding to the decrease in entropy on heating. Indeed, Elliot [26] has reported that medium-range order polyhedron and cage-like molecular clusters in covalently bonded amorphous solids are responsible for symmetrical stretching vibrational and rotational diffusional motions, respectively (Fig. 17.3b). The metal–metalloid amorphous $Pd_{40}Cu_{30}Ni_{10}P_{20}$ alloy has two typical polyhedra:trigonal prism capped with three-half octahedral and a tetragonal dodecahedron (Fig. 17.3b, insert) [3]. The glassy alloy is to the four-body correlation with torsion motion what the rubber is to the paraffin or polyethylene bonds, permitting free rotation of c–c

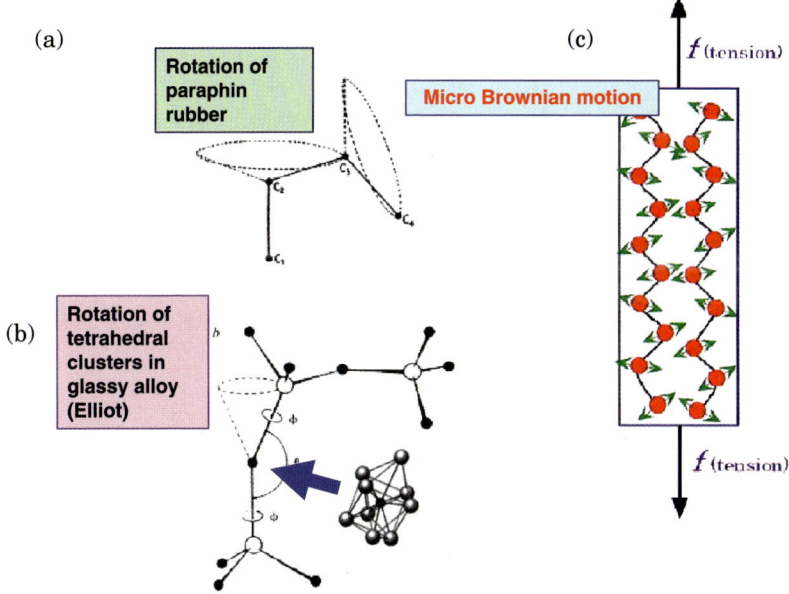

Fig. 17.3. Rotation of paraffin rubber (**a**) and tetrahedral clusters (**b**) and a model (**c**) of entropy elasticity in the glassy alloy

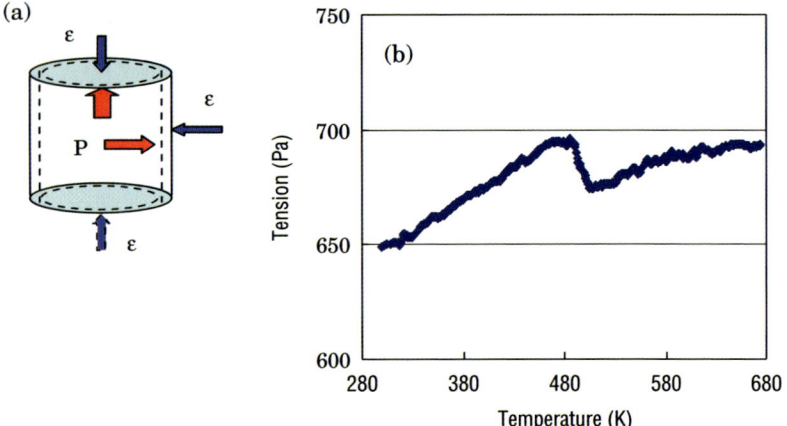

Fig. 17.4. Schematic figure (**a**) for calculation of tension and temperature-induced tension (**b**)

bonds (Fig. 17.3a, insert).To determine the rubbery elasticity, we calculate the temperature-dependent force of the circular cylinder glassy alloy, by using the definition of compressibility, $-\Delta V/V = \Delta F\beta$, where F is load. If we assume $\Delta V/V = \varepsilon_1 + \varepsilon_2 + \varepsilon_3$ for small ΔV and $\varepsilon_2 = \varepsilon_3 = 0.02$ for lateral surface [27], the rubber-like tension as longitudinal force can be obtained (Fig. 17.4b). The tension between 398 and 480 K can be plotted in a linear form as

$$F \text{ (Pa)} = 0.282T + 562. \tag{17.1}$$

The intercept, 562 Pa, is energy elasticity, $(\partial U/\partial \varepsilon)_T$, which is repulsive stress against the binding (inner) energy U between atoms in the alloy. The tension force (650 Pa) at room temperature is considerably lower than that (29 kPa) of vulcanized rubber [28], indicating the total amount of entropy contribution for elasticity. The rubbery-like entropy elasticity model based on the stretching rotational and vibrational motions of polyhedral clusters [26] or locally collective atomic motion, which corresponds to a hinge-like motion between trigonal prismatic structure units connected by shearing edges [29], is presented in Fig. 17.3c.

17.3.2 Gough–Joule Effect

Finally, to certify the entropy elasticity of the glassy alloy, we looked for a possible Gough–Joule effect by using the glassy alloy bolt. The change in heat transfer as a function of tensile stress is shown in Fig. 17.5. When tension up to 400 MPa is applied to the glassy alloy, temperature increases in a zigzag pattern and then decreases with holding time, showing the occurrence of the Gough–Joule effect. The zigzag behavior may result from a friction problem between the thread and nut or delay of heat transfer for stress value response.

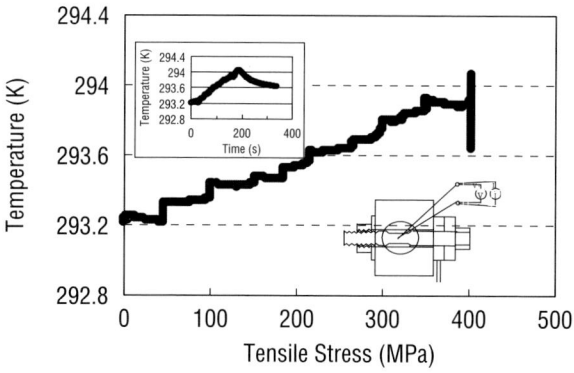

Fig. 17.5. Gough–Joule of glassy alloy $Pd_{40}Cu_{30}Ni_{10}P_{20}$

17.4 Conclusion

Characteristic elastic and thermoelastic behaviors in $Pd_{40}Cu_{30}Ni_{10}P_{20}$ glassy alloy were measured as a function of temperature. The increase of porosity in the glassy alloy reduces E and G linearly and K and λ parabolic. The order of K, λ, E, and G, which is monopolistically character of glassy alloys, changes to the order of E, K, λ, and G, being crystalline-like deformation mode at 58% porosity. Deformation mode shifts from dilational and shear (uniaxial volume-preserving) deformation to volumetric (three-dimensional volume-nonpreserving) one with mixed mode-II–mode-III, as porosity increases. Poisson's ratio and G/K ratio decreases and increases, respectively, showing a crystalline-like deformation mode. From dilational, shear, and volumetric dynamic viscosities, the viscoelasticity of the glassy alloys is predominated by shear motion, regardless of porosity. Contrast to continuous increment in noncomplex dilational damping, the first general increment in shear one could be elucidated by accumulation of strains induced by formation of pores. These elastic parameters are sensitive probes for evaluating the deformation mode and viscoelasticity of porous glassy alloys, associated with complex waves.

Acknowledgement. This work was supported by a Grant-In-Aid for Science Research in a priority Area' Research and Development Project on Advanced Metallic Glasses, Inorganic Materials and Joining Technology' from the Ministry of Education, Science, Sports and Culture, Japan, and by JSPS Asian CORE Program.

References

1. M. Fukuhara, A. Inoue, N. Nishiyama, Appl. Phys. Lett. **89**, 101903 (2006)
2. P. Chaudhari, D. Turnbull, Science **199**, 11 (1978)
3. C. Park, M. Saito, Y. Waseda, et al., Mater. Trans. Jpn. Inst. Met. **40**, 491 (1999)

4. A. Van Den Beule, J. Sietsma, Acta. Metall. Mater. **38**, 383 (1990)
5. A. Makino, A. Inoue, T. Mizushima, Mater. Trans. Jpn. Inst. Met. **41**, 1471 (2000)
6. Y. Kawamura, T. Nakamura, A. Inoue, Scripta Mater. **39**, 301 (1998)
7. M. Fukuhara, I. Yamauchi, J. Mater. Sci. **28**, 4681 (1993)
8. M. Fukuhara, A. Sanpei, J. Polym. Sci. Part. B: Polym. Phys. **34**, 1579 (1596)
9. N. Nishiyama, A. Inoue, Mater. Trans. Jpn. Inst. Met. **37**, 1531 (1997)
10. M. Fukuhara, T. Tsubouchi, Chem. Phys. Lett. **371**, 184 (2003)
11. N. Nishiyama, A. Inoue, J.Z. Jiang, Appl. Phys. Lett. **78**, 1985 (2001)
12. M. Fukuhara, A. Sanpei, Philo. Mag. Lett. **80**, 325 (2000)
13. M. Fukuhara, M. Yagi, A. Matsuo, Phys. Rev. B **65**, 224210 (2002)
14. M. Fukuhara, A. Sampei, Jpn. J. Appl. Phys. **39**, 2916 (2000)
15. M. Fukuhara, A.H. Matsui, M. Takeyama, Chem. Phys. **258**, 97 (2000)
16. F.E. Wang, B.F. DeSavage, W.J. Buehler, J. Appl. Phys. **39**, 2166 (1968)
17. M. Fukuhara, A. Sampei, Phys. Rev. B **49**, 13099 (1994)
18. H.D. Landau, E.M. Lifshitz, *Statistical Physics* (Pergamon, London, 1959), p. 438
19. T. Kojima, Dissertation, Department of Material Science, Osaka University, 1999
20. O. Haruyama, Private Communication, University of Tokyo
21. M. Fukuhara, A. Sampei, Jpn. J. Appl. Phys. **35**(Part 1, No. 5B), 3218 (1996)
22. R. Kubo, *Elasticity in Rubber* (Japanese) (Shokabo, Tokyo, 1996)
23. R. Houwink, Trans. Faraday Soc. **32**, 131 (1936)
24. S. Spinner, G.W. Cleek, J. Appl. Phys. **31**, 1407 (1960)
25. H.M. James, E. Guth, J. Chem. Phys. **11**, 455 (1943)
26. S.R. Elliot, Nature **354**, 445 (1991)
27. S.P. Timoshenko, J.N. Goodier, *Theory of Elasticity*, 3rd edn. (Tokyo, McGraw-Hill, Kogakusha, 1970), p. 244
28. R.L. Anthony, R.H. Caston, E. Guth, J. Phys. Chem. **46**, 826 (1942)
29. K. Suzuki, K. Shibata, H. Mizuseki, J. Non Cryst. Solids, **156–158**, 58 (1993)

Formation and Mechanical Properties of Bulk Glassy and Quasicrystalline Alloys in Zr-Al-Cu-Ti System

J.B. Qiang, W. Zhang, G.Q. Xie, and A. Inoue

Summary: The glass-forming ability (GFA) of $Zr_{65}Al_{7.5}Cu_{27.5}$ alloy can be improved effectively by a certain amount of Ti addition. Bulk glassy samples of 3 mm in diameter were obtained in the compositions containing 3–7 at% Ti by copper mold casting, while further Ti additions will deteriorate GFA and lead to the formation of quasicrystals at $(Zr_{65}Al_{7.5}Cu_{27.5})_{90}Ti_{10}$. Meanwhile, the addition of Ti in the $Zr_{65}Al_{7.5}Cu_{27.5}$ alloy was found to alter the crystallization behavior of the metallic glasses; thus bulk quasicrystalline alloy can also be achieved by glass devitrification of Ti-containing BMGs. Room-temperature compression testing showed that bulk quasicrystalline alloy, fabricated either by mold casting or by glass devitrification method, exhibited higher fracture strength and Young's modulus than that of the relative monolithic BMGs.

18.1 Introduction

An icosahedral short-range order (ISRO) has been suggested to be the intrinsic local structure of metallic glasses [1]. It creates a thermodynamic barrier to the formation of periodic crystals. Therewith, the enhanced stability of such a local atomic configuration leads to improved glass-forming ability (GFA) of an alloy [2–4]. Meanwhile, icosahedral atomic clusters are the building blocks of icosahedral quasicrystals (I-phases) and their crystalline approximants [5]. Under favorable preparation conditions, the icosahedral order could be extended orderly in real space to form an I-phase. And the formation of I-phase is available during the alloy melt cooling down path or on the subsequent heating of these Zr-based BMGs [6, 7]. In most cases, nanometer scaled I-phases revealed as the primary precipitation phases when annealing the BMGs within their undercooled liquid regions. These I-phases are metastable and their grains can not grow easily into a large size [8, 9]. Recently, Kühn et al. found that, in the Zr-Ti-Nb-Cu-Ni-Al BMG-forming alloy system, a large-grained I-phase was formed directly from the melt at low cooling rates [10]. The above experimental evidence implies certain relations in the formation of BMGs and I-phases in these alloys. Generally, I-phases have strict structures and a limit compositional stability as well [11, 12]. BMG formation is also compositional sensitive [4]. Thus in a given system, it is worthwhile to systematically examine the compositional effect on their formations. In a previous work, we reported that the addition of early transition metals as Ti, Nb, or Ta in the $Zr_{65}Al_{7.5}Cu_{27.5}$ alloy triggered the I-phase formation via crystallization [13]. In the present work, we will make a thorough investigation on the compositional dependence of BMG and I-phase formation in the Zr-Al-Cu-Ti alloy system. The mechanical properties of the bulk glassy and quasicrystalline alloys are also studied.

18.2 Experimental

Alloy ingots of compositions $(Zr_{65}Al_{7.5}Cu_{27.5})_{100-x}Ti_x$ (x = 0, 2, 3, 5, 7, 10, 12, and 15 at%) were prepared by arc melting the constituent elements under a Ti-gettered high purity argon atmosphere. The purities of metals are 99.9 wt% for Zr and Ti, 99.999 wt% for Al, and 99.99 wt% for Cu. The ingots were remelted four times to improve homogeneity. The alloy ribbons with a cross section of about $0.05 \times 1.0\,mm^2$ were prepared by using a single roller melt-spinning apparatus with a wheel surface velocity of $40\,m\,s^{-1}$. Alloy rods with different diameters were made by means of copper mold suction casting.

X-ray diffraction (XRD) for phase identification was conducted on a Rigaku RINT-ultima IIIsp diffractometer with Cu $K\alpha$ irradiation (λ = 0.15406 nm). TA-DSC Q100 type differential scanning calorimetry (DSC) and TA-STD Q600 type differential thermal analysis (DTA) were employed to study the thermal stability. The constant heating rates for DSC and DTA measurements were 40 and $20\,K\,min^{-1}$, respectively. Isothermal annealing treatment was done in the DSC sample chamber under a continuous flow of Ar.

Rod specimens with a length/diameter ratio of 6/3 were adopted for the uniaxial compression testing. Quasi-static loadings (strain rate of $5 \times 10^{-4}\,s^{-1}$) were conducted on an Instron at room temperature, and the strain was measured by using a strain gauge. The fracture morphology of the broken samples was observed by scanning electron microscopy (SEM).

18.3 Results and Discussion

18.3.1 Formation of BMG and QCs by Liquid Cooling

XRD analysis indicated that all the melt-spun $(Zr_{65}Al_{7.5}Cu_{27.5})_{100-x}Ti_x$ (x = 0, 2, 3, 5, 7, 10, 12, and 15 at%) samples were metallic glasses. The constant heating rate DSC traces of them were shown in Fig. 18.1. Ti-addition is found to lead to a shift of the onset crystallization temperature (T_x), while the change glass transition temperature (T_x) is very subtle. The under-cooled

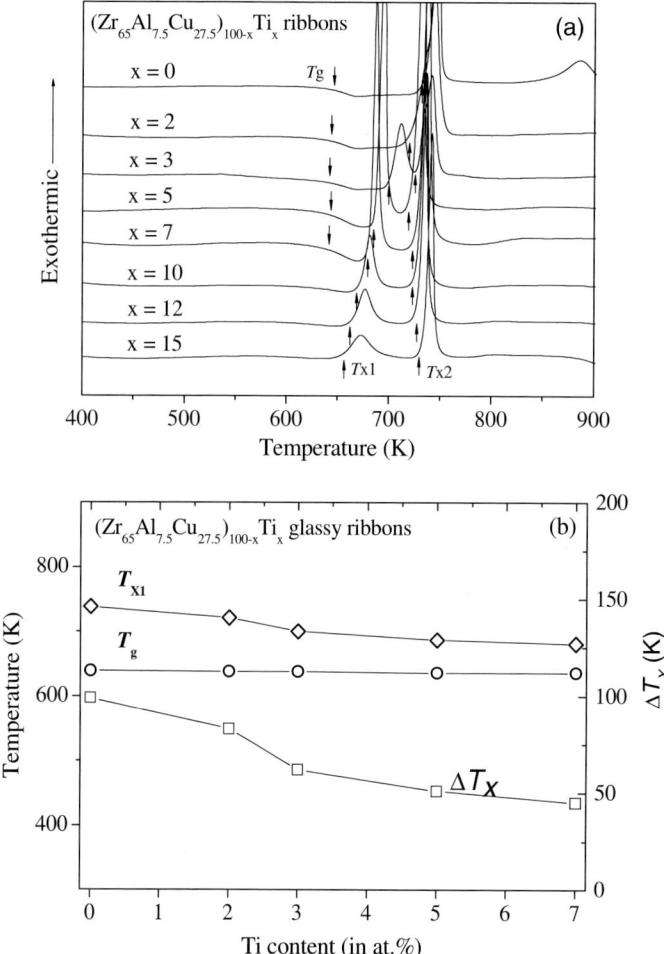

Fig. 18.1. (a) DSC traces of the melt-spun $(Zr_{65}Al_{7.5}Cu_{27.5})_{100-x}Ti_x$ alloys at the heating rate of $40\,K\ min^{-1}$; (b) composition dependence of glass transition temperature T_g, onset temperature of crystallization T_x, and under-cooled liquid region $\Delta T_x = T_x - T_g$ of these melt-spun $(Zr_{65}Al_{7.5}Cu_{27.5})_{100-x}Ti_x$ glassy alloys

liquid span, $\Delta T_x = T_x - T_g$, reduced gradually with the increase of Ti content. The DTA curves of the melt-spun samples were shown in Fig. 18.2, from which the onset melting temperature (T_m) and the liquidus temperature (T_l) was obtained. The GFAs of the serial compositions were assessed by the well recognized indicators: T_g/T_m [14], T_g/T_l [15], ΔT_x [4], and $\gamma = T_x/(T_g + T_l)$ [16]. To avoid ambiguities, the calculation was made on those compositions showing a distinct glass transition. As shown in Fig. 18.3, T_{rg} increased with Ti-addition signal and enhanced GFA, while the indicators γ (Fig. 18.3) and

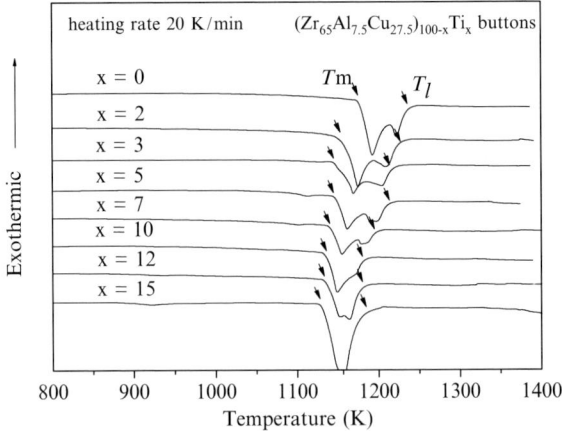

Fig. 18.2. DTA traces of the $(Zr_{65}Al_{7.5}Cu_{27.5})_{100-x}Ti_x$ alloys

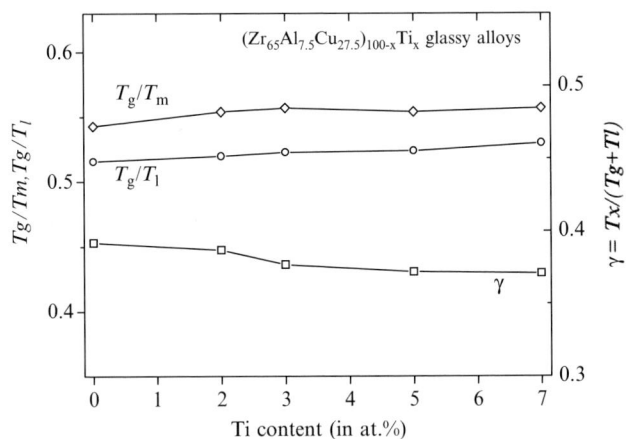

Fig. 18.3. Composition dependence of reduced glass transition temperature parameters T_g/T_m, T_g/T_l, and $\gamma = T_x/(T_g + T_l)$ of the $(Zr_{65}Al_{7.5}Cu_{27.5})_{100-x}Ti_x$ glassy alloys

ΔT_x (Fig. 18.1b) showed an opposite variation tendency of GFAs vs. Ti-content. To clarify the validity of the above three GFA indicators, alloy rods with 3 mm in diameter at these compositions were made by copper mold casting. The XRD patterns of these as-cast samples were shown in Fig. 18.4. BMG formation was observed within the composition range 3–7 at% Ti. At compositions with the Ti content less than 3 at%, the bulk samples were partially crystallized. In the 10% Ti alloy, I-phase in half-micrometer scale formed and coexisted with minor retaining glassy phase. At even higher Ti contents, the Zr_2Cu type phase was found to be the major precipitant. The information

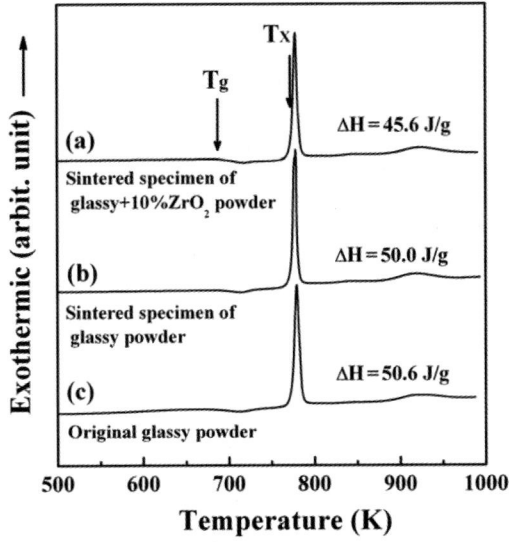

Fig. 19.6. DSC curves of the sintered $Zr_{55}Cu_{30}Al_{10}Ni_5$ glassy matrix composite containing 10 vol.% ZrO_2 powder (**a**), the sintered monolithic $Zr_{55}Cu_{30}Al_{10}Ni_5$ BMG specimen (**b**), and the original $Zr_{55}Cu_{30}Al_{10}Ni_5$ glassy powder (**c**)

The monolithic gas-atomized $Zr_{55}Cu_{30}Al_{10}Ni_5$ glassy powder shows only a diffuse diffraction pattern typical for a glassy phase and no diffraction peak corresponding to crystalline phases is observed, while the XRD pattern of the sintered metallic glassy matrix composite specimen shows sharp peaks diffracted from the ZrO_2 superimposed on a broad halo peak, indicating that the matrix of the sintered composite consists of a fully glassy phase.

Figure 19.6 shows the DSC curves of the sintered glassy matrix composite specimen (Fig. 19.6a) and the monolithic gas-atomized $Zr_{55}Cu_{30}Al_{10}Ni_5$ glassy powder (Fig. 19.6c) taken at a heating rate of $0.67\,\mathrm{K\ s^{-1}}$. The features of the DSC curve of the sintered glassy matrix composite specimen are similar to that of the original powder, namely, an endothermic reaction due to the glass transition, followed by a large supercooled liquid region and two exothermic reactions due to crystallization events. The sintered composite specimen shows that glass transition temperature (T_g) of 685 K, the onset temperature of the first-stage crystallization (T_X) of 774 K, and the extent of the supercooled liquid region ($\Delta T = T_X - T_g$) of 89 K are the same for those from the original powder. It indicates that no crystallization occurred during the SPS process. Furthermore, the crystallization enthalpy (ΔH) of the sintered glassy matrix composite specimen is $45.6\,\mathrm{J\ g^{-1}}$. It is around 90% of that of the monolithic gas-atomized $Zr_{55}Cu_{30}Al_{10}Ni_5$ glassy powder, which is $50.6\,\mathrm{J\ g^{-1}}$, as indicated in Fig. 19.6. This further demonstrated that the crystallization of the $Zr_{55}Cu_{30}Al_{10}Ni_5$ glassy matrix in the sintered composite specimen did not take place during the SPS process.

19.3.3 Mechanical Properties of Metallic Glassy Matrix Composite

Figure 19.7 shows nominal compressive stress–strain curves of the sintered glassy matrix composite specimen (Fig. 19.7a), together with that of the as-cast $Zr_{55}Cu_{30}Al_{10}Ni_5$ bulk glassy alloy rod sample (Fig. 19.7c). The results of the compression tests are summarized in Table 19.1. For the monolithic $Zr_{55}Cu_{30}Al_{10}Ni_5$ as-cast rod sample, the compressive fracture strength is 1,640 MPa, but with no plastic deformation. In contrast, for the sintered glassy matrix composite specimen, the ultimate compression stress reaches

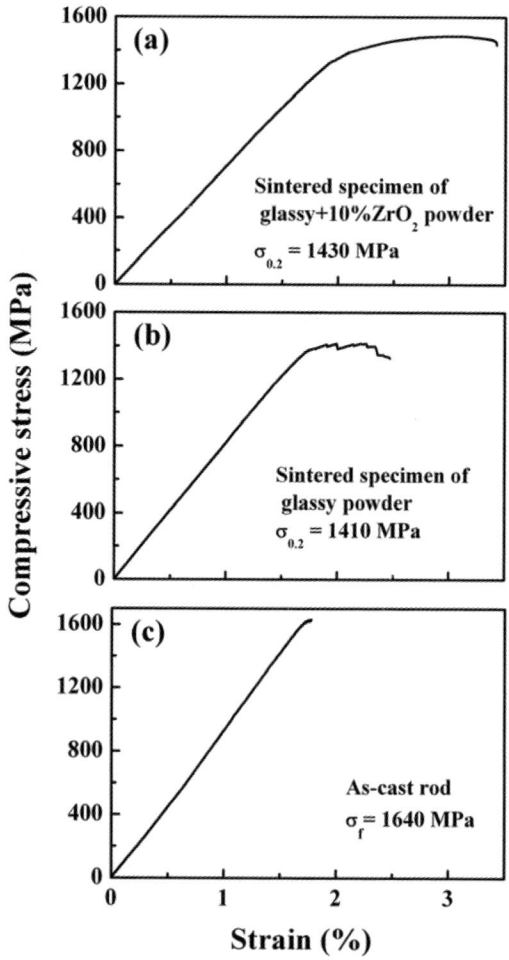

Fig. 19.7. Compressive stress–strain curves of the sintered $Zr_{55}Cu_{30}Al_{10}Ni_5$ glassy matrix composite containing 10 vol.% ZrO_2 powder (**a**), the sintered monolithic $Zr_{55}Cu_{30}Al_{10}Ni_5$ BMG specimen (**b**), and the as-cast $Zr_{55}Cu_{30}Al_{10}Ni_5$ glassy alloy rod specimen (**c**)

Table 19.1. Densities and some measured mechanical properties of the sintered monolithic $Zr_{55}Cu_{30}Al_{10}Ni_5$ BMG specimen and the ZrO_2 particulate reinforced composite as well as the as-cast $Zr_{55}Cu_{30}Al_{10}Ni_5$ bulk glassy alloy rod specimen

Specimens	SPS conditions	ρ (g cm^{-3})	ρ/ρ_0 (%)	Yield stress, $\sigma_{0.2}$ (MPa)	Strain at offset yield, ε_y (%)	Ultimate compression stress, σ_{max} (MPa)	Fracture strain, ε_f (%)
Glass +10% ZrO_2	623 K, 600 MPa, 10 min	6.39	95.2	1,430	2.07	1,500	3.42
Monolithic glass	623 K, 30 MPa, 10 min	6.51	95.3	1,410	1.93	1,420	2.47
Glass casting rod	–	6.83	100	–	1.78	1,640	1.78

1,500 MPa and the sample exhibits macroscopic compressive plastic deformation after the elastic strain.

Although the compressive strength of the sintered $Zr_{55}Cu_{30}Al_{10}Ni_5$ glassy matrix composite specimen was slightly lower than that of the as-cast rod specimen, it is noted that the sintered glassy matrix composite specimen exhibited larger plastic ductility than that of the as-cast $Zr_{55}Cu_{30}Al_{10}Ni_5$ glassy alloy rod specimen, as indicated in Fig. 19.7. The plastic deformation of the sintered glassy matrix composite specimen with the ZrO_2 ceramic particles exceeded 1.55% under compression, while it was lower than 0.1% for the as-cast $Zr_{55}Cu_{30}Al_{10}Ni_5$ glassy alloy rod specimen.

19.4 Discussion

It is known that the ductility enhancement of the BMGs is mainly achieved by introducing crystalline phase and free volume into the metallic glassy matrix. To clarify the reinforced mechanical effect of the ZrO_2 particulates in the metallic glassy matrix composite, we specially prepared the sintered $Zr_{55}Cu_{30}Al_{10}Ni_5$ glassy alloy specimens without the ZrO_2 particulates by the SPS process at a sintering temperature of 623 K, with a loading pressure of 30 MPa, and a holding time of 10 min. It exhibits similar relative density to that of the sintered $Zr_{55}Cu_{30}Al_{10}Ni_5$ glassy matrix composite with 10% ZrO_2 particulates, as shown in Table 19.1. This indicates that there is similar porosity, or same free volume in the two sintered specimens. The glassy features of the sintered monolithic $Zr_{55}Cu_{30}Al_{10}Ni_5$ glassy alloy specimens (without the ZrO_2 particulates) have been identified by XRD and DSC analyses, which are shown in Figs. 19.5b and 19.6b, respectively, while the compressive stress–strain curve and the data of the compressive test are given in Fig. 19.7b and Table 19.1, respectively. Compared with the sintered mono-lithic $Zr_{55}Cu_{30}Al_{10}Ni_5$ BMG specimen, both the compressive strength and the plastic ductility in the sintered $Zr_{55}Cu_{30}Al_{10}Ni_5$ glassy matrix compos-ite were enhanced by adding the ZrO_2 particulates into the metallic glassy alloy. The compressive strength increased from 1,420 MPa for the sintered monolithic $Zr_{55}Cu_{30}Al_{10}Ni_5$ BMG specimen to 1,500 MPa for the sintered $Zr_{55}Cu_{30}Al_{10}Ni_5$ glassy matrix composite specimen. Meanwhile, the compres-sive plastic ductility after the elastic strain increased from 0.74% for the sin-tered monolithic $Zr_{55}Cu_{30}Al_{10}Ni_5$ BMG specimen to 1.55% for the sintered glassy matrix composite specimen. It is a very large improvement in compar-ison with that of the as-cast $Zr_{55}Cu_{30}Al_{10}Ni_5$ metallic glassy rod specimen, which is almost zero, as shown in Fig. 19.7c and Table 19.1. Because of the similarity of free-volume in the sintered $Zr_{55}Cu_{30}Al_{10}Ni_5$ glassy matrix com-posite and in the sintered monolithic $Zr_{55}Cu_{30}Al_{10}Ni_5$ BMG specimen, the effect of free-volume on the plastic ductility in the sintered composite spec-imen can be determined. Therefore, except for the effect of the free-volume, the improvement of the compressive strength and plastic ductility in the sin-tered $Zr_{55}Cu_{30}Al_{10}Ni_5$ glassy matrix composite should mainly originate from the reinforcement ZrO_2 particulates. Conner et al. [22] have reported that particulates of W, WC, Ta, or SiC can restrict the propagation of shear bands and promote the generation of multiple shear bands. In the present study, the existence of the ZrO_2 particulates, on the one hand, makes shear band prop-agation more difficult, which leads to an increase of compressive strength. On the other hand, the ZrO_2 particulates also lead to the generation of multiple shear bands [10, 23]. Choi-Yim et al. [23] have demonstrated that the forma-tion of multiple shear bands is initiated by particles, which are blocking the propagation of the single shear band.

Based on the aforementioned analysis and discussion, it is clear that the improvement of the compressive strength and the plastic ductility originates from the structural inhomogeneity caused by the microparticles inclusion. The ZrO_2 ceramic particulates act as a resisting media causing deviation, branching, and multiplication of shear bands.

19.5 Conclusions

$Zr_{55}Cu_{30}Al_{10}Ni_5$ bulk metallic glassy matrix composites have been fabricated by the SPS process using a mixed powder of gas-atomized $Zr_{55}Cu_{30}Al_{10}Ni_5$ glassy alloy powders blended with 10 vol.% ZrO_2 powders. The structure, thermal stability, and mechanical properties of the sintered bulk metallic glassy matrix composite were investigated. No crystallization of the glassy matrix is demonstrated during the SPS process. The crystallization behavior of the sintered glassy matrix composite is similar to that of the glassy alloy powder. The ZrO_2 particles are homogeneously dispersed in the $Zr_{55}Cu_{30}Al_{10}Ni_5$ glassy matrix. Because of the addition of the ZrO_2 particulates into the metallic glassy alloy, the plastic ductility of the sintered $Zr_{55}Cu_{30}Al_{10}Ni_5$ glassy matrix composite is enhanced, which originates from the structural inhomogeneity caused by the microparticles inclusion.

Acknowledgements. The authors express their gratitude to Mr. A. Okubo and Mr. Kanomata at Institute for Materials Research of Tohoku University for their help in the experiments. This work was supported by a Grant-In-Aid for Science Research in a priority Area "Research and Development Project on Advanced Metallic Glasses, Inorganic Materials, and Joining Technology" as well as Grant-In-Aid "Priority Area on Science and Technology of Microwave-Induced, Thermally Non-Equilibrium Reaction Field" No. 18070001 (2006–2010) from the Ministry of Education, Science, Sports and Culture, Japan. A part of this work was carried out at Advanced Research Center of Metallic Glasses, Institute for Materials Research, Tohoku University, Japan.

References

1. A. Inoue, Acta Mater. **48**, 279 (2000)
2. H.A. Bruck, A.J. Rosakis, W.L. Johnson, J. Mater. Res. **11**, 503 (1996)
3. D.V. Louzguine-Luzgin, A. Inoue, J. Nanosci. Nanotech. **5**, 999 (2005)
4. H. Kato, K. Yubuta, D.V. Louzguine, A. Inoue, H.S. Kim, Scripta Mater. **51**, 577 (2004)
5. H. Kato, T. Hirano, A. Matsuo, Y. Kawamura, A. Inoue, Scripta Mater. **43**, 503 (2000)
6. J. Eckert, M. Seidel, A. Kübler, U. Klement, L. Schultz, Scripta Mater. **38**, 595 (1998)
7. H. Choi-Yim, W.L. Johnson, Appl. Phys. Lett. **71**, 3808 (1997)
8. H. Kato, A. Inoue, Mater. Trans. **38**, 793 (1997)
9. C.C. Hays, C.P. Kim, W.L. Johnson, Phys. Rev. Lett. **84**, 2901 (2000)
10. H.M. Fu, H. Wang, H.F. Zhang, Z.Q. Hu, Scripta Mater. **54**, 1961 (2006)
11. V. Mamedov, Powder Metall. **45**, 322 (2002)
12. M. Tokita, Mater. Sci. Forum **308–311**, 83 (1999)
13. Z.A. Munir, U. Anselmi-Tamburini, M. Ohyanagi, J. Mater. Sci. **41**, 763 (2006)
14. G.Q. Xie, O. Ohashi, N. Yamaguchi, M. Song, K. Mitsuishi, K. Furuya, T. Noda, J. Mater. Res. **19**, 815 (2004)
15. G.Q. Xie, O. Ohashi, N. Yamaguchi, M. Song, K. Mitsuishi, K. Furuya, T. Noda, Jpn. J. Appl. Phys. **42**, 4725 (2003)
16. K. Yamazaki, S.H. Risbud, H. Aoyama, K. Shoda, J. Mater. Processing Technol. **56**, 955 (1996)
17. O. Yanagisawa, T. Hatayama, K. Matsugi, Materia Jpn. **33**, 1489 (1994)
18. M. Omori, Mater. Sci. Eng. A **287**, 183 (2000)
19. G.Q. Xie, O. Ohashi, M. Song, K. Furuya, T. Noda, Metall. Mater. Trans. A **34**, 699 (2003)
20. A. Inoue, T. Zhang, Mater. Trans. **37**, 185 (1996)
21. A. Castellero, S. Bossuyt, M. Stoica, S. Deledda, J. Eckert, G.Z. Chen, D.J. Fray, A.L. Greer, Scripta Mater. **55**, 87 (2006)
22. R.D. Conner, H. Choi-Yim, W.L. Johnson, J. Mater. Res. **14**, 3292 (1999)
23. H. Choi-Yim, B. Busch, U. Köster, W.L. Johnson, Acta Mater. **47**, 2455 (1999)

Nucleation and Growth of Thin Pentacene Films Studied by LEEM and STM

J.T. Sadowski, A. Al-Mahboob, Y. Fujikawa, and T. Sakurai

Summary: The application of the low-energy electron microscopy (LEEM) to the in-situ, real-time investigations of the nucleation and growth of organic pentacene (Pn) thin films is discussed. The relation between the substrate surface reactivity and the Pn film growth mechanisms is evidenced. We also discuss the anisotropy in the Pn film growth under kinetically limited growth conditions, which results in preferential growth of Pn in the direction of the b-axis, of the in-plane unit cell. We show how this kinetic preference can dramatically affect the crystallinity of the organic film, on the example of self-polycrystallization in epitaxial Pn films on H-Si(111) surface.

20.1 Introduction

The microelectronics industry has long revolved around inorganic materials such as silicon and gallium arsenide, but recently there is a significant interest in the application of organic semiconductors to microelectronic devices. Among other materials, pentacene (Pn, $C_{22}H_{14}$) appears to be particularly promising, since it has been successfully used in organic thin-film transistors with field-effect mobilities matching or even surpassing that of amorphous silicon [1]. Pentacene is relatively easy to grow on a wide range of substrates, ranging from insulators [2–5] and semiconductors [6] to metals [7–10]. Furthermore, fabrication of Pn thin film organic field effect transistors (FETs) on plastic substrates has also been demonstrated [11], which makes Pn an important material for constructing low-cost, large area electronic devices, flexible displays, etc.

The record carrier mobility achieved in a Pn single crystal-based devices is as high as $35 \, \text{cm}^2 \, \text{V}^{-1} \, \text{s}^{-1}$ [12], whereas mobility of Pn thin-film FETs is at best in the order of $\sim 1 \, \text{cm}^2 \, \text{V}^{-1} \, \text{s}^{-1}$ [11, 13], and reproducibility is still not very good. Origin of this discrepancy is related to the thin film crystal quality, since the defects and impurities associated with domain boundaries may induce localized electronic states that can inhibit the charge transport [14]. The electrical conduction in organic thin-film transistors (OTFTs) is also strongly affected by defects and the structure of the interface [15, 16]. In fact, the Pn even becomes almost nonconducting if OTFT is made of amorphous Pn film [17]. Therefore, there are at least two main reasons why the understanding and controlling of the nucleation and growth of thin organic films is crucial for the successful device fabrication: (1) properties of the interface between the organic film and a substrate determine the efficiency of the charge carrier injection and thus the device stability; (2) crystallinity and the size of the crystalline domains influence the device performance. As the organic thin film growth is usually performed in conditions being far from the thermodynamic equilibrium, the kinetics controlling self-organized organic film growth requires comprehensive understanding, before the organic-based electronics can fully materialize.

20.2 Experimental Setup

In-situ observations of thin pentacene films were performed in the ultra-high vacuum (UHV) low-energy electron microscope (Elmitec LEEM III) with a base pressure of 5×10^{-10} Torr. LEEM [18, 19] is a relatively new and powerful technique for studying the dynamic and static properties of surfaces and thin films, including growth and decay [20, 21], phase transitions [22–24], surface reactions [25], surface structure [26, 27], and morphology [28, 29]. It utilizes low energy, elastically backscattered electrons to image surfaces with few nanometers lateral resolution and atomic (molecular) depth resolution. Its real-time imaging capability and several unique contrast mechanisms for image formation makes it invaluable tool for surface science.

The photo of the Elmitec LEEM III system used in the experiments described in this paper is shown in Fig. 20.1a. The structure of LEEM system can be divided into three major parts: (1) illumination (electron gun) and imaging columns (except objective lens), (2) main chamber with sample manipulator, and (3) sample preparation chamber with air lock-load facility. Imaging column (together with illumination system) and main chamber are connected through a gate valve. Illumination column consists of LaB_6 electron gun with a Wehnelt cylinder for controlling the electron emission, three magnetic condensor lenses with magnetic deflection coils and illumination-aperture manipulator having three apertures allowing reducing the size of the beam spot on the sample surface to 20, 8, and 2 μm respectively. Without any illumination aperture the beam spot size on the sample is 80 μm in diameter. The magnetic prism (beam separator) is a part of both illumination and imaging systems. Objective lens hosted with deflectors and stigmators is placed in main chamber, where molecular beam deposition on substrate specimen can be done. Imaging system consists of five magnetic lenses, four deflectors, two stigmators, one contrast aperture manipulator (with three apertures), and a chevron channel plate array with a fluorescent screen. The imaging system is

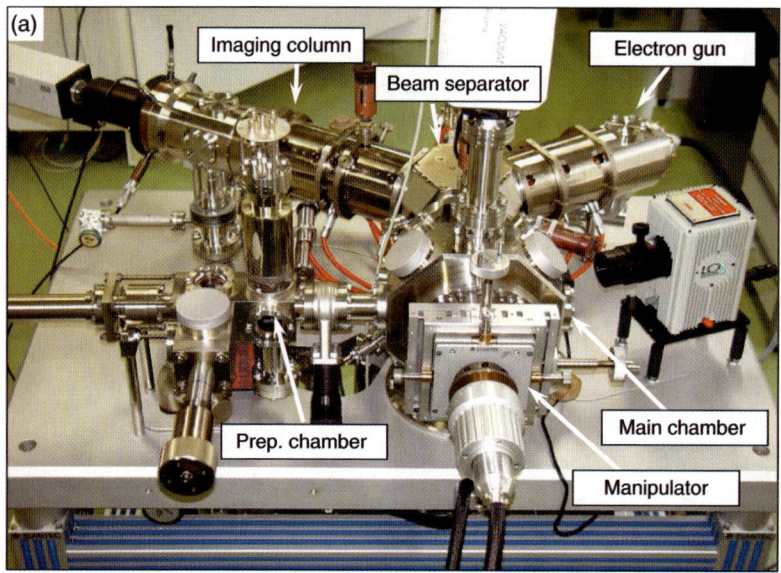

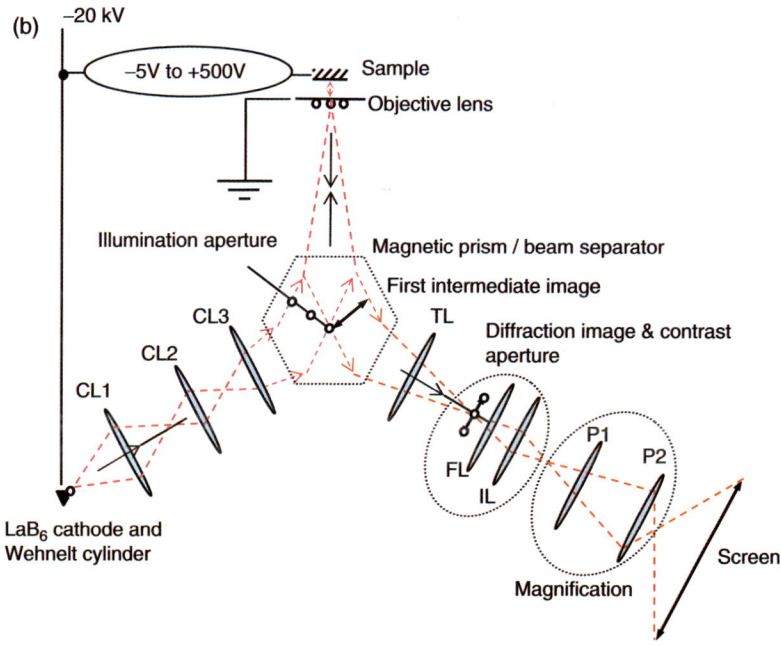

Fig. 20.1. (a) A photo of the Elmitec LEEM III system used in the experiments described in this paper; the main components are labeled in the photo. (b) Schematic presentation of the electron optics of LEEM III system: *CL* condensor lens, *TL* transfer lens, *FL* field lens, *IL* intermediate lens, *PL* projective lenses; all CLs, objective lens, FL, IL, and PLs are hosted with XY deflectors; the objective lens and TL also contain stigmators

equipped with high resolution CCD camera and controlled by extensive control and image processing software. Sample manipulator has three translations (±4 mm for X, Y and 50 mm travel along manipulator axis) and two nearly eucentric tilts $\pm2°$, exchangeable sample cartridge with W/Re-thermocouple, electron bombardment heating of a sample up to $2,000°C$, radiation heating up to $650°C$, and cooling by liquid nitrogen flow through spiral coil down to $-125°C$.

A schematic of the LEEM III electron optics is shown in Fig. 20.1b. Intensity of the electron beam is controlled by electrostatic voltage applied to Wehnelt cylinder. The beam is collimated through three magnetic condensor lenses CL1, CL2, and CL3, and then it passes through a beam separator. A beam separator is an essential part in LEEM, which allows the physical separation between incident and outgoing beams. It is a magnetic prism, which uses a magnetic field to bend the electron beam through a desired angle in moving from one arm of the microscope to another. LEEM III uses 60° prism deflector. At the center of the beam separator, the illumination aperture can be placed at the path of beam. After passing through the aperture, beam reaches to the sample via objective lens. Sample is kept at the potential of -20 kV. The effective kinetic energy of the imaging electrons is controlled by applying small changes to this potential. Objective lens is grounded. Objective lens and sample together form a cathode lens where the objective lens places the sample in a strong electrostatic field (up to 10 kV mm^{-1} for optimum resolution). The electrons traverse the focusing field of the objective lens at a relatively high electron beam energy of 20 keV, and are decelerated between objective lens and sample to a final low energy determined by sample potential. In this first pass, the "objective lens" functions as the final condenser, controlling the angle (illumination deflection), location of illumination (focusing), and stigmatism. The magnetic lens field inside the objective lens acts both as the final condenser lens and as the first image forming lens. The magnetic field does not extend to sample itself.

The strong interaction of low-energy electron with matter limits the depth-profile information, which yields extreme surface sensitivity even under normal incidence. The low-energy electron beam is back-scattered, reflected, or diffracted from the sample surface upon incidence, and accelerated by strong uniform electrostatic field between the sample and objective lens to acquire the same kinetic energy of 20 kV as the illumination beam. Instead of using illumination electron beam, photo-emitted electrons, excited by an external UV light source, can also be used for imaging surface, where contrast is achieved due to difference in local work-function at the surface. In any case, the electron beam coming from the sample surface passes through the beam separator, behind which the TL relays the first intermediate image, formed by objective lens and deflected by 120° with respect to beam originated from the gun, towards IL and FL that are coupled together. A contrast aperture is placed at the back focal plane of the combination of TL and objective lens, where parallel rays coming from the sample plane are converged and

LEED patterns (or Fourier transformed pattern of a surface) are formed. The position of the contrast aperture can be manipulated at this plane for choosing/cutting off the beam having a particular diffraction/reflection angle. The images are formed at Gaussian focal plane behind the contrast aperture, at which nonparallel rays coming from a point of the sample surface are converged. FL and IL allow the switching between imaging of back-focal plane (diffraction plane/Fourier transformation) and Gaussian focal-plane (image plane/inversed Fourier transformation of chosen diffraction spots by the aperture). PLs project the image on a channel plate (MCP) and control the magnification. MCP relay highly magnified images on a fluorescent screen for video rate image acquisition.

Nucleation and growth of Pn films were observed in real time in LEEM, with imaging electron energy typically as low as 0.2 eV in order to avoid electron irradiation-induced damage to the organic film. The μ-beam low-energy electron diffraction (μ-LEED) patterns were also recorded in the LEEM experiments with electron energies typically in the range of 8–17 eV. In this technique, an aperture placed in the path of electron beam allows one to limit the beam size to approximately 2 μm in diameter on the surface of interest and to choose the probing area, for example, including both the edge of the organic island and the bare substrate at once. In a separate experiment, the STM images were obtained from the 1 ML thick Pn film deposited at RT onto Bi(0001) surface,

To evaluate the molecular-scale crystallinity of the Pn films and to confirm the in-plane molecular packing and epitaxial relations indicated by the diffraction experiments, additional experiments were carried out in the UHV field-ion scanning tunneling microscope (FI-STM) [30], with base pressure of 2×10^{-11} Torr. In both systems, Pn was thermally evaporated from the tantalum crucible at the deposition rates of 0.02–0.3 ML min^{-1}, where 1 ML corresponds to the molecular density of the Pn(001) plane [31].

20.3 Results and Discussion

Pentacene (Pn, $C_{22}H_{14}$) is a flat, elongated molecule, consisting of five π-conjugated, aromatic, benzene-like rings fused sideways along the long molecular axis (LMA), as shown schematically in Fig. 20.2. Pn bulk phase has a triclinic crystal structure [31] with two molecules per unit cell and it forms a layered structure along ⟨001⟩ direction, with the herringbone arrangement of molecules in the (001) plane (called also *ab*-plane). Northrup et al. have shown by the first-principles density-functional theory (DFT) calculations [32] that the (001) plane has much lower surface energy than other low index planes, and the steps along short axis (*a*-axis) of the in-plane unit cell have the lowest step energy. The bonds between molecules in the same layer (*ab*-plane) are several times stronger than those between molecules in adjacent layers. This anisotropy in the bonding is a driving force for the film to exhibit

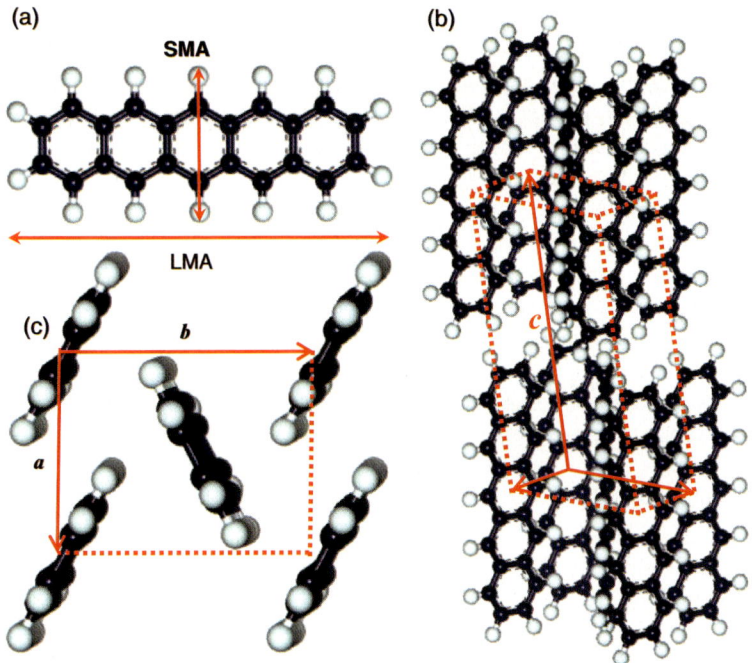

Fig. 20.2. (**a**) Sketch of the Pn ($C_{22}H_{14}$) molecule, long molecular axis (LMA), and short molecular axis (SMA) are outlined by arrows; (**b**) Tilted, side view of Pn unit cell; (**c**) Molecular packing of Pn in the ab-plane viewed along direction normal to the surface; the nonequivalent molecule of a herringbone basis is located at (1/2, 1/2) with respect to the equivalent molecule at (0, 0)

self-organized (001)-"standing up" orientation, when deposited on a substrate with which Pn interacts weakly. In most of the cases of Pn deposition, a flat-laying first layer (in the case of metallic substrates) [7–10] or often disordered wetting layer, or interfacial layer with poor crystallinity [6] is formed before an ordered, standing up molecular structure eventually appears.

The growth of the pentacene thin films is governed by the balance of two interactions: weak Van der Waals-like pentacene–pentacene interaction and the interaction between pentacene molecule and the substrate. The latter one depends to certain extent on the surface electronic structure [33]. On the insulating substrates, like SiO_2, the second interaction is again Van der Waals-like, which allows pentacene to crystallize into the bulk-like film with the most energetically favorable (001) face parallel to the growth front [2, 3]. From the other side, when the density of states at or near the Fermi level increases, the π-type pentacene system starts to interact more strongly with these valence electrons, forcing the molecules down onto the surface, as in the case of the Pn films on metallic substrates [7–10].

Theoretical works [34] predict a wide spectrum of polymorphism in Pn and indeed there are several different bulk polymorphs of pentacene. Moreover, in thermally evaporated pentacene thin films, deposited for example on oxides, so called thin film phase with a larger interplane spacing between (001) planes (up to 15.5 Å) is observed frequently [35]. Such polymorphism opens the possibility of growing epitaxial Pn films having standing-up geometry on substrates with wide variation in lattice parameters, even if there is no lattice commensuration between bulk Pn and a substrate, if the interaction between the Pn molecules is weak enough in comparison with the Pn–Pn intermolecular interactions.

20.3.1 Highly Ordered Pentacene Films on Bi(0001)

The question arises of whether an appropriate substrate exists, which allows the Pn films to have good electrical connectivity to the substrate, while at the same time the Pn–substrate interaction is small enough for the film structure to be predominantly determined by Pn–Pn interactions. We have found that thin, well-ordered bismuth (Bi) thin films are a possible candidate having an electronic structure that is intermediate between the metallic and insulating substrates commonly used for Pn growth.

Bi has the A7 (arsenic) crystal structure, with two atoms per rhombohedral unit cell. This structure can be also understood as pseudo-cubic [36]. The natural cleaved (111) plane of rhombohedral Bi is called in this work the (0001) plane for hexagonal crystal indexing, with the simplest trigonal in-plane coordinates $|\mathbf{a}| = |\mathbf{b}| = 4.5332$ Å and out-of-plane trigonal axis $|\mathbf{c}| = 11.7967$ Å. As the LEED patterns represent the surface lattice and the interface relations are described explicitly, indexing of plane (hkil) (i $= -($h $+$ k$))$ or directions $\langle$hkil$\rangle$ for epitaxy will be denoted here for simplicity as (hk) or $\langle$hk$\rangle$, respectively, in terms of trigonal lattice vectors. Here, a bilayer (BL) is the smallest unit in the $\langle$0001$\rangle$ direction, with bilayer (BL) step height of 3.932 Å [37].

In the recent work [38], it has been determined that during the UHV Bi deposition on Si(111)-7 × 7 at room temperature (RT), the film undergoes an unique and unexpected structural phase transformation from the pseudocubic (01$\bar{1}$2) (in terms of hexagonal indexing) phase into a trigonal Bi(0001) phase – the same as the bulk terminated Bi. The morphology of the Bi(0001) film can be further improved by moderate annealing [39], as shown in Fig. 20.3b.

Highly ordered Bi(0001)/Si(111) substrates, kept at RT, were used for subsequent pentacene deposition [40, 41]. Real-time LEEM observations of Pn growth on Bi(0001) surface show that the nucleation density of Pn islands is very low ($<10^{-5} \, \mu m^{-2}$) and Pn is still growing for a while even after stopping the deposition, due to material available on the substrate surface. This implies that Pn molecules are highly mobile on Bi(0001) surface at RT. The growth mode of first Pn layer on well annealed Bi(0001) appears to be a step-flow-like, resulting in compact islands, as it is shown in Fig. 20.4a. At the growth front some holes initially appear in the film, but they are gradually filled-up

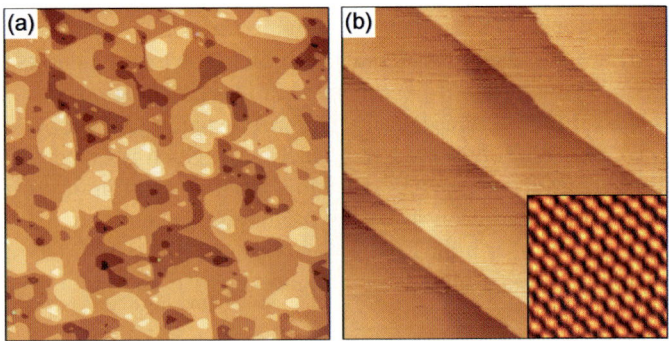

Fig. 20.3. (a) Topographic STM image ($200 \times 200\,\text{nm}^2$, sample bias $V_S = -2.0\,\text{V}$, tunneling current $I = 20\,\text{pA}$) of the as-deposited, 20 BL-thick Bi(0001) film grown on Si(111)-7 × 7; **(b)** STM image ($200 \times 200\,\text{nm}^2$, $V_S = -2.0\,\text{V}$, $I = 20\,\text{pA}$) of the surface shown in (a), after annealing at $\sim$400 K for 30 min; in the inset: high-resolution STM image ($4 \times 4\,\text{nm}^2$, $V_S = -0.03\,\text{V}$, $I = 500\,\text{pA}$) showing hexagonal arrangement of the Bi atoms on the annealed surface

by the molecules supplied to the surface. Examining the shape of the holes at the moment of their formation reveals that they probably form at natural defects such as a BL-high Bi islands. Nevertheless, the STM images (Fig. 20.4) confirm very high crystallinity of the Pn film and perfect molecular ordering, starting from the first Pn layer, directly adsorbed on Bi(0001).

The μ-LEED patterns taken from the first Pn layer on Bi(0001) correspond to lattice parameters of ab-plane, indicating standing-up orientation of molecules. LEED patterns taken before formation of crystalline Pn layer and taken from bare substrates near the Pn islands do not show any change in the structure of the substrate. This fact together with self-organized, standing-up Pn orientation implies lack of formation of the wetting layer and chemical bonds between Pn and Bi(0001) substrate, resulting in sharp and smooth interface. Thus we can conclude that the growth is not kinetically limited, unless there are defects with unsaturated dangling bonds locally present. High mobility of Pn molecules and a very low nucleation density on atomically flat Bi(0001) result in the sizes of the first-layer Pn islands on Bi(0001) surface exceeding $\sim$0.2 mm [40].

LEED pattern taken from the Bi(0001) surface partially covered by the first Pn layer is shown in Fig. 20.5a. Spots originating from the noncovered Bi(0001) surface are marked by arrows, while those coming from the overlapping Bi and Pn patterns are marked by circles. The Pn (02) diffraction spot exactly overlaps with the Bi(01) spot of the annealed Bi(0001) film, which means that $\langle 10 \rangle$ directions of both Bi(0001) and Pn(001) are identical. The distance between equivalent (01) Pn molecular lines – spacing between molecular rows that describe the intersections of Pn(010) planes by Pn(001) surface – is $d(|\boldsymbol{b}|) = \sqrt{3}|\boldsymbol{a}_{\text{Bi}}|$, where $|\boldsymbol{a}_{\text{Bi}}| = 4.5532$ Å is the inter-atomic

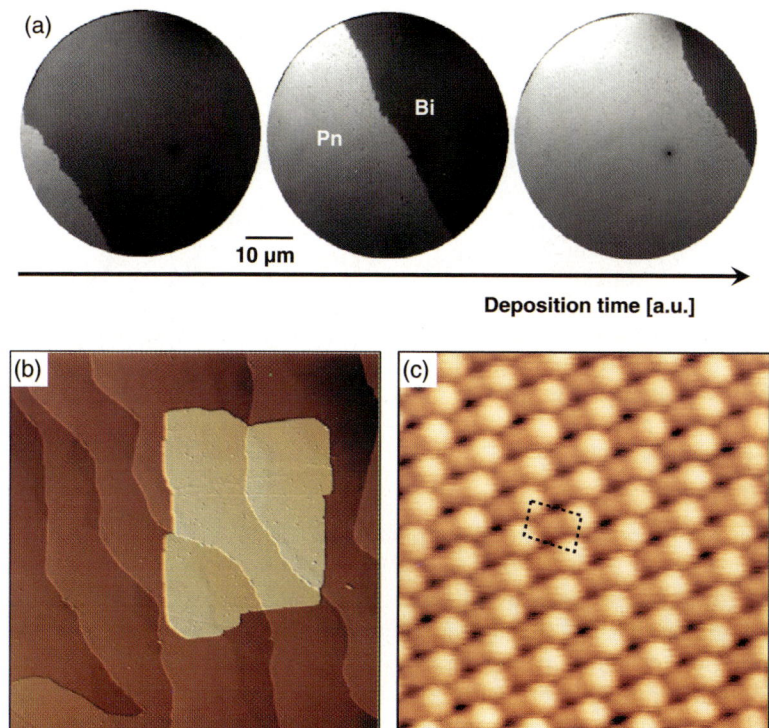

Fig. 20.4. (a) Time series (from left to right in arbitrary time scale) of the bright-field LEEM images ($E = 0.2\,\mathrm{eV}$) obtained during Pn deposition on a Bi(0001) surface at RT – step flow growth of first pentacene layer (bright contrast) having standing up bulk-like structure; single nucleation occurred within observed several hundreds micrometer region of the sample; (b) Topographic STM image ($500 \times 500\,\mathrm{nm}^2$, sample bias $V_S = -2.3\,\mathrm{V}$, tunneling current $I = 20\,\mathrm{pA}$) showing Pn-covered Bi(0001) surface, with second-layer Pn island (brighter) visible on the top of fist layer – lines running from the top to the bottom of the image are the underlying, bilayer high Bi steps; (c) High-resolution STM image ($6 \times 6\,\mathrm{nm}^2$, $V_S = +2.2\,\mathrm{V}$, $I = 50\,\mathrm{pA}$) of the first Pn layer on Bi(0001) showing molecular arrangement in the Pn(001) plane – the unit cell with two inequivalent molecules is outlined

distance in the Bi(0001) plane. The subscripts Bi denote Bi in-plane trigonal lattice vectors, and the Pn surface lattice (ab-plane) is described without subscripts.

The in-plane unit cell parameters for the Pn(001) plane (ab-plane) obtained from the analysis of the LEED diffraction patterns are $|\boldsymbol{a}| = 6.1\pm0.2\,\text{Å}$, where $\boldsymbol{a}\|\boldsymbol{a}_{\mathrm{Bi}}$, $|\boldsymbol{b}| = 7.871 \pm 0.005\,\text{Å}$, and $\gamma = 86° \pm 0.5°$, respectively. Diffraction data imply a "point-on-line" commensurate adsorption geometry of the first Pn layer on Bi(0001), schematics of which is shown in Fig. 20.5b. The important observation is that the in-plane lattice parameter of Pn/Bi(0001) is very close to that first reported for bulk Pn [31]. The in-plane spacing

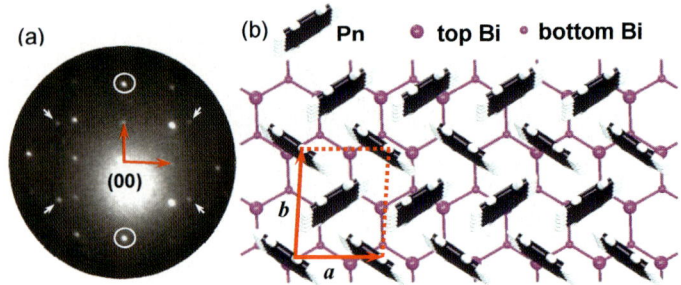

Fig. 20.5. (a) μ-LEED pattern ($E = 16.4$ eV) taken from the edge of the Pn island grown on the Bi(0001), diffraction spots from both Pn layer and Bi(0001) substrate are visible simultaneously; (b) Adsorption geometry (view from the direction normal to Pn(001) plane) for the first Pn layer adsorbed on Bi(0001) surface, showing proposed commensurate "point-on-line" alignment with the Bi surface

between alternate molecular rows parallel to Pn a-axis is the same in both the bulk Pn and Pn/Bi(0001), and it is also same as in plane spacing of Bi atomic rows. In the cases of all reported self-organized thin Pn films of few monolayers other than this system, the obtained thin films are so-called Pn thin film polymorphs, having compressed in-plane lattice with respect to bulk. Observed here, monolayer of Pn on Bi(0001) is the only example of the thin film with lattice parameters having the values close to that as in bulk, as well as it is the first reported standing-up epitaxial Pn thin film.

The above discussion concerns a single-domain Pn grain. Since the Bi(0001) surface has a threefold symmetry, and we detected an epitaxial relation between Pn lattice and Bi(0001) surface, we should expect to find three equivalent Pn domains. They are indeed observed in our experiments, as it is shown in Fig. 20.6, when the nonannealed (more rough) Bi(0001) surface is used. The LEEM observation of the Pn film at electron energy of approximately 2 eV reveals a distinctive contrast levels within single Pn grain (Fig. 20.6a), when the incident electron beam is slightly tilted from the surface normal (this is so-called tilted bright-field imaging mode). Each individual contrast observed in the LEEM image can be assigned to the specific type among the three LEED patterns, characterized by its a-axis (shorter real-space axis) aligning to each of three equivalent directions of the trigonal substrate surface, $\langle 10 \rangle$, $\langle 01 \rangle$, and $\langle \bar{1}\bar{1} \rangle$, respectively.

Pn crystal lattice [11] is described as having $P\bar{1}$ symmetry. Since molecule has tilted orientation with respect to normal to ab-plane, interface further lowers the symmetry that results in its surface lattice (ab-plane) having no symmetry at all (**P1** surface lattice). Mirroring of surface lattice (for an example, twining by mirror parallel to a-axis) results in non-superimposable Pn molecular orientation due to asymmetry or chirality induced by interface. The μ-LEED pattern (Fig. 20.7a) taken on Pn film on Bi(0001) at incident electron energy of 4.9 eV also revealed the asymmetry (chirality) in the intensity

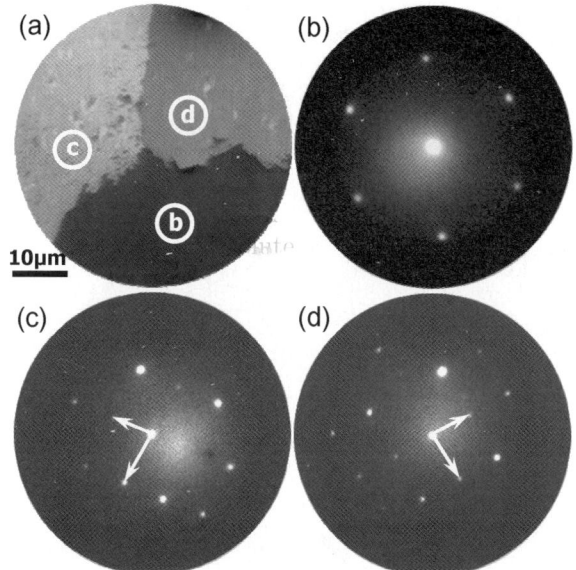

Fig. 20.6. (a) Tilted bright-field LEEM image showing distinctive contrasts within a single Pn grain together with corresponding μ-LEED patterns recorded from the (b) bare Bi(0001) surface and (c, d) two Pn domains having point-on-line epitaxial relations with two of the three equivalent trigonal directions, $\langle 10 \rangle$, $\langle 01 \rangle$, and $\langle \bar{1}\bar{1} \rangle$ of the substrate surface, respectively; additional contrast in the Pn domains shown in (a) is associated with holes in Pn layer (black) and a second Pn layer (brighter) nucleated on top of the first one

of particular diffraction spots. Such asymmetry depends on the wavelength of electron beam used and the orientation of c^*-axis (direction of reciprocal vector normal to ab-plane) of the Pn film. The inversion of c^*-axis will change Pn ab-plane from right-handed to left-handed (ba-plane) due to oblique unit cell of surface lattice. The asymmetry in LEED also can be reversed by 180° rotation of Pn unit cell orientation. Using asymmetry of (01) Pn diffraction spots in the dark-field imaging mode, where contrast of sub-domains depends on the presence and intensity of diffracted beam within the imaging aperture, we can directly observe this reversing of asymmetry between two sub-domains in a two-dimensional thin film, which is visible as a contrast reversal between Pn domains shown in Fig. 20.7b,c.

In our epitaxial Pn thin films grown on Bi(0001) surface we frequently detect twinning. This is likely associated with secondary nucleation at substrate defects. The twinning in Pn domains can be generally described by twin operation in terms of either Pn lattice itself or the substrate lattice, or both lattices together. However, in our case, because of strong commensuration between Pn lattice and Bi(0001) substrate surface, only a twining realized by mirroring a Pn unit cell by having a mirror axis described in surface lattice

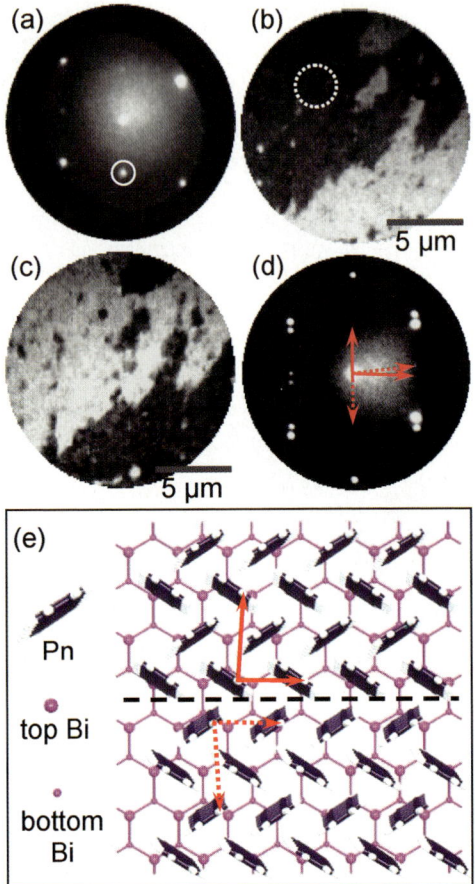

Fig. 20.7. (a) μ-LEED pattern (4.9 eV) recorded from the region marked by dotted circle in (b) showing asymmetry (chirality) in intensity of the diffraction spots
(asymmetric spot is marked by a circle); (b, c) asymmetric dark-field LEEM images taken with electron beam of 4.0 and 4.9 eV, respectively, using the diffraction
spot marked by circle in (a) – reversed contrast between twins is associated with
mirroring molecular tilt at interface by the reversal of the c^*; (d) μ-LEED pattern ($E = 10.0$ eV), recorded in a separate experiment, showing twinning in the
Pn/Bi(0001) island; (e) schematic adsorption geometry for the twinned Pn layer,
twinning plane is marked by a dashed line

vectors of ab-plane has been observed. The mirror plane is parallel to the Pn
a-axis (which is also a direction of the point-on-line commensuration) and
results in the inversion of c^*-axis, as well as the reversal of the molecular tilt
between the twins. An example on such twinning in Pn grown on Bi(0001) is
shown in the asymmetric dark-field imaging in Fig. 20.7b,c, and in the LEED
pattern in Fig. 20.7d, in which two sets of diffraction spots, originating from

twinned domains, are visible. Using same principles for the analyzing of diffraction patterns as we did above, we can obtain the molecular arrangement in the Pn(001) plane for the twinned layer, shown in Fig. 20.7e.

20.3.2 Anisotropic Growth of Pentacene

Pentacene is known to grow non-epitaxially having fractal islands' shapes on chemically active (having dangling bonds) semiconductor surfaces, like Si(001)-2 × 1. Meyer zu Heringdorf et al. have concluded [6] from the analysis of islands' shapes that in such cases the growth proceeds under the regime of diffusion-limited aggregation (DLA) [42]. Our LEEM data confirm that the nucleation of monolayer-high, (001)-oriented Pn islands on the clean Si(111)-7 × 7 surface follows the formation of a disordered wetting layer (Fig. 20.8a).

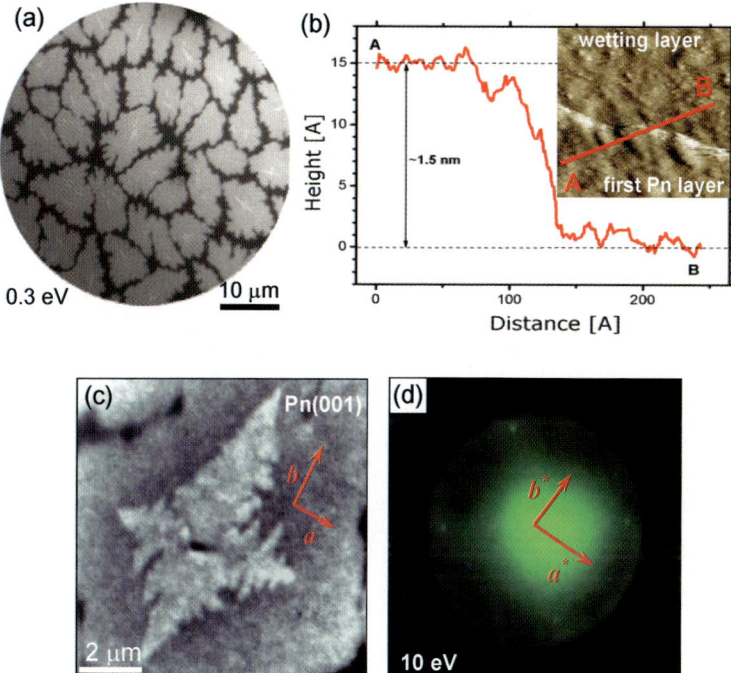

Fig. 20.8. (a) Bright-field LEEM image showing first and second layer Pn islands (brighter contrast) nucleated on top of the disordered wetting layer (*dark*) in Pn deposition on Si(111)-7 × 7 at RT; (b) STM height profile taken from the edge of the first crystalline Pn layer (shown in the STM image in the inset: $25 \times 25 \, nm^2$, $V_S = -1.7 \, V$, $I = 20 \, pA$) on top of the wetting layer; (c) LEEM image of the Pn island (second layer) – the primitive vectors constituting the in-plane unit cell are outlined in the image, demonstrating that under kinetically limited growth conditions the fast growth direction aligns to the b-axis in this anisotropic system; (d) μ-LEED pattern evidencing the relation between island shape and Pn in-plane unit cell lattice vectors

Random adsorption and possible dissociation of Pn molecules on the Si dangling bonds makes the wetting layer very rough (as depicted in the STM image, Fig. 20.8b). Local roughness of the substrate (interface) reduces the mobility of the molecules, resulting in DLA-like growth of Pn islands. The nucleation density of Pn islands is rather high – at the flux of 0.5 ML min^{-1} the typical nucleation density is about $3 \times 10^{-2}\,\mu m^{-2}$.

The Pn in-plane lattice parameters obtained from μ-LEED patterns, taken from the second crystalline Pn layer, correspond to thin-film phase with unit cell parameters as follows: $|a| = 5.71 \pm 0.07$ Å, $|b| = 7.39 \pm 0.16$ Å, and $\gamma = 91.0 \pm 0.7°$. The diffraction patterns also indicate rather poor crystallinity of Pn up to first few layers. Most interestingly, our real-time LEEM observations of the Pn islands evolution on Si(111)-7 × 7 reveal a kinetic preference in their development. Analysis of the diffraction patterns clearly indicates a preferential growth along the b-axis (Fig. 20.8c), with anisotropy in island shape apparent until the diffusion fields of neighboring islands overlap. This preference is in contrast with theoretically predicted surface energies or attachment energies [43] and experimental Pn crystal shape [44]. According to these reports, the step energy is lowest for step parallel to a-axis and highest for step parallel to b-axis; thus island-shapes are not expected to be elongated along the b-axis, which contradicts the experimental LEEM observation discussed here. The electronic energy of step formation for a Pn having the in-plane lattice parameter obtained from LEEM experiments has been calculated [45]. The results of calculation together with experimentally observed growth rate for the Pn on Si(111)-7 × 7 are summarized in the Table 20.1. It is also observed that initial anisotropy in growth of first and second layers is identical. We propose that not only the fractal growth itself, but also the observed overall anisotropy in the evolution of fractal Pn islands is associated with a kinetics, rather than thermodynamics.

To examine the generality of kinetic preference of self-organized thin Pn film, we grew a highly ordered epitaxial Pn on a flat, well ordered $\alpha\sqrt{3}$-Bi-Si(111) surface [46]. In this case, we observed very low nucleation density ($<10^{-5}\,\mu m^{-2}$) with heterogeneous nucleation usually occurring on defects at the surface. LEED patterns taken before formation of crystalline Pn layer

Table 20.1. DFT-calculated step energies and experiments growth anisotropy of Pn islands grown on Si(111)-7 × 7

Step parallel	E_{step} (meV Å^{-1})	Growth direction (normal to step)	Growth rate (relative)	
			Calculated (equilibrium)	Observed (kinetic)
a	90.93	b^*	0.74	1
b	122.37	a^*	1	0.47 ± 0.07
$a - b$	97.27	$(a + b)^*$	0.79	0.28 ± 0.04
$a + b$	98.28	$(a - b)^*$	0.80	0.32 ± 0.03

and taken from the bare substrate near the Pn islands do not show change in
the structure of the substrate upon the Pn deposition. LEED patterns also
show that the film terminates with (001) surface. These observations imply
lack of formation of the wetting layer or lack of formation of chemical bonds
between Pn and $\alpha\sqrt{3}$-Bi-Si(111) substrate, resulting in a sharp and smooth
interface. The LEED patterns taken from the edge of Pn island, which contain
diffraction spots originated from both substrate surface and Pn layer (example
shown in Fig. 20.9b), show always a definite relation between substrate and
Pn film. The b-axis of Pn is rotated by a small angle with respect to one of
the three equivalent directions ($\langle 1\bar{1}0\rangle_{Si}$, $\langle 01\bar{1}\rangle_{Si}$, and $\langle \bar{1}01\rangle_{Si}$) of the trigonal

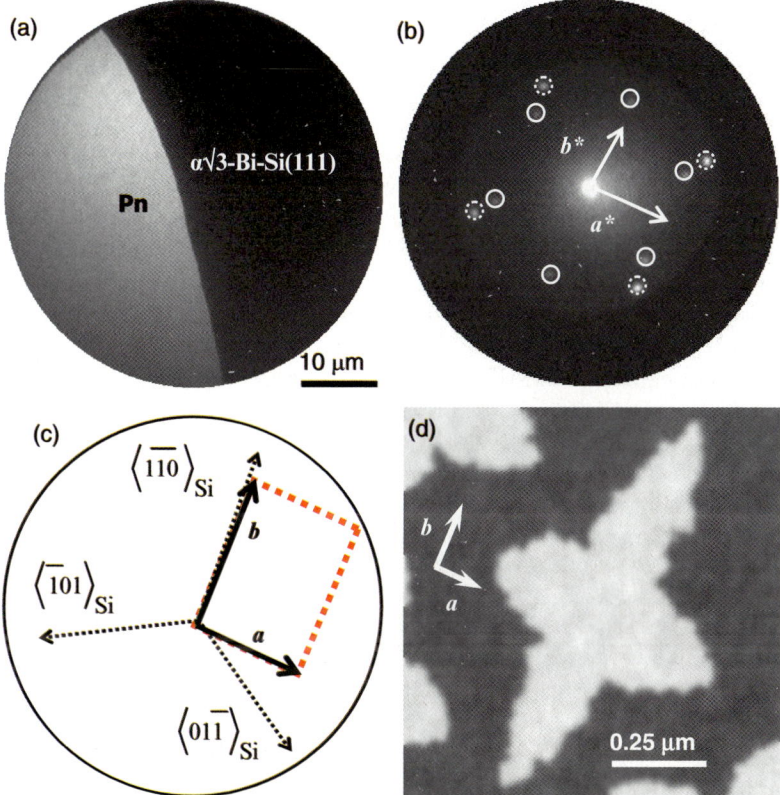

Fig. 20.9. (a) Bright-field LEEM image (at $E = 0.5\,\mathrm{eV}$) showing the front growth
of the first Pn layer (*bright*) nucleated on the $\alpha\sqrt{3}$-Bi-Si(111) surface (*dark*); (b)
μ-LEED pattern ($E = 8.3\,\mathrm{eV}$) taken from the edge of the Pn island – spots from
both substrate (*solid circles*) and Pn (*dotted circles*) are visible and Pn reciprocal
lattice vectors are outlined; (c) Sketch of the orientation of Pn unit cell in relation
to substrate lattice vectors; (d) Bright-field LEEM image (at $E = 1.8\,\mathrm{eV}$) of the
second layer island showing the kinetic preference in growth in the direction of the
b-axis

substrate surface, such that, the directions $\langle 2\bar{1}\rangle$ of Pn are matching with another direction of these equivalent surface lattice directions, as shown in Fig. 20.9c. The Pn in-plane lattice parameters obtained from LEED patterns correspond to thin-film phase (similar to that in case of monolayer Pn on Si(111)-7×7)): $|a| = 5.86\pm0.11$ Å, $|b| = 7.34\pm0.17$ Å, and $\gamma = 90.6\pm0.5°$. The formation of a first Pn layer in this case is governed by a step-flow growth with a compact growth front (Fig. 20.9a), and sizes of the grains in the first layer are exceeding hundreds of micrometers in diameters. The second layer islands have higher density (typically $\sim 0.1\,\mu m^{-2}$) due to stronger Pn–Pn interaction between molecules in adjacent layers. From combined LEEM/LEED study it is apparent that the second layer islands are elongated along their b-axes (Fig. 20.9d) [45].

20.3.3 Self-Polycrystallization in Epitaxial Pentacene Films on H-Si(111)

In the previous paragraphs we have discussed the kinetic preference in Pn film growth. Now, we will demonstrate how this growth anisotropy can dramatically affect the crystal structure of the thin molecular film. We use hydrogen-terminated Si(111) wafers, prepared by wet etching [47, 48], as the substrates for the Pn growth. In atomic scale, the H-terminated Si(111) surface is smooth and practically defect-free, and it is characterized by regular array of silicon steps, as it is shown in Fig. 20.10. However, since the wet chemical etching is used for the preparation of the H-Si(111) surface, a certain number of defects (pits, islands, rough steps, possibly metal contaminations) is always present on this surface.

Upon the deposition of pentacene on H-Si(111), under typical deposition conditions (substrate kept at RT; Pn flux, $0.2\,ML\ min^{-1}$) the nucleation is heterogeneous, but the nucleation density is still very low, typically

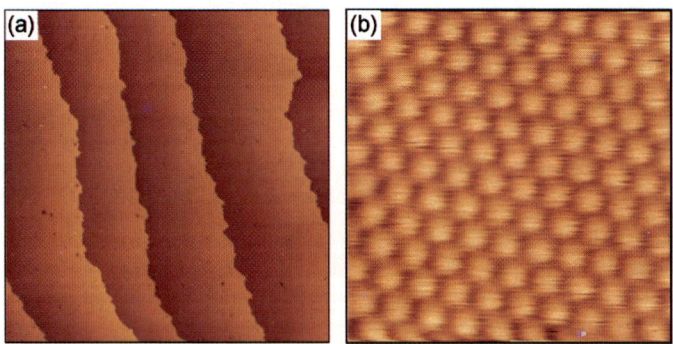

Fig. 20.10. (a) Topographic STM image ($200 \times 200\,nm^2$, sample bias $V_S = -1.3\,V$, tunneling current $I = 20\,pA$) of the H-terminated Si(111) surface prepared by wet etching; (b) High-resolution STM image ($4 \times 4\,nm^2$, $V_S = -1.3\,V$, $I = 100\,pA$) showing atomic structure of the H-Si(111) surface

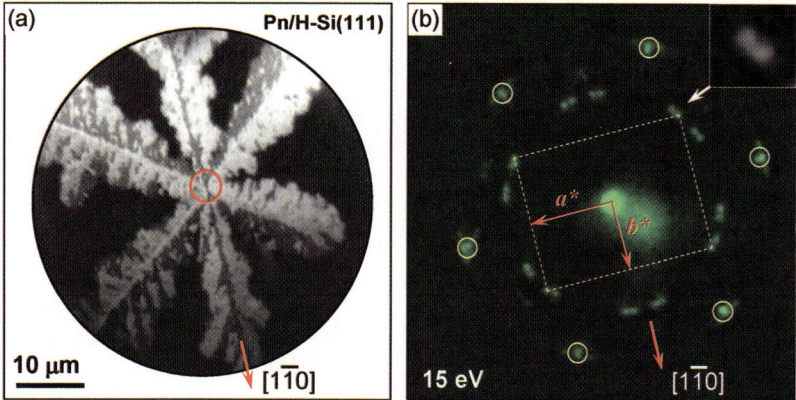

Fig. 20.11. (a) Tilted-beam LEEM image of the pentacene island grown on the H-Si(111) surface (imaging electron energy of 2.2 eV) – at these imaging conditions a complicated domain pattern is visible within the island; (**b**) LEED diffraction pattern obtained from the island shown in (**a**); spots originating from Si(111)-1 × 1 lattice are marked by yellow circles and a unit cell of one of the Pn rotational domains is outlined by a rectangle; magnified image of the pair of spots defining the "paired" Pn domains, rotated with respect to each other by an angle of 4.6 ± 0.3°, is shown in the inset

$1.7 \times 10^{-4}\,\mu m^{-2}$. LEEM images show single-monolayer high Pn islands with fractal shapes, typical of the DLA-like growth. Surprisingly, a quite complicated pattern of different contrasts levels is observed in the individual branches of the Pn islands (Fig. 20.11a) when the incident electron beam direction is tuned by tilting slightly away from the surface normal in the LEEM experiment. The LEED pattern obtained from the island in Fig. 20.11a is shown in Fig. 20.11b. There are several distinctive features in this pattern. First, the diffraction spots originating from the bare H-Si(111)-1 × 1 surface are still visible (marked by yellow circles in Fig. 20.11b), showing that nucleation of the Pn islands starts immediately without formation of a wetting layer. Second, careful analysis of the LEED pattern shows six pairs of double spots, corresponding to six different rotational Pn domains grown epitaxially on the H-Si(111)-1 × 1 surface. The unit cell of one such domain is marked as an example by a yellow rectangle in Fig. 20.11b. The above observation makes immediately clear that the different contrast levels in Fig. 20.11b originate from different rotational Pn domains and demonstrates a polycrystalline structure within the single Pn grain, which was observed to develop from a single nucleus [49], as it is illustrated in Fig. 20.12.

The crystal structure of the Pn layer grown on H-Si(111) and its relationship with the substrate are determined from the micro-beam LEED (μ-LEED) patterns [50]. It is found that each individual contrast level (Fig. 20.13a) can be assigned to one of the three μ-LEED patterns (one is shown as an example in Fig. 20.13b), corresponding to the Pn b-axis (longer real-space axis)

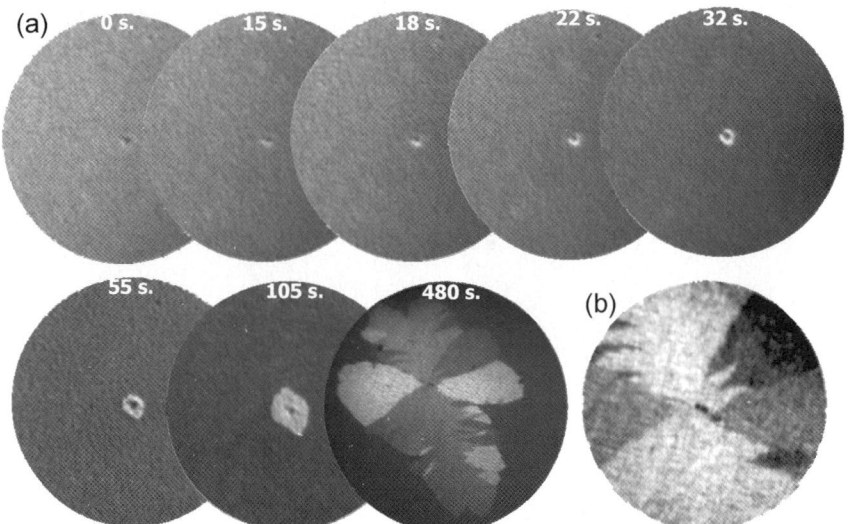

Fig. 20.12. (**a**) A time-series (time is denoted in each frame) of still frames taken from the movie recorded during bright-field LEEM imaging of the nucleation and development of a single, isolated Pn island grown on H-Si(111). Nucleation starts at the defect visible as a darker spot in the center of a frame. There is clearly a single nucleation site, which subsequently develops into an asymmetric Pn island. Last frame shows internal polycrystalline structure of the same island, recorded under tilted bright-field LEEM imaging conditions – different contrasts correspond to different Pn rotational domains; Field-of-view for all frames is 15 μm; (**b**) LEEM image (tilted bright-field with different tilt angle), showing vicinity of the nucleation site – on the basis of the single, bright contrast in the vicinity of the nucleation site we can conclude that the grain nucleated a single-crystalline domain, which further developed into a polycrystalline structure

aligning almost exactly (rotated by an angle of $2.3 \pm 0.3°$) to each of three equivalent $<1\bar{1}0>$ directions of the Si substrate surface (Fig. 20.13d). A simple model illustrating the epitaxial relations and lattice parameters determined for the in-plane 2D Pn unit cell: $|a| = 6.02 \pm 0.02$ Å, $|b| = 7.62 \pm 0.02$ Å, and $\gamma = 90.0 \pm 0.4°$, respectively, are shown in Fig. 20.13c. The values of the in-plane lattice parameters are verified through LEED patterns obtained from dozens of islands, showing that the Pn molecules grow in the "standing up" orientation but the Pn crystal lattice is somewhat distorted from its bulk structure.

This "point-on-line" coincidence is realized between the molecular lattice and surface crystal lattice along this axis. Because the H-Si(111) surface has a threefold symmetry and the Pn(001) plane in our thin film consists of a rectangular unit cell, there are three, 120°-rotated, equivalent epitaxial orientations of the 2D Pn lattice on H-Si(111), each of them consisting of *paired* domains misaligned by $2.3 \pm 0.3°$ from one of $<1\bar{1}0>$ axes. The rotated domains yield

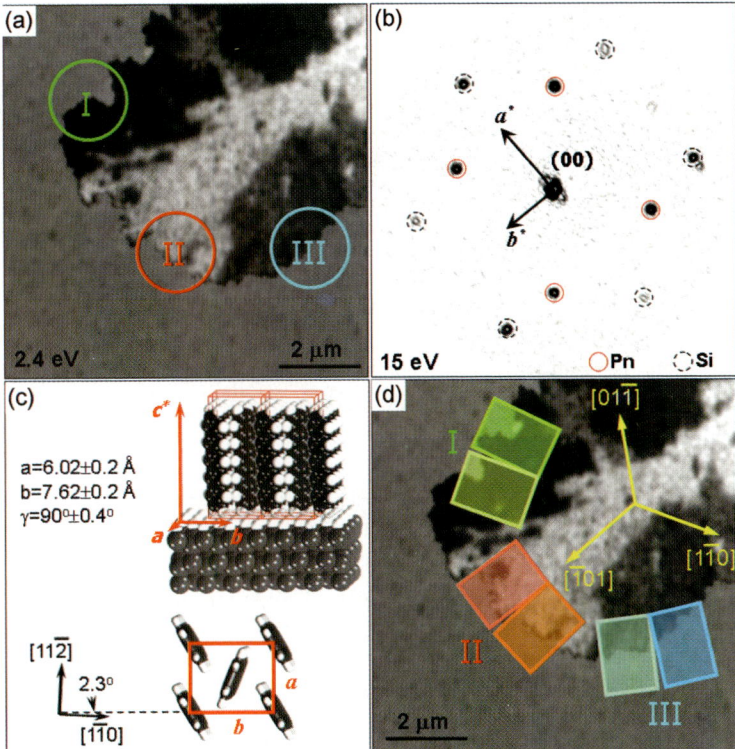

Fig. 20.13. (a) Tilted-beam LEEM image showing tip of the one of the branches in Pn island on H-Si(111) with three distinctive contrasts (I, II, III) visible; (b) The μ-LEED diffraction pattern obtained from an individual contrast segment II in (a) – spots in black circles are from the Si(111) lattice and spots marked by red solid circles are from the Pn(001) lattice; (c) A simple model outlining the epitaxial relations between Pn lattice and H-Si(111) substrate; (d) Schematic of Pn 2D crystal alignment reflecting the contrast difference in the domain structure shown in (a); shorter side of each rectangle corresponds to *a*-axis while longer side corresponds to *b*-axis of the 2D unit cell in a particular domain

different contrasts when the incident electron beam is slightly tilted from the surface normal. Since the paired domains (with their respective lattice orientations shown schematically in Fig. 20.13d) have just a small relative rotation of $4.6 \pm 0.3°$ between directions of their *b*-axes, the molecular arrangements within them are similar and result in almost identical image intensity. Pn islands with another type of epitaxial relation have been also observed in our experiments, with *b*-axes rotated by an angle of $3.6 \pm 0.3°$ to each of three equivalent <11$\bar{2}$> directions of the Si substrate surface, albeit less frequently. Moreover, the individual Pn islands have exclusively one type of epitaxial orientations (i.e., $b||$<1$\bar{1}$0> or $b||$<11$\bar{2}$>): the mixture to epitaxial structures was never observed [50].

For this polycrystallization to be realized in spite of the energetic disadvantage of forming lengthy domain boundaries, an overriding driving force must exert itself. Popescu and coworkers have proposed recently a theoretical model for the anisotropic (in the sense of the existence of privileged radial directions for growth) diffusion-limited aggregation in two dimensions [51]. The roughly sixfold shapes of Pn islands observed in our experiments (Fig. 20.10a) have remarkable similarities to shapes predicted by this model, if we assume that there are three preferred directions for the molecule attachment in Pn/H-Si(111) system. We can also clearly see that the fastest growth direction of every individual domain constituting the island aligns with the direction of the b-axis of its in-plane unit cell, which happens also to be an approximate direction of the point-on-line epitaxial relation between Pn and substrate lattices. Although this epitaxial relation appears to be an important factor in determining the preferential growth directions, we cannot exclude other possibilities such as intrinsic anisotropy (in bonding or kink formation energies) in the Pn crystal lattice, as demonstrated in the previous paragraph that under kinetic growth conditions, the crystal structure anisotropy results in the preferential growth along the b-axis of the Pn in-plane unit cell.

The existence of preferential growth directions is important for the elucidation of the branching mechanism of Pn islands on H-Si(111). The observed complicated, dendritic shape of the Pn islands (Fig. 20.14a) implies that their growth is mainly controlled by Pn surface diffusion. According to the Berg effect [52], the supersaturation on the surface is highest at the protrusions along the island perimeter. Such a configuration of the diffusion field allows the tips of the grain to grow faster, leading to the formation of fractal islands,

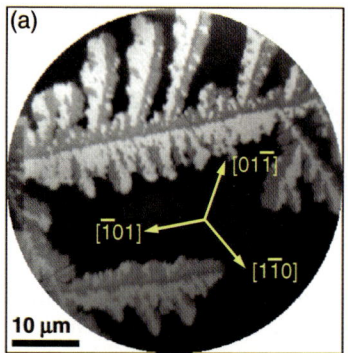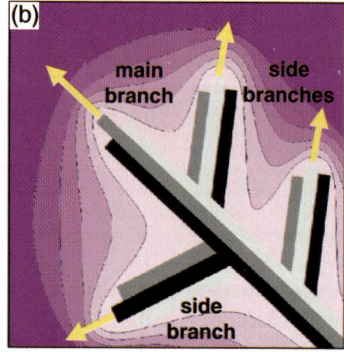

Fig. 20.14. (**a**) Tilted-beam LEEM image showing the internal structure of the Pn island grown on the H-Si(111) surface; the arrows outline the major crystallographic directions of the substrate, which are also a direction of development of the branches; (**b**) Sketch illustrating the branching mechanism observed in (**a**); the equipotential lines of the density gradient of diffusing molecules are shown schematically (yellow arrows indicate the fast growth directions of the branches, which align to the density gradient)

with the distribution of the branches governed by the surface diffusion coefficients [42] and branching directions generally following the density gradients of the diffusing molecules. However, in the case of anisotropic growth favoring certain crystallographic directions, the growth rate can also be affected by the crystal lattice orientation.

From the LEEM images we have found that the growth and evolution of the Pn islands into multiple domains are governed by a well-defined set of rules: (1) Branch growth begins with the development of the *main* domain (which becomes the center domain), whose *b*-axis aligns to the direction selected from one of the <1$\bar{1}$0> directions of the substrate surface; (2) *Outer* domains grow at each side of the main domain with their *b*-axis rotated by $\pm 60°$, thus each branch consists of three different domains displaying different contrasts I, II, and III in the LEEM images; (3) New (*side*) branches grow from the side walls of the outer domains replicating procedures (1) and (2), as shown schematically in the Fig. 20.14b. Even if domains (branches) with different orientations are created in random secondary nucleation events, as suggested by the rather disordered domain boundaries observed in our experiments, such domains (branches) are promptly suppressed by the favored domain (with proper *b*-axis alignment), developing a well defined, polycrystalline structure. Therefore, under kinetics-dominated conditions with a significant dendritic growth mode, the formation of side domains whose *b*-axis aligns to the density gradient and develops into a new branch is preferred over the continuous growth of the center domain.

This experimentally observed branching mechanism also sheds light on the origin of dislocations in Pn islands on H-Si(111). In previous report [53], the presence of deep holes, which are often located close to the center/top of the multilayered Pn grains, was attributed to screw dislocations with an empty core. From our experimental results it is quite apparent that these defects should be rather attributed to the local disorder in the first Pn monolayer, originating from the presence of long domain boundaries between Pn rotational domains.

20.4 Conclusions

In conclusion, we have demonstrated that low-energy electron microscope is a powerful tool for studying the nucleation and growth of thin organic films, especially when it is combined with scanning tunneling microscope for molecular-scale surface investigation. Using LEEM we were able to follow in real time the development of the Pn films and correlate their growth mechanisms and morphology with the crystal structure. This enabled us to propose adsorption structures of Pn molecules on substrates with various types of chemical and electronic activity. We have also demonstrated that the growth kinetics is extremely important in case of pentacene growth, when it can result in an intrinsic, self-driven polycrystallization within a single grain, reflecting the dispersed directions of density gradient around the Pn island. The observed mechanism can be applied to various systems, in particular crystallization of organic films, where both kinetic growth conditions and anisotropic effect are present.

References

1. C.D. Dimitrakopoulos, J. Kymissis, S. Purushothaman, D.A. Neumayer, P.R. Duncombe, R.B. Laibowitz, Adv. Mater. **11**, 1372 (1999)
2. C.D. Dimitrakopoulos, A.R. Brown, A. Pomp, J. Appl. Phys. **80**, 2501 (1996)
3. R. Ruiz, B. Nickel, N. Koch, L.C. Feldman, R.F. Haglund Jr., A. Kahn, F. Family, G. Scoles, Phys. Rev. Lett. **91**, 136102 (2003)
4. W. Kalb, P. Lang, M. Mottaghi, H. Aubin, G. Horowitz, M. Wuttig, Synth. Met. 146, **279** (2004)
5. G. Yoshikawa, T. Miyadera, R. Onoki, K. Ueno, I. Nakai, S. Entani, S. Ikeda, D. Guo, M. Kiguchi, H. Kondoh, T. Ohta, K. Saiki, Surf. Sci. **600**, 2518 (2006)
6. F.J. Meyer zu Heringdorf, M.C. Reuter, R.M. Tromp, Nature **412**, 517 (2001)
7. L. Casalis, M.F. Danisman, B. Nickel, B. Bracco, T. Toccoli, S. Ianotta, G. Scoles, Phys. Rev. Lett. **90**, 206101 (2003)
8. J.Z. Wang, K.H. Wu, W.S. Yang, X.J. Wang, J.T. Sadowski, Y. Fujikawa, T. Sakurai, Surf. Sci. **579**, 80 (2005)
9. C. Baldacchini, C. Mariani, M.G. Betti, L. Gavioli, M. Fanetti, M. Sancrotti, Appl. Phys. Lett. **89**, 152119 (2006)
10. M. Satta, S. Iacobucci, R. Larciprete, Phys. Rev. B **75**, 155401 (2007)
11. C.D. Dimitrakopoulos, S. Purushothaman, J. Kymissis, A. Callegari, J.M. Shaw, Science **283**, 822 (1999)
12. O.D. Jurchescu, J. Baas, T.T.M. Palstra, Appl. Phys. Lett. **84**, 3061 (2004)
13. C.D. Dimitrakopoulos, D.J. Mascaro, IBM J. Res. Dev. **45**, 11 (2001)
14. S.H. Jin, K.D. Jung, H. Shin, B.G. Park, J.D. Lee, Synth. Met. **156**, 196 (2006)
15. M. Shtein, J. Mapel, J.B. Benziger, S.R. Forrest, Appl. Phys. Lett. **81**, 268 (2002)
16. I. Yagi, K. Tsukagoshi, Y. Aoyagi, Appl. Phys. Lett. **86**, 103502 (2005)
17. R.G. Della Valle, A. Brillante, E. Venuti, L. Farina, A. Girlando, M. Masino, Org. Electron. **5**, 1 (2004)
18. E. Bauer, Rep. Prog. Phys. **57**, 895 (1994)
19. R.M. Tromp, IBM J. Res. Develop. **44**, 503 (2000)
20. P. Sutter, M.G. Lagally, Phys. Rev. Lett. **81**, 3471 (1998)
21. J.B. Hannon, M. Copel, R. Stumpf, M.C. Reuter, R.M. Tromp, Phys. Rev. Lett. **92**, 216104 (2004)
22. R.M. Tromp, M.C. Reuter, Phys. Rev. Lett. **68**, 820 (1992)
23. E. van Vroonhoven, H.J.W. Zandvliet, B. Poelsema, Phys. Rev. Lett. **91**, 116102 (2003)
24. R. van Gastel, R. Plass, N.C. Bartelt, G.L. Kellogg, Phys. Rev. Lett. **91**, 055503 (2003)
25. R.M. Tromp, M.C. Reuter, Phys. Rev. Lett. **68**, 954 (1992)
26. S. Kodambaka, S.V. Khare, W. Święch, K. Ohmori, I. Petrov, J.E. Greene, Nature 429, **49** (2004)

27. J.B. Hannon, J. Sun, K. Pohl, G.L. Kellogg, Phys. Rev. Lett. **96**, 246103 (2006)
28. H. Hibino, Y. Watanabe, C.W. Hu, I.S.T. Tsong, Phys. Rev. B **72**, 245424 (2005)
29. W.L. Ling, N.C. Bartelt, K.F. McCarty, C.B. Carter, Phys. Rev. Lett. **95**, 166105 (2005)
30. T. Sakurai, T. Hashizume, I. Kamiya, Y. Hasegawa, N. Sano, H.W. Pickering, A. Sakai, Prog. Surf. Sci. **33**, 3 (1990)
31. R.B. Campbell, J.M. Robertson, Acta Crystallogr. **15**, 289 (1962)
32. J.E. Northrup, M.L. Tiago, S.G. Louie, Phys. Rev. B **66**, 121404 (2002)
33. G.E. Thayer, J.T. Sadowski, F. Mayer zu Heringdorf, T. Sakurai, R.M. Tromp, Phys. Rev. Lett. **95**, 256106 (2005)
34. R.G. Della Valle, E. Venuti, A. Brillante, A. Girlando, J. Chem. Phys. **118**, 807 (2003)
35. C.C. Mattheus, G.A. de Wijs, R.A. de Groot, T.T.M. Palstra, J. Am. Chem. Soc. **125**, 6323 (2003)
36. F. Jona, Surf. Sci. **8**, 57 (1967)
37. P. Hofmann, The Structure of Bismuth: Structural and Electronic Properties, University of Aarhus, Denmark, November 2005
38. T. Nagao, J.T. Sadowski, M. Saito, S. Yaginuma, Y. Fujikawa, T. Kogure, T. Ohno, Y. Hasegawa, S. Hasegawa, T. Sakurai, Phys. Rev. Lett. **93**, 105501 (2004)
39. S. Yaginuma, T. Nagao, J.T. Sadowski, A. Pucci, Y. Fujikawa, T. Sakurai, Surf. Sci. **547**, L877 (2003)
40. J.T. Sadowski, T. Nagao, S. Yaginuma, Y. Fujikawa, A. Al-Mahboob, K. Nakajima, G.E. Thayer, R.M. Tromp, T. Sakurai, Appl. Phys. Lett. **86**, 073109 (2005)
41. A. Al-Mahboob, J.T. Sadowski, T. Nishihara, Y. Fujikawa, Q.K. Xue, K. Nakajima, T. Sakurai, Surf. Sci. **601**, 1304 (2007)
42. T.A. Witten Jr., L.M. Sander, Phys. Rev. Lett. **47**, 1400 (1981)
43. J.E. Northrup, M.L. Tiago, S.G. Louie, Phys. Rev. B **66**, 121404 (2002)
44. L.F. Drummy, P.K. Miska, D. Alberts, N. Lee, D.C. Martin, J. Phys. Chem. B **110**, 6066 (2006)
45. A. Al-Mahboob, J.T. Sadowski, Y. Fujikawa, K. Nakajima, T. Sakurai, Phys. Rev. B **77**, 035426 (2008)
46. R. Shioda, A. Kawazu, A.A. Baski, C.F. Quate, J. Nogami, Phys. Rev. B **48**, 4895 (1993)
47. S. Suto, R. Czajka, S. Szuba, A. Shiwa, S. Winiarz, H. Nagashima, H. Kato, T. Yamada, A. Kasuya, Acta Phys. Pol. A **104**, 289 (2003)
48. T. Takahagi, S. Shingubara, H. Sakaue, Solid State Phenomena **76–77**, 105 (2001)
49. J.T. Sadowski, G. Sazaki, S. Nishikata, A. Al-Mahboob, Y. Fujikawa, K. Nakajima, R.M. Tromp, T. Sakurai, Phys. Rev. Lett. **98**, 046104 (2007)
50. S. Nishikata, G. Sazaki, J.T. Sadowski, A. Al-Mahboob, T. Nishihara, Y. Fujikawa, S. Suto, T. Sakurai, K. Nakajima, Phys. Rev. B **76**, 165424 (2007)
51. M.N. Popescu, H.G.E. Hentschel, F. Family, Phys. Rev. E. **69**, 061403 (2004)
52. W.F. Berg, Proc. R. Soc. A **164**, 79 (1938)
53. B. Nickel, R. Barabash, R. Ruiz, N. Koch, A. Kahn, L.C. Feldman, R.F. Haglund, G. Scoles, Phys. Rev. B **70**, 125401 (2004)

Mechanism of Chiral Growth
of 6,13-Pentacenequinone Films
on Si(111)

A. Al-Mahboob, J.T. Sadowski, T. Nishihara, Y. Fujikawa, and T. Sakurai

Summary: Thin film growth of 6,13-pentacenequinone, $(C_{24}H_{12}O_2, PnQ)$, on Si(111)-7 × 7 at room temperature (RT) was studied by low-energy electron microscopy (LEEM) and ab initio density functional theory (DFT) calculations. Our experiments yielded direct microscopic observation of enantiomorphic evolution mechanism in the initial stage of the chiral-like growth of PnQ islands under kinetic growth conditions. We observed that the faster growth direction aligns with the direction of easier molecule incorporation, or lowest kink formation energy, rather than along the lowest energy step. Real time observation of the growth and subsequent relaxation of island shape revealed that kinetically stiff direction differs from the thermodynamic one. This feature together with anisotropic mass incorporation determines the enantiomorphic evolution and rotational arrangement of crystallites during the growth of elongated organic molecules, like PnQ.

21.1 Introduction

Recent progress in the application of organic molecular materials to the microelectronic devices signifies big advantages of the organic electronics, such as flexibility, low-cost fabrication, and biodegradability. Among currently investigated organic materials, pentacene $(C_{22}H_{14}, Pn)$ has been studied widely [1, 2]. Pn oxidizes in the presence of oxygen, forming 6,13-pentacenequinone $(C_{24}H_{12}O_2, PnQ)$, and pentacene-based field-effect transistors (FETs) show increased saturation current in oxygen atmosphere [3]. The decrease in mobility and on/off ratio induced by long exposure to air has also been reported [4]. Recently, the p-type electronic transport and yellow luminescence in PnQ thin film has been demonstrated [5]. Here we report thin film growth of PnQ studied by LEEM.

Organic materials are usually highly anisotropic, either in their crystal structures, or in bonding energies, or both. Such anisotropy may lower the crystal symmetry of the film in comparison to that of bulk [6] and result in frequently observed crystallization of achiral molecules in the chiral space groups [7–9]. Moreover, a molecular tilt, often present in organic films of elongated molecules such as Pn and PnQ, has significant impact on crystal morphology. There has been observed in case of Pn that the molecular tilt results in lack of mirror symmetry at the surface [10], which consequently may induce a surface chirality in 2D space [6]. In this study we grew self-assembled, chiral thin films of 6,13-pentacenequinone on Si(111)-7×7 surface through the kinetic processes; we point out that here both template and components are nonchiral. The film growth was studied by real-time microscopy and ab initio density functional theory (DFT) calculations.

21.2 Experimental and Calculation Procedures

6,13-Pentacenequinone (PnQ) has been thermally evaporated from a tantalum crucible in situ in the low-energy electron microscope (LEEM) on a clean Si(111)-7 × 7 kept at room temperature (RT). The deposition rate was in the range of 0.01–0.3 ML min^{-1}, where 1 ML corresponds to the molecular density of the PnQ(001) plane [11]. Growth of PnQ was observed in real time in LEEM, with imaging electron energy as low as 0.2–0.5 eV in order to avoid desorption of the PnQ caused by the electron irradiation. Micro-beam low energy electron diffraction (μ-LEED) patterns were acquired in the LEEM experiments [12] from various positions on the PnQ grains, in order to determine local crystallographic orientations and lattice parameters of surface unit cell. To support our experimental results we performed ab initio DFT calculations of total electronic energies in generalized gradient approximation (GGA) functional (p91) [13] for geometry optimization and molecular packing. We employed the quantum mechanical material modeling software DFT Electronic Structure Program "Materials Studio DMol^3 version 3.1" [14]. The total electronic energy calculations were done in the local density approximation (LDA) functional (pwc) [13]. To minimize random errors, we either considered the same k-spacing in calculations for ensuring the same integration grid, or we used the same supercell parameters if possible.

21.3 Results and Discussion

21.3.1 Growth and Thin Film Phase of PnQ on Si(111)

6,13-Pentacenequinone is an oxidation product of pentacene. It crystallizes in monoclinic structure, having two molecules per unit cell. The molecules in the alternate (002) planes of bulk crystal have a herringbone arrangement [11], while these are equivalent within the same layer. Interlayer spacing (d_{002}) of bulk is 8.93 Å [11], while that for recently observed thin film phase is ~12.7 Å [5].

Upon the deposition of PnQ on the clean Si(111)-7 × 7, initially the dangling bonds on Si surface were saturated, leading to the formation of a disordered wetting layer. Its formation was evidenced by a gradual disappearance

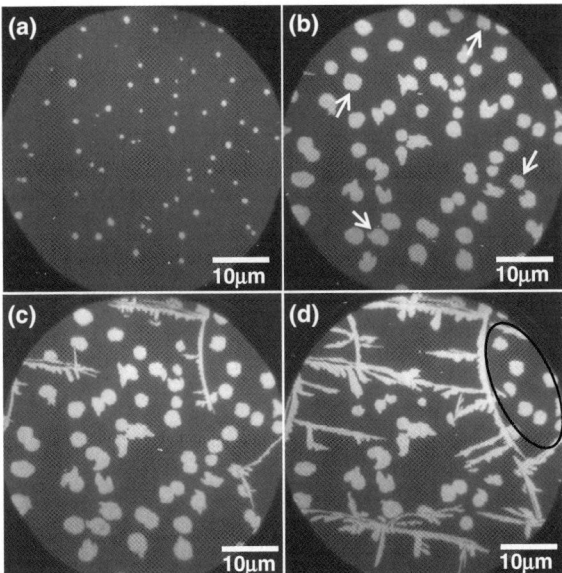

Fig. 21.1. Successive LEEM images of PnQ growth on Si(111)-7 × 7 surface at RT: (**a**) Nucleation of compact islands (type-I, *bright contrast*) on top of disordered wetting layer (*dark*); (**b**) Onset of the nucleation of some of crystalline islands (type-II) on compact islands, nucleation sites are indicated by arrows; (**c**) Evolution of type-II islands by absorbing material from the compact islands; (**d**) Mass from type-I island has diffused into a fractal side of a bigger islands (type-II), while at its concave side the compact islands (marked by a black ellipse) remain for long time until they dissolve after being approached by needle-like branches

of the LEED spots that were originating from a clean Si surface. Subsequently, two stages of the islands' nucleation were observed in the LEEM experiment. In the first stage, the isotropic islands (hereafter compact islands, or type-I) were formed, as shown in Fig. 21.1a,b. No apparent crystal structure of these islands could be detected by LEED, suggesting that they were amorphous. The density of the type-I islands saturated soon after the onset of their nucleation and the linear increase in the average island area has been observed. Practically all the molecules delivered to the surface were being attached along boundaries of the compact islands, indicating presence of large terrace diffusion. The saturation density of type-I islands and their critical size were dependent on the deposition rate.

After the compact islands grew big enough – typically a few hundreds to several thousands nanometers in diameters – the crystalline islands (hereafter type-II) nucleated (Fig. 21.1c) at the type-I islands by a phase transition or by stabilizing a critical type-II island. The nucleation of type-II islands on a bare substrate was never observed. The LEED data show that the type-II islands are crystalline. They are randomly oriented on the substrate, but all are having

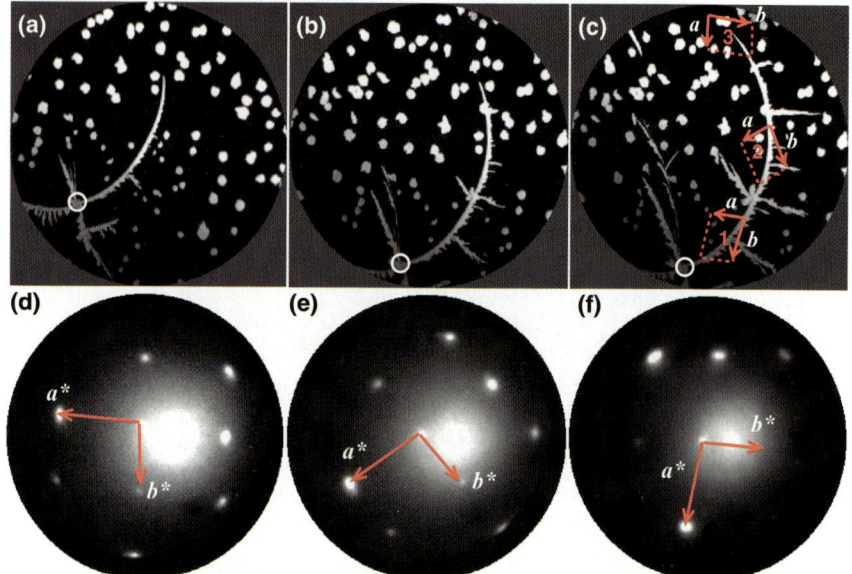

Fig. 21.2. Set of LEEM images and LEED patterns showing evolution of ring-like island: (**a–c**) Successive LEEM images taken at beam energy of 0.4 eV and separated by ∼50 s; White circle in the images is shown as a reference to the same area of the substrate; (**d–f**) μ-LEED patterns at beam energy of 15.0 eV, taken from the area marked by representative unit cell orientations 1, 2, and 3, respectively, in the image (c); The gradual rotation of LEED patterns from (**d–f**) is associated with local orientation of crystallites in the curved island

the same in-plane lattice parameters. The type-II islands grow in feather-like shapes, with some of them showing either a left-handed or right-handed, gradual change in the direction of the tips of their main branches (Figs. 21.2 and 21.3a). One side of the particular type-II island appeared to be smooth, while the opposite edge usually exhibited complicated, wavy morphology often accompanied by pronounced branching (Fig. 21.1d). LEEM data taken with different optimal focusing conditions at opposite edges of the elongated islands suggest that the island height gradually increased from the wavy edge towards the straight edge. We also observed anisotropic mass transport from compact islands to needle-like islands. The mass transport was always faster if any wavy side or growth tip of type-II island was present even within few micrometer of a compact island (Fig. 21.1d). At a flux of 0.05 ML min^{-1} and surface kept at RT, the mass transport from ∼2.2 μm distant islands was clearly visible.

All LEED patterns indicated that surface of the type-II islands terminated with the (002) crystal plane, which means that PnQ molecules in the type-II island are aligned in a tilted, standing-up fashion. Therefore, the in-plane lattice of the thin film corresponds to the *ab* crystal plane, while *c*-axis is directed

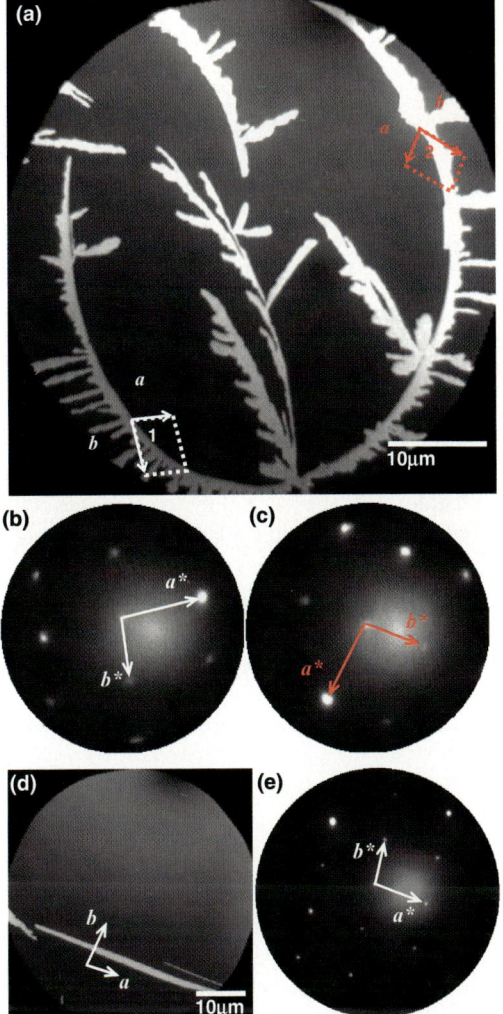

Fig. 21.3. (a) A LEEM image of a type-II island showing chiral branch and its enantiomorph; local in-plane unit cell orientations are marked, the straight (smooth) edge of concave side is locally parallel to the long diagonal of the unit cell; (**b** and **c**) μ-LEED patterns (at 15.0 eV) from the positions marked 1 and 2 in (**a**) respectively; (**d**) Relaxed type-II grain after 20 h from the end of the deposition – the elongated edge is perpendicular to b-axis; (**e**) μ-LEED pattern (at 17.0 eV) taken at the grain shown in (**d**)

either towards the substrate or out of the substrate. In the investigated thin film, the in-plane lattice parameters were $|a| = 5.3 \pm 0.1$ Å, $|b| = 7.0 \pm 0.1$ Å, and $\gamma = 82 \pm 0.5°$, while in the bulk PnQ, the corresponding values are 4.935 Å, 8.18 Å, and 90°, respectively [11]. In relation to bulk, the thin film is

Table 21.1. Lattice parameters of PnQ surface lattice

	Unit cell parameters	
	Thin film PnQ	Bulk PnQ [11]
$a(\text{Å})$	5.3 ± 0.1	4.935
$b(\text{Å})$	7.0 ± 0.1	8.18
$\gamma(°)$	82 or 98 ± 0.5	90
Ratio $|b|/|a|$	1.3 ± 0.02	1.66
Compression of ab-plane vs. bulk	9%	–
	Layer spacing normal to ab-plane	
d_{002} (Å)	12.7 [5]	8.93

compressed in the direction of b-axis (14.4% compression) and expanded in the direction of a-axis (7.4% expansion), resulting in overall compression of ab-plane of 9% (see Table 21.1). Since in our experiments only the single phase has been observed, we expect that this is the thin film phase, characterized by the layer spacing $d_{002} = 12.7$ Å [5]. Even though the PnQ film in [5] was grown on Si(001)-2 × 1 surface, PnQ grows on both Si(001)-2 × 1 and Si(111)-7 × 7 substrates through the amorphous wetting layer formation, which makes the subsequent island growth mechanism on both substrates similar.

After stopping the deposition, the type-II islands slowly relaxed without change in their in-plane lattice parameters, having long edges parallel to the a-axis (Fig. 21.3d).

21.3.2 Growth Anisotropy

There are two distinct variations of type-II islands and growth of both of them is highly anisotropic. One grows preferentially along the long diagonal of the surface unit cell (the $(-a-b)$ direction in terms of the in-plane primitive lattice vectors), having pronounced ring-like shape with either clockwise or anticlockwise rotation of the growth tip (Fig. 21.3a). The other one, with needle-like shape (Fig. 21.1d), grows preferentially in the direction of short diagonal of in-plane unit cell (the $(-a + b)$ direction). It also exhibits preference in the branching direction, which is related to the c-axis orientation. In the case of ring-like island, one of the side edges is very smooth and parallel to the preferential growth direction, which suggests that this particular edge is very stiff under kinetic growth conditions. Being locally aligned with the long diagonal of the 2D unit cell (the $(-a - b)$ direction in Fig. 21.3a), the smooth edge direction is different than the thermodynamically stiff directions (along a), and it is also different from the direction of long edge (parallel to a-axis) of relaxed chiral island (Fig. 21.3d).

To understand the mechanism of the faster growth along the directions of in-plane unit cell diagonals, we performed the DFT calculations of the

Table 21.2. Molecule incorporation, kink and crystal bonding energies in PnQ bilayer structure

Layer	Step parallel to	Incorporation and bonding direction	Incorporation energy (meV)	Kink energy (meV)	Bonding energy (meV)
L1	b	$-a$	−842.44	118.12	−768.85
	b	a	−842.79	117.77	
	a	b	−246.54	764.32	−146.22
	a	$-b$	−246.48	764.38	
	$(a+b)$	$(-a+b)$	−759.48	217.97	−696.01
	$(a+b)$	$(a-b)$	−759.42	218.02	
	$(a-b)$	$(a+b)^{\mathrm{a}}$	−954.37	21.43	−842.36
	$(a-b)$	$(-a-b)^{\mathrm{a}}$	−957.72	18.08	
L2	b	$-a^{\mathrm{b}}$	−950.77	9.79	−920.99
	b	a^{b}	−951.56	9.00	
	a	b	−292.14	718.72	−228.42
	a	$-b$	−291.48	719.38	
	$(a+b)$	$(-a+b)^{\mathrm{c}}$	−978.87	−3.07	−924.75
	$(a+b)$	$(a-b)^{\mathrm{c}}$	−976.14	−0.34	
	$(a-b)$	$(a+b)$	−663.23	314.22	−626.59
	$(a-b)$	$(-a-b)$	−664.06	313.39	

[a] Favorable asymmetric incorporation direction in the layer L1
[b] Second-most favorable asymmetric incorporation direction in the layer L2
[c] Most favorable asymmetric incorporation direction in L2; Note that the calculated energies may contain an offset error, the values are given only for the indication of favorable incorporation and asymmetry directions

molecule incorporation, kink, and crystal bonding energies in PnQ bilayer structure (Table 21.2). For the structure optimization, our experimental values of surface unit cell parameters of PnQ (Table 21.1) and value of the inter-plane spacing for the thin film phase ($d_{002} = 12.7$ Å) from [5] were used as constraints. By the analogy with the bulk PnQ, we also considered that the PnQ thin film has a bilayer structure, with unit cell incorporating two molecular layers: L1 and L2, respectively. For the calculations we assumed that the layer L1 is the bottom layer and L2 is the alternate layer; the molecules in the first layer, L1, have an arbitrary direction of the tilt with short molecular axis (SMA) closely aligned along in-plane diagonal direction (in bulk-like molecular arrangement), for instance along the long diagonal. The molecules in the second (alternate) layer, L2, are in herringbone arrangement with respect to molecules in L1, but all the molecules have the same orientation within the same layer, similar as in the bulk. Various models were then constructed with the same molecular orientation as it was found by the geometry optimization of the unit cell. Possibility of PnQ film formation by the stacking of just one type of layers (L1–L1 or L2–L2) has been excluded, since DFT

Table 21.3. DFT-calculated surface energies[a] of PnQ thin film polymorph

Surface	Surface energy per unit cell (eV)	Surface energy per molecule (meV)	Surface energy per area (meV $\mathring{A}^{-2}$)	Step energy[b] (meV $\mathring{A}^{-1}$)
(100)	0.90567	452.8	5.1	129.4
(010)	0.34446	172.2	2.5	65.0[c]
(001)	0.08576	85.8	2.3	–
(110)	0.945175	472.6	4.5	115.7
(−110)	1.15352	576.8	4.8	123.4

[a]Electronic energy at $0\,K$ (vibrational energy and correction for finite temperature was not calculated)

[b](001) plane is a surface plane; Energies are calculated for bilayer-high steps

[c]Step parallel to the *a*-axis

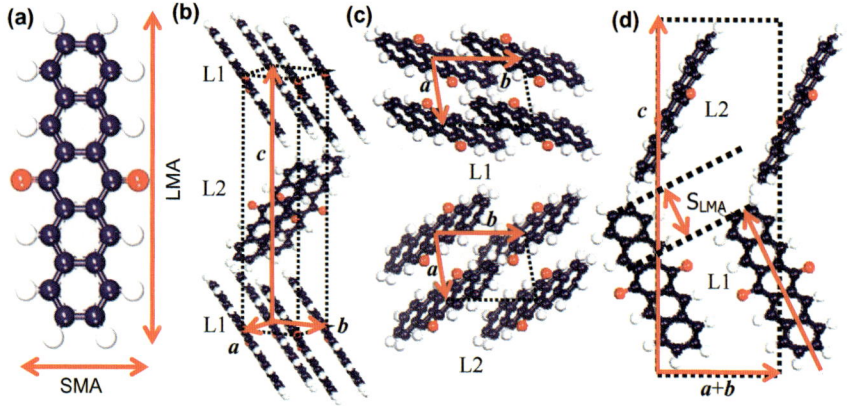

Fig. 21.4. DFT optimized molecular packing of thin film structure of 6,13-pentacenequinone ($C_{22}H_{12}O_2$): (**a**) An isolated PnQ molecule, the directions of long molecular axis (LMA) and short molecular axis (SMA) are shown; (**b**) Outline of thin film unit cell; (**c**) Top view of molecular arrangement in L1 and L2 layers viewed along (002) planes; (**d**) Side view of molecular arrangement in thin film, the shift in LMA (S_{LMA}), between two adjacent molecules arranged along the long diagonal of surface unit cell is shown

total energy calculations show energetically unstable nature of such stacking sequence. Additionally, the DFT-calculated surface energies for PnQ thin film polymorph (summarized in Table 21.3) show that PnQ(001) plane has the lowest surface energy, and thus the PnQ thin film is expected to grow with the (001) plane parallel to the surface, which we indeed see in our experiment.

Anticipated molecular packing in a PnQ unit cell, with the geometry optimized by ab initio DFT calculations is shown in Fig. 21.4. The anisotropy in bonding and molecular tilt in the bottom layer (L1) and alternate layer (L2) are found to be different. In L1 the molecules interact sideways along the long diagonal, with σ-bonds of neighboring molecules facing each other, while in L2 the molecules interact also sideways, but along the short diagonal of the in-plane unit cell. As PnQ molecules are arranged in a tilted, standing-up geometry in the surface unit cell, two adjacent molecules in a molecular row are shifted with respect to each other along long molecular axis (LMA), giving a relative shift (S_{LMA}) of two adjacent molecules along in-plane diagonal by an amount close to twice the length of a benzene ring (see Fig. 21.4d). It is to be noted that in case of Pn molecules, they interact more strongly when they are somewhat shifted sideways, rather than in $\pi - \pi$ facing configuration [15]. By analogy, in case of PnQ the strongest interaction is thus expected along the long diagonal of L1 and along the short diagonal of L2, due to the sideway ($\sigma - \sigma$ facing) configuration.

Our experimental data show that the faster growth of type-II islands occurs along the diagonal directions of the in-plane unit cell, which is consistent with the DFT calculations, showing that this is the direction of the strong interaction between adjacent molecules in the first (bottom) layer of an island. The observed preferential growth of ring-like island is consistent with the lowest kink formation and molecule incorporation energies found from DFT calculations (Table 21.2), if the sequence in bilayer is L1–L2 (L1 nucleates as a first, bottom layer). Since the inclusion of the molecules takes place initially in the bottom layer of the L1–L2 bilayer structure, we believe that the growth anisotropy (preferential growth and kinetically stiff direction) is induced in the L1. The energy of a kink formed at the sites along the long diagonal of in-plane unit cell in L1 is the lowest, while along other directions the kink formation energy is significantly higher (in comparison with RT thermal energy), which results in forming a growth tip along long diagonal of in-plane unit cell.

For the domain of inversed stacking sequence, L2–L1, the preferential growth will be along the short diagonal of in-plane unit cell. The L2 has two comparable faster asymmetric incorporation directions: along in-plane short diagonal direction and along the direction of the a-axis, respectively. Along other directions the incorporation is negligible. The edge parallel to the growth direction in this case is less stiff than that for the domain with L1–L2 stacking. In principle, both sequences, L1–L2 and L2–L1, can be detected in our experiment by mapping unit cell orientation from µ-LEED patterns and observation of preferential growth direction in real space-time, and we have also observed both types of preferential growth orientations, as it is shown in Fig. 21.5. Here, crystallites with c-axis pointing downwards (marked by crossed circles in Fig. 21.5) and c-axis pointing upwards (dotted circles) can be observed, and both stacking sequences L1–L2 (dotted arrows) and L2–L1 (solid arrows) are distinguished. It is also to be noted that the side branches are elongated always along the short diagonal of L1–L2 grain, whereas DFT-calculated energies show that the fastest incorporation in L2 is also along this branching direction. The sideway branching, associated with the layer stacking reversal, may be thus associated with a stacking fault at a bottom layer induced by secondary nucleation at defect sites like surface steps.

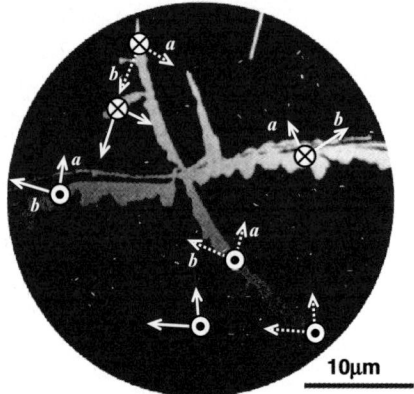

Fig. 21.5. LEEM image of the PnQ film on Si(111) showing a chiral evolution of islands growing preferentially along the long diagonal (L1–L2 stacking, *dotted arrows*) and short diagonal (L2–L1 stacking, *solid arrows*) axes of the in-plane unit cell, respectively; PnQ crystallites with *c*-axis pointing downwards (*crossed circles*) and *c*-axis pointing upwards (*dotted circles*) are observed

21.3.3 Chiral Evolution of the Islands

Considering only the kink energy anisotropy in the bottom layer, we cannot fully explain the chiral shape evolution of the ring-like islands. To fully grasp this problem, we should consider the presence of intermediate states that can enhance anisotropy in molecular incorporation; however, calculations that would reveal such states are currently beyond our capabilities. The μ-LEED patterns show gradual rotation of in-plane unit cells (two-dimensional chirality) following the shape evolution of ring-like structure (as shown in Fig. 21.2). A clockwise rotation of the island tip occurs if *c*-axis is pointed downwards (a PnQ island has (001) surface plane) and anti-clockwise rotation is realized when *c*-axis is directed upwards (PnQ island has a (001) surface plane), as it is shown in Fig. 21.3b. Although the PnQ crystallizes in an achiral space group ($P2_1c^{-1}$), the (001) and ($00\bar{1}$) faces are enantiomorphs because they lack a centre of symmetry [11]. Reversing of the direction of the lattice vector ***c*** reverses the molecule tilt with respect to a vertical plane (out-of-plane mirror), resulting in mirroring of asymmetry and thus favorable mass incorporation at one side of the island. The specific direction of the preferential growth is resulting mainly from the molecular assembly enhancing favorable mass incorporation in the bottom layer.

Gradual rotation of specific growth tip of this ring-like island is related to a successive rotation of small crystallites, which should be associated with gradual inclusion (or exclusion) of extra partial arrays of molecules, as we indeed see the gradual rotation of reciprocal lattice in μ-LEED patterns (Figs. 21.2d–f). These inclusions in the ring-like PnQ island take place only at one side of the chiral structure (convex side of a ring). This also results in

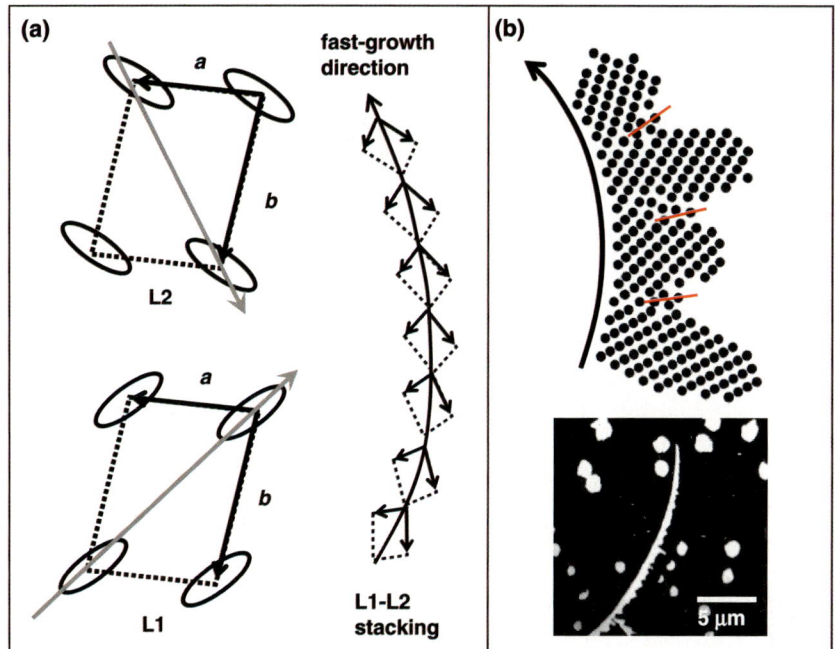

Fig. 21.6. Schematic presentation of a kinetic growth model of ring-like, chiral PnQ island (for simplicity only one – counter-clockwise rotation of growth tip is shown); *c*-axis is pointing outward, long axis of ellipses represents the cross-sectional orientation of PnQ molecules: (**a**) Relation between the *ab* unit cell orientations and the preferential growth directions in the L1 (*bottom layer*) and L2 (*alternate layer*); the gradual rotation of crystallites in L1–L2 stacking is represented by outlining the orientation of cell vectors: curved arrow shows the motion of growth tip consistent with observation, grey arrows show the fast growth directions in L1 and L2, respectively; (**b**) Illustration of the gradual rotation of growth tip (*arrow*) due to assymmetric molecule incorporation at the convex side that produces additional partial rows of molecules at this side; thin lines schematically mark the inclusions of extra molecular rows. Inset: LEEM image showing corresponding ring shape of the PnQ island

wavy shape of the edge or occasional branching on convex side associated with secondary nucleation at the defect sites. Schematic presentation of the evolution of ring-like chiral island (having L1–L2 stacking) under kinetic growth is shown in Fig. 21.6. Here, *c*-axis is pointing up, and long axes of ellipses represent the cross-sectional orientation of PnQ molecules (see Fig. 21.4). The gradual rotation of crystallites in L1–L2 stacking is represented by outlining the orrientation of the 2D unit cell vectors. The assymmetric molecule incorporation at convex side produces additional partial rows of molecules at this side.

21.4 Conclusions

At the initial stage of PnQ growth on Si(111), noncrystalline islands or clusters (type-I) nucleate on top of a disordered wetting layer. Crystalline islands (type-II) nucleate subsequently by the phase transition or by stabilizing a nucleus of critical size on the type-I islands. PnQ thin film growth mechanism at RT is found to be kinetic rather than thermodynamic one. The observed anisotropy in the growth of type-II islands is related to (1) asymmetric (chiral) mass incorporation and (2) mass transport from noncrystalline islands via a large diffusion length. The left-handed or right-handed chirality (and preferential branching at only one side of both, needle-shaped and ring-shaped islands) can be related to the orientation of c-axis (or inversion of the right-handed unit cell, ab, to left handed unit cell, ba, for fixed orientation of c) and associated with the selective mass incorporation at the opposite side of island having a straight (smooth) edge. In the case of ring-shaped island, the direction of this smooth edge is parallel to the long diagonal of in-plane unit cell, which is the easiest direction of kink formation at bottom layer, and also the fastest growth direction. The ring shape of the island originates from the gradual rotation of crystallites associated with asymmetric (chiral) arrangement of dislocations. The inversion of c-axis reverses the tilt of molecule, resulting in mirroring of asymmetry and thus favorable mass incorporation at one side. After long-time relaxation the elongated islands are formed, with their straight edges perpendicular to the b-axis. The results of DFT calculations support the presence of anisotropy in molecule incorporation and kink formation, showing a stiff direction parallel to the growth direction under kinetic condition. From the observation of growth and subsequent relaxation, and from the DFT calculations, it is clear that kinetically stiff direction (here one diagonal of the in-plane unit cell) differs from the thermodynamic one. This feature together with the asymmetric mass incorporation associated with molecular tilt with respect to surface normal and bond anisotropy determines the chirality (enantiomorphic evolution) of organic growth.

Acknowledgements. Authors thank the Foundation Advanced Technology Institute (ATI) for the financial support; and Center for Computational Materials Science, IMR, Tohoku University for the assistance with the DFT calculations.

References

1. C.D. Dimitrakopoulos, S. Purushothaman, J. Kymissis, A. Callegari, J.M. Shaw, Science **283**, 822 (1999); A.R. Brown, A. Pomp, C.M. Hart, D.M. de Leeuw, Science **270**, 972 (1995)
2. S.F. Nelson, Y.Y. Lin, D.J. Gundlach, T.N. Jackson, Appl. Phys. Lett. **72**, 1854 (1998); C.D. Dimitrakopoulos, J. Kymissis, S. Purushothaman, D.A. Neumayer, P.R. Duncombe, R.B. Laibowitz, Adv. Mater. **11**, 1372 (1999)

3. R. Ye, M. Baba, K. Suzuki, Y. Ohishi, K. Mori, Thin Solid Films **464–465**, 437 (2004)
4. Y. Qiu, Y. Hu, G. Dong, L. Wang, J. Xie, Y. Ma, Appl. Phys. Lett. **83**, 1644 (2003)
5. D.K. Hwang, K. Kim, J.H. Kim, D. Jung, E. Kim, S. Im, Appl. Surf. Sci. **244**, 615 (2005)
6. S.M. Barlow, R. Raval, Surf. Sci. Rep. **50**, 201 (2003)
7. M. More, G. Odou, J. Lefebvre, Acta Crystallogr. B **43**, 398 (1987)
8. H. Aoyama, T. Hasegawa, Y. Omote, J. Am. Chem. Soc. **101**, 5343 (1979)
9. D. Rabinovich, H. Hope, Acta Crystallogr. A **36**, 670 (1980)
10. A. Al-Mahboob, J.T. Sadowski, T. Nishihara, Y. Fujikawa, Q.K. Xue, K. Nakajima, T. Sakurai, Surf. Sci. **601**, 1304 (2007)
11. C.C. Mattheus, Dissertation, University of Groningen, The Netherlands, 2002; After transformation (Mattheus) of data [A.V. Dzyabchenko, V.E. Zavodnik, V.K. Belsky, Acta Crystallogr. **35** 2250 (1979)] into conventional unit cell: $|a| = 4.951$ Å, $|b| = 8.170$ Å, and $|c| = 17.784$ Å, $\gamma = 93.26°$, respectively
12. E. Bauer, Rep. Prog. Phys. **57**, 895 (1994)
13. J.P. Perdew, Y. Wang, Phys. Rev. B **45**, 13244 (1992); J.P. Perdew, Y. Wang, Phys. Rev. B **33**, 8800 (1986); J.P. Perdew, J.A. Chevary, S.H. Vosko, K.A. Jackson, M.R. Pederson, D.J. Singh, C. Fiolhais, Phys. Rev. B **46**, 6671 (1992)
14. B. Delley, J. Chem. Phys. **92**, 508 (1990); B. Delley, J. Chem. Phys. **113**, 7756 (2000)
15. K. Lee, J. Yu, Surf. Sci. **589**, 8 (2005)

GaN Integration on Si via Symmetry-Converted Silicon-on-Insulator

Y. Fujikawa, Y. Yamada-Takamura, Z.T. Wang, G. Yoshikawa, and T. Sakurai

Summary: Integration of metals and semiconductors having three- or sixfold symmetry on device-oriented (i.e., (001)) silicon wafers, which have fourfold symmetry, has been a longstanding challenge. We demonstrate that, by using symmetry-converted (111) silicon-on-insulator, we can integrate wurtzite-structure gallium nitride, which has threefold symmetry, with Si(001). The stability of the symmetry-converted Si(111) layer makes this technique appealing to the commercial integration of wide-ranging important materials onto Si (001) base wafers.

22.1 Introduction

Silicon serves as the fundamental material for the semiconductor industry because of its superior processability in the fabrication of various device structures, and especially because of its ability to form a high-quality oxide. Si(001), with a square surface lattice, has been used for actual device fabrication primarily because of the lower interface state density. A major and important class of materials having three- or sixfold symmetries has intrinsic difficulty growing on Si(001), because the symmetry mismatch at the interface induces polycrystalline-film formation and roughness, factors that seriously degrade electronic and optoelectronic performance. For this reason, many important and interesting systems, such as Pb quantum wells, where the quantum size effect controls superconductivity [1], organic network structures [2], and GaN films [3–6] with promising applications for optoelectronics and power devices, have been studied using Si(111) substrates; thus at least avoiding the symmetry mismatch issue. However, the use of Si substrates with indices other than (001) in technology has been limited by the poor quality of oxide/Si interface mentioned above.

A general solution for this long-standing problem has been given by utilizing a silicon-on-insulator (SOI) structure where a thin Si(111) layer is bonded to Si(001) via the oxide layer [7]. We can thus use the Si(111) template layer in those regions where we need to integrate a three- or sixfold symmetric material, while using the Si(001) wafer in other regions to create CMOS electronic devices. This SOI structure provides a uniform Si(111)-7 × 7 clean surface using the surface treating method recently developed for the Si(001)-SOI surface [8]. The thin Si layer with the clean surface is stable up to 800°C, enabling the growth of a broad range of materials tested previously on Si(111)-7 × 7. A GaN film is directly grown on this SOI structure to demonstrate the validity of our approach. The ultrathin Si(111) layer, with a thickness of 14 nm, is stable against the irradiation of Ga and N-plasma fluxes at the growth temperature, which results in the formation of uniform N-polar GaN film on the SOI structure.

22.2 Interface Control of GaN Film Growth on Si(111)

GaN growth has been extensively studied due to the applications of GaN-based materials as short-wavelength light emitting diodes and laser diodes. Different substrates, such as c-plane sapphire [9], 6H-SiC(0001) [10], GaAs(111) [11], and Si(111) [3–6], have been used for wurtzite GaN growth due to the lack of large size single-crystal GaN substrates. Among these substrates, Si is considered the most attractive because of its mature electronic industry and availability of high quality, low cost, and large diameter wafers. The wurtzite structure of GaN has a freedom in its polarity, which in turn has direct influence on surface structure, growth process, and its quality [12, 13]. The polarity is largely influenced by the choice of substrate and buffer layers used for the initial film nucleation [10, 12]. In case of nonpolar substrate like Si, it is especially important to control the nucleation stage to accomplish mono-polar film with atomically smooth surface.

Polarity control of direct GaN growth on Si(111) by radio frequency plasma-assisted molecular beam epitaxy (PAMBE) has been reported [5]. The initial nucleation and growth stages of GaN are studied in detail, applying N-rich and Ga-rich growth conditions during the nucleation. Successive film growth is always carried out under Ga-rich growth conditions to obtain smooth films, making a good use of surfactant effect of Ga [13]. Figure 22.1 shows the scanning tunneling micrographs of the surface structures observed on the GaN films grown with the N-rich nucleation condition, which exhibit 3 × 3, 6 × 6, and c(6 × 12) reconstructions, typical of the surface structures for N-polar GaN films [14]. The step height observed in Fig. 22.1a is about 0.26 nm, half the lattice constant along c-axis of wurtzite GaN unit cell, corresponding to a bilayer step. These STM results confirm that the surface is covered uniformly with N-polar related reconstructions and the terrace size is in the range of 50–250 nm. Increasing the amount of Ga on this surface does not induce further reconstruction or fluid surface, and only the formation of clusters is observed in STM images.

On the other hand, the nucleation under Ga-rich conditions results in formation of a rougher surface after the completion of the film growth, as is seen in its STM image (Fig. 22.2a). Small size reconstructed domains, indicated by

Fig. 22.1. STM images of the GaN films grown on Si(111)-7×7 with N-rich nucleation conditions. (**a**) Surface in the size of 200 nm square is covered by c(6 × 12) reconstruction, (**b**) 3 × 3 in the size of 10 nm square, (**c**) 6 × 6, (**d, e**) c(6 × 12) reconstructions in the size of 15 nm square. Sample bias voltages are −0.45, −0.2, +0.70, −3.8, and +2.1 V, respectively. Tunneling currents are in the range of 60–200 pA. Reused with permission from Ref. 5. Copyright 2005, American Institute of Physics

the black arrow, are observed frequently in between the rough areas. Almost all of these small reconstructed domains are c(6 × 12) rather than 3 × 3, in spite of faint 3 × 3 patterns given by reflection high-energy electron diffraction (RHEED) observations. Occasionally, very flat areas without any atomic feature coexisting with c(6 × 12) reconstructed areas can be seen in the STM images (Fig. 22.2b). The flat, featureless area is likely due to Ga-fluid over Ga-polar crystal [15, 16]. Bright feature at the edge of the fluid island [17] is also seen in these flat areas in Fig. 22.2b. The observed boundary plane between c(6 × 12) and featureless region is {1$\bar{1}$00}, which is typical of inversion domain (ID) boundaries. All these facts suggest that this surface is mixed in polarity with N-polar regions (I and III) and Ga-polar regions (II, IV, and V). These results indicate that N-rich condition in the nucleation layer growth stage is critical for achieving the growth of mono-polar GaN films on Si(111) without inversion domain boundaries.

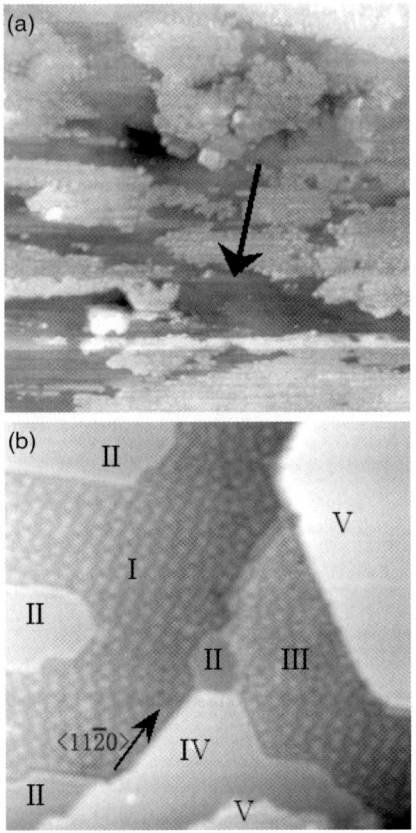

Fig. 22.2. STM images of the GaN films grown on Si(111)-7 × 7 with Ga-rich nucleation conditions. (**a**) The small size c(6 × 12) domain with rough feature, and (**b**) the fluid structure surrounded by c(6 × 12) reconstruction in the size of 30 nm square. Sample bias voltages are +1.5 and −4.0 V, respectively. Tunneling current is 100 pA. Observed area is classified into regions *I–V* according to their reconstructed features and terrace heights. Reused with permission from Ref. 5. Copyright 2005, American Institute of Physics

22.3 Integration of Wurtzite GaN on Si(001) via Symmetry-Converted SOI

A SOI wafer with a 100-nm-thick Si(111) template layer and 200-nm-thick oxide layer on a Si(001) handle wafer, fabricated by the Smart Cut^{TM} technique, was obtained from SOITEC. The Si(111) layer has a resistance of 13–22 Ω cm. This layer was oxidized at 1,100°C for 80 min. The oxidized surface layer was removed by HF to leave 14 nm of unoxidized Si(111), as measured using ellipsometry. This SOI (Fig. 22.3) was used for the symmetry-converted growth of GaN. The ex situ cleaning of the sample was achieved by treating it in HCl and H_2O_2 solutions (HCl (35% aq.):H_2O_2 (30% aq.):purified water = 4:5:5) at

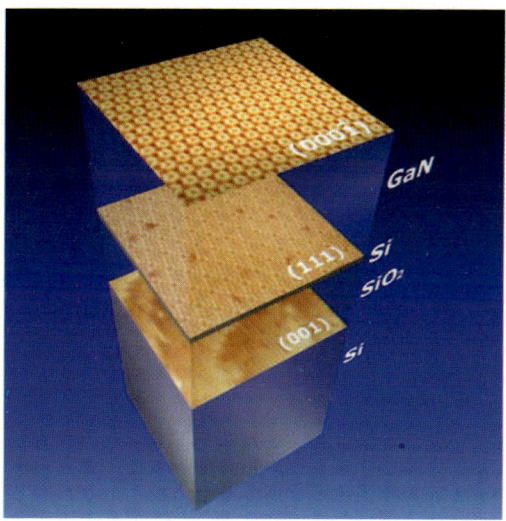

Fig. 22.3. Schematics of symmetry converted growth of GaN on Si(001) with superimposed STM images ($25 \times 25\,\text{nm}^2$, tunneling current: 30 pA) observed in this work. GaN is grown on a 7×7 clean surface formed on a Si(111) template layer (sample bias: $+1.3$ V) bonded with Si(001) handle wafer via SOI structure. An ordered 6×6 reconstructed surface appears as a consequence of excess Ga termination of grown film, showing uniform formation of a N-polar GaN film with $(000\bar{1})$ surface (sample bias: $+1.1$ V). Removal of the SOI structure by heating at $1,300°$C results in the formation of the Si(001) dimer-row structure (sample bias: $+1.2$ V), having square-base surface lattice

$80°$C for 10 min twice, removing surface oxide with 5% HF solution after the first treatment. The thickness decrease by this treatment is estimated as 1 nm or less, which is the typical thickness of the native-oxide layer. After a degas at $600°$C for 12 h, $1\,\text{ML min}^{-1}$ of Si flux from a resistively heated Si wafer was applied for 5 min to the specimen surface held at $750°$C to remove the surface oxide layer that had built up during the transfer and degas. Finally, the specimen was annealed at $800°$C for 2 min to obtain the Si(111)-7×7 reconstruction. GaN was grown applying the N-rich nucleation condition described in the previous section. The SOI substrate is nitrided by N plasma (rf power, 300 W; N_2 pressure, 2×10^{-3} Pa) at substrate temperature of $700°$C for 3 min to form a silicon nitride layer prior to the growth. After that, the N plasma condition is set at rf power of 300 W and N_2 pressure of 3×10^{-3} Pa for the growth. GaN is nucleated and grown at a fixed substrate temperature of $750°$C, and the Ga K-cell temperature is kept at $980°$C for the nucleation layer growth, to give a relatively lower beam equivalent pressure (BEP) of 1.3×10^{-4} Pa at the Ga flux monitor. These parameters assure a N-rich condition to allow uniform N-polar nucleation at the initial growth stage [5]. After 10–20 min of nucleation layer growth, the K-cell temperature is increased to

1,050°C (BEP, 3.5×10^{-4} Pa) to change the growth condition to a Ga-rich one, in order to grow a flat GaN film [5]. All growth processes are monitored in real time by RHEED. GaN films approximately 300 nm thick were grown.

The superimposed image on the Si(111) template layer in Fig. 22.3 shows an STM micrograph of the symmetry-converted SOI surface prepared by cleaning with in-situ Si deposition and successive annealing. A uniform surface with flat terraces is observed, accompanied by typical Si(111)-7 × 7 atomic structures. The Si(001)-2 × 1 dimer-row structure can be observed after flash heating this specimen at 1,350°C (lower image in Fig. 22.3). At this temperature, the oxide decomposes and the Si(111) template layer will disappear with it. The fact that we observe first the clean-Si(111) surface structure and later, after oxide decomposition, the Si(001) surface structure confirms that the surface Si(111) layer was bonded to Si(001). The stable tunneling current to the thin Si(111) layer assures that the Si(111) template layer, which is the only possible source of the conductivity on the insulating oxide layer, maintains connectivity to the end of the specimen even after the surface treatment. Resistance measurements of the point contact of the STM tip to the sample surface give stable values in the range of 10^7 Ω. Because point contacts to bulk-Si(111) surfaces always give resistances of 10^6 Ω or less, the higher resistance measured for the thin template layer of Si(111) on the surface of SOI(111)/Si(001) must be dominated by the sheet resistance of the Si(111) template layer with its clean Si(111) surface. The carrier density in the 14 nm Si(111) template layer is estimated to be 10^{15} cm^{-3} from the resistance of the original wafer [17], giving an areal density of 10^9 cm^{-2} in the Si(111) template layer. This density is much smaller than the adatom density on the Si(111)-7 × 7 surface (10^{14} cm^{-2}) or the oxide/Si(111) interface trap density ($10^{11} - 10^{12}$ cm^{-2}eV^{-1} for the Si(111)/SiO$_2$ interface, distributed uniformly across the 1.2 eV-wide bandgap [17]). Therefore, with this thickness of template layer, the carriers in the bulk band are totally depleted [8], because it is known that the Si(111)-7 × 7 surface pins the Fermi level almost exactly at the middle of the band gap [18]. Therefore, the charge carrier transport through the thin Si(111) membrane observed here, which enables the STM observation, cannot be via conventional bulk conduction but must be via the Si(111) surface state bands directly.

The resistance of the same SOI surface increases to 10^{10} Ω by depositing a fraction of a monolayer of Ga to form the Si(111)-$\sqrt{3} \times \sqrt{3}$-Ga reconstruction. This reconstruction eliminates the in-gap surface bands of the clean surface [19] and thus eliminates the surface conduction channel. This resistance change is completely reversible with the removal of Ga at 750°C, a result that allows us to exclude the possibilities of permanent changes in morphology and doping status. The observations above strongly support our conclusion on the mechanism of conduction in these films.

GaN is grown on this well-defined Si(111)-7 × 7 reconstructed structure. RHEED observations of the clean substrate (Fig. 22.4a) give a 7×7 streak pattern with no evidence of possible agglomeration of Si 3D islands in the Si(111)

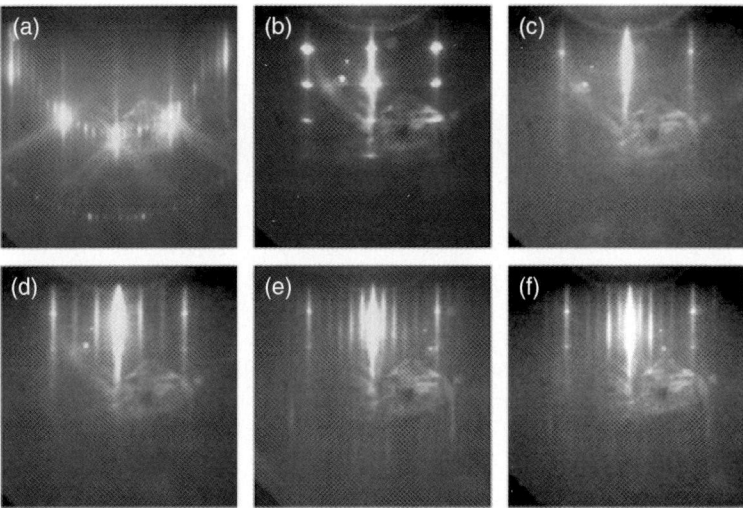

Fig. 22.4. (*Upper row*) Sequential change of RHEED patterns during the growth of GaN films on symmetry-converted SOI, observed with an electron beam parallel to Si<1$\bar{1}$0> showing (**a**) the 7 × 7 reconstruction of the SOI(111) surface, (**b**) nucleation of wurtzite GaN, and (**c**) after film growth. (*Lower row*) Sequential change of surface reconstruction with the deposition of additional Ga on the GaN film at room temperature observed with an electron beam parallel to GaN<$\bar{1}\bar{1}$20>, which aligns to Si<1$\bar{1}$0>. The patterns correspond to GaN(000$\bar{1}$)-(**d**) 3 × 3, (**e**) 6 × 6, and (**f**) c(6 × 12) reconstructions with higher intensity in 3× streaks, which are all known for the N-polar GaN surface. Reused with permission from Ref. 7. Copyright 2007, American Institute of Physics

template layer (spots in the RHEED pattern) before growth. Initial nitridation gives a diffuse RHEED pattern, indicating the existence of a disordered nitride layer on the surface [5]. Because the initial nucleation is performed under N-rich conditions to allow initial N-polar nucleation [5], the lack of a Ga surfactant effect [13] causes the growth to proceed in the three-dimensional mode. At this nucleation stage of GaN, we obtain a RHEED pattern (Fig. 22.4b) consisting of transmission diffraction spots, without any streaky component from possible regions of flat growth. From this transmission RHEED pattern, GaN is determined to be wurtzitic, having the following epitaxial relationship with the Si(111) layer: GaN <0001> // Si <111> and GaN <$\bar{1}\bar{1}$20> // Si<1$\bar{1}$0>, similar to earlier results on conventional bulk Si(111) [5]. After changing the growth condition to a Ga-rich one, the RHEED pattern recovers its streaky nature (Fig. 22.4c), indicating improvement to a smoother surface morphology under these conditions. The GaN in-plane lattice constant, derived from the RHEED analysis, is 0.319 ± 0.002 nm. This value, close to that of pure GaN, suggests that the lattice mismatch stress between the GaN film and Si(111) is mostly relaxed via the generation of dislocations, within experimental error.

Ga deposition on the GaN film formed on the SOI(111) structure results in clear and high-intensity 3×3, 6×6, and $c(6 \times 12)$ RHEED patterns (Figs. 22.4d–f) depending on the amount of Ga deposited on the 1×1 surface, which are typical of the N-polar film [15]. STM observations of these Ga covered surfaces demonstrate uniform imaging of atomically resolved features (example shown as the upper image in Fig. 22.3), which has been reported for N-polar reconstructions [15]. Anti-polar domains, typically covered by a Ga-fluid surface after Ga termination [5], were never found in our observations, confirming that N-polar GaN was selectively grown on the SOI(111) surface. Even at the end of the growth, the conductivity of the SOI structure was not damaged.

The growth of GaN on conventional SOI(001) [20] and conventional SOI(111) [21] (template and handle wafer both Si(111) and also a much larger template layer thickness) has been investigated with the goal of enhancing strain relaxation between the GaN and the Si. The improved relaxation of residual strain in a thick GaN layer has already been reported on a 200 nm-thick (111) SOI layer without symmetry conversion [21]. The stability of ultrathin SOI(111) layers down to at least 10 nm widens the application of this symmetry conversion technique by allowing the completion of strain relaxation in an earlier stage of growth.

22.4 Conclusions

The wurtzite GaN growth on symmetry converted SOI with a (001) handle and a (111) oriented template layer has potential technological importance by enabling integrated GaN devices intermingled with Si controlling circuits on a single chip. Direct growth of single-domain wurtzite GaN on vicinal Si(001) via an AlN seed layer [22] aims at the same goal. The growth of GaN on SOI reported here will be more suitable for integration of GaN with Si devices, because it is possible to avoid direct contact of other elements with Si(001) and there is no need for a large miscut angle of the Si(001) substrate. The Si(111) template layer is stable under a nitrogen flux for nitridation and a Ga flux for successive growths at temperatures between 700 and 800°C. This stability of a thin Si(111) layer on the SOI structure enables most of the heteroepitaxial growths reported on Si(111)-7×7 that can be performed at lower growth temperatures. The stability of the orientation converted SOI membrane on Si(001) observed here would enable general integration of the intriguing class of materials with three- or sixfold surface symmetries [1, 2], as well as functional high-index surfaces [23], with Si devices.

References

1. Y. Guo, Y.F. Zhang, X.Y. Bao, T.Z. Han, Z. Tang, L.X. Zhang, W.G. Zhu, E.G. Wang, Q. Niu, Z.Q. Qiu, J.F. Jia, Z.X. Zhao, Q.K. Xue, Science **306**, 1915 (2004)

2. J.A. Theobald, N.S. Oxtoby, M.A. Phillips, N.R. Champness, P.H. Beton, Nature **424**, 1029 (2003)
3. T. Takeuchi, H. Amano, K. Hiramatsu, N. Sawaki, I. Akasaki, J. Cryst. Growth **115**, 634 (1991)
4. T.D. Moustakas, T. Lei, R.J. Molnar, Phys. B **185**, 36 (1993)
5. Z.T. Wang, Y. Yamada-Takamura, Y. Fujikawa, T. Sakurai, Q.K. Xue, Appl. Phys. Lett. **87**, 032110 (2005)
6. Y. Yamada-Takamura, Z.T. Wang, Y. Fujikawa, T. Sakurai, Q.K. Xue, J. Tolle, P.L. Liu, A.V.G. Chizmeshya, J. Kouvetakis, I.S.T. Tsong, Phys. Rev. Lett. **95**, 266105 (2005)
7. Y. Fujikawa, Y. Yamada-Takamura, G. Yoshikawa, T. Ono, M. Esashi, P.P. Zhang, M.G. Lagally, T. Sakurai, Appl. Phys. Lett. **90**, 243107 (2007)
8. P.P. Zhang, E. Tevaarwerk, B.N. Park, D.E. Savage, G.K. Celler, I. Knezevic, P.G. Evans, M.A. Eriksson, M.G. Lagally, Nature **439**, 703 (2006)
9. S. Nakamura, Jpn. J. Appl. Phys. **30**, L1705 (1991)
10. T. Sasaki, T. Matsuoka, J. Appl. Phys. **64**, 4531 (1988)
11. H. Okumura, S. Misawa, S. Yoshida, Appl. Phys. Lett. **59**, 1058 (1991)
12. E.S. Hellman, MRS Internet J. Nitride Semicond. Res. **3**, 11 (1998)
13. G. Mula, C. Adelmann, S. Moehl, J. Oullier, B. Daudin, Phys. Rev. B **64**, 195406 (2001); N. Gogneau, E. Sarigiannidou, E. Monroy, S. Monnoye, H. Mank, B. Daudin, Appl. Phys. Lett. **85**, 1421 (2004)
14. A.R. Smith, R.M. Feenstra, D.W. Greve, J. Neugebauer, J.E. Northrup, Phys. Rev. Lett. **79**, 3934 (1997); Appl. Phys. A **66**, S947 (1998)
15. A.R. Smith, R.M. Feenstra, D.W. Greve, M.S. Shin, M. Skowronski, J. Neugebauer, J.E. Northrup, J. Vac. Sci. Technol. B **16**, 2242 (1998)
16. Q.Z. Xue, Q.K. Xue, R.Z. Bakhtizin, Y. Hasegawa, I.S.T. Tsong, T. Sakurai, T. Ohno, Phys. Rev. B **59**, 12604 (1999)
17. S.M. Sze, *Physics of Semiconductor Devices*, 2nd edn. (Wiley, New York, 1981)
18. R.I.G. Uhrberg, G.V. Hansson, J.M. Nicholls, P.E.S. Persson, S.A. Flodström, Phys. Rev. B **31**, 3805 (1985); F.J. Himpsel, Th. Fauster, G. Hollinger, Surf. Sci. **132**, 22 (1983)
19. J.M. Nicholls, B. Reihl, J.E. Northrup, Phys. Rev. B. **35**, 4137 (1987)
20. J. Cao, D. Pavlidis, Y. Park, J. Singh, A. Eisenbach, J. Appl. Phys. **83**, 3829 (1998)
21. S.Q. Zhou, A. Vantomme, B.S. Zhang, H. Yang, M.F. Wu, Appl. Phys. Lett. **86**, 081912 (2005)
22. F. Schulze, A. Dadgar, J. Bläsing, A. Diez, A. Krost, Appl. Phys. Lett. **88**, 121114 (2006)
23. Y. Fujikawa, T. Nagao, Y. Yamada-Takamura, T. Sakurai, T. Hashimoto, Y. Morikawa, K. Terakura, M.G. Lagally, Phys. Rev. Lett. **94**, 086105 (2005)

Functional Probes for Scanning Probe Microscopy

K. Akiyama, T. Eguchi, M. Hamada, T. An, Y. Fujikawa, Y. Hasegawa, and T. Sakurai

Summary: Functional probes for scanning probe microscopy (SPM) were fabricated with focused ion beam (FIB) method. Metal-tip cantilevers were fabricated for Kelvin probe force microscopy (KFM) and glass-coated tungsten tips were fabricated for scanning tunneling microscopy under irradiation of synchrotron-radiation light (SR-STM). Here we report the fabrication process and the characterization of those functional probes.

23.1 Introduction

SPM is a powerful technique in nanotechnology. It can be used not only in nanoscale characterization of materials but also in manipulation and lithography of nanomaterials. In spite of the remarkable progress of SPM, one of its key elements, a probe has not been investigated extensively. Excellent SPM results often came out by chance after repetitive experiments using probes prepared by very primitive methods. If we design and fabricate probes desired for each experiment, we believe that we can perform SPM experiments more efficiently and we can even open up new possibilities of SPM.

Here we report fabrications of original functional probes such as a metal-tip cantilever for KFM and lithography with atomic force microscope (AFM) and a glass-coated tungsten tip for SR-STM. A sharp metal probe attached to a regular Si cantilever works well in precise KFM measurements and AFM lithography. A glass-coated tungsten tip is suitable for reducing background noises in the SR-STM experiments. The fabrication processes and performances of the functional probes are described in this paper.

23.2 Fabrication of Functional Probes

23.2.1 Fabrication of a Metal-Tip Cantilever for KFM and AFM Lithography

The metal-tip cantilever (Fig. 23.1) was fabricated by attaching a thin tungsten wire to a regular Si cantilever and milling it into a sharp probe using FIB [1].

A thin tungsten wire (diameter, 5 μm; length, ~1 mm) is attached to a regular Si cantilever with a silver epoxy using a thin glass tube whose end is slightly bent so that it can hold the wire. The motion of the glass tube is controlled precisely by a micromanipulator under a high-resolution optical microscope. After the attachment, the cantilever is preheated at 80°C on the stage of the optical microscope and baked in an oven at 150°C for 1 h.

After attachment of the wire to the Si cantilever, the cantilever is transferred into a FIB chamber for fabrication of a sharp tip. In our FIB system, we can design a shape and size of an area on which the ion beam is focused to mill a specimen. For example, the milling area is set to a rectangular shape when we cut the wire and to a doughnut shape when we sharpen the wire to form a tip.

At first, the wire is cut from its side into ~40 μm in length by using the rectangular milling area (Fig. 23.2a). In this process, the ion current and the beam size are set at 0.8 μA and 0.5 μm, respectively. The cutting process is easy but often the cut piece remains on the wire hanging or bridging with another part of the Si cantilever. In this case, we shake the cantilever several times by hand outside the FIB chamber to drop off the piece.

After cutting, the wire attached to a Si cantilever is set parallel to the ion beam in the FIB chamber so that it can be milled from its top. The wire is first milled with a doughnut-shape milling pattern slightly smaller than the

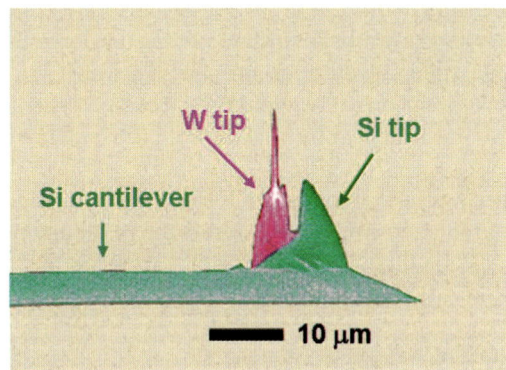

Fig. 23.1. A SEM image of the tungsten-tip cantilever. A tungsten tip, whose diameter is 5 μm before sharpened by FIB, is attached on a silicon cantilever near its silicon tip

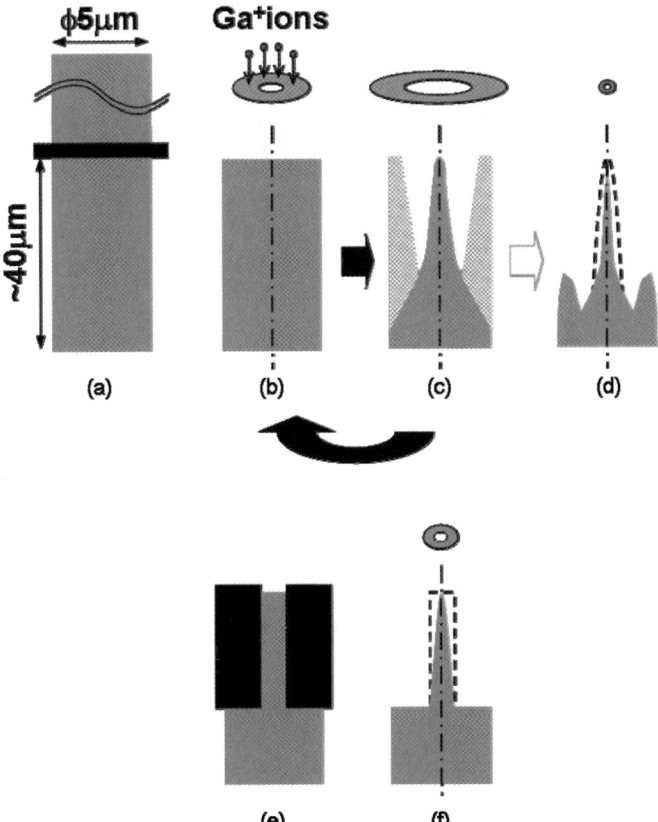

Fig. 23.2. Schematics showing a fabrication process of a sharp tip using focused ion beam

diameter of the wire (Fig. 23.2b). After digging a doughnut-shape hole on the wire, an outer rim is milled off by using a doughnut-shape milling pattern slightly larger than the diameter of the wire. After repeating this procedure, forming of a tip shape (Fig. 23.2b) and milling the outer rim (Fig. 23.2c) for a few times, we get a thin and long tip. Then the apex of the tip is milled using a milling area of tiny doughnut shape patterns (Fig. 23.2d). With gradual decrease in the inner diameter of the doughnut shape patterns, the apex of the tip becomes sharp. The inner and outer diameter of the final doughnut shape pattern is 0.4 and 0.9 μm, respectively. After milling with this pattern, the diameter of the tip apex is usually less than 5 nm in the case of tungsten wire. After each milling process we monitor the tip shape by taking an image of secondary electrons emitted by ion irradiation. At the last milling process of the tip, however, we do not observe the image since otherwise the apex would be milled out. We characterize the tip apex by TEM observation.

Various materials other than tungsten can also be used as a tip of the metal-tip cantilever as far as their thin wires are available, since the FIB sharpening can be performed on any conductive wires. So far we have fabricated metal-tip cantilevers with Au and PtIr wires. Au-tip cantilevers are used for AFM lithography and PtIr-tip cantilevers are useful for AFM observation in air and KFM.

It is, however, not easy to handle Au and PtIr wires, since the wires are sticky to a thin glass tube in which the wires are inserted and manipulated, compared with tungsten. Despite their sticky nature, we can somehow manage to attach wires of both materials to Si cantilevers. The FIB process of the Au wire is just the same as tungsten. But in the case of PtIr, it is hard to mill its wire. PtIr wire was not easily milled comparing to Si cantilever resulting in making a hole on Si cantilever before fabrication of sharp PtIr tip. Because of this reason, the same method cannot be used. We first mill sidewalls of a PtIr wire using a rectangular milling area to make the wire thin (Fig. 23.2e). Then we form a tip shape with doughnut-shape patterns as in Fig. 23.2f. These processes correspond to a milling with the larger doughnut-shape patterns in the case of W and Au as shown in Fig. 23.2b,c. After the rough milling processes, the same fine milling with tiny doughnut-shape patterns was performed to make a sharp tip as shown in Fig. 23.2d.

23.2.2 Fabrication of a Glass-Coated Tungsten Tip for SR-STM

A glass-coated tungsten tip is fabricated in a process of glass coating of a electrochemically etched tungsten tip and milling of its apex with FIB to remove the glass layer and sharpen the tip apex [2].

The tungsten wire (ϕ 0.2 mm) is etched in conventional electrochemical method with KOH solution. After the etched wire is inserted into the thin glass tube, the glass tube is pulled under heating with a Bunsen burner so that an unused part of the glass tube is removed (Fig. 23.3a). Then, the glass layer covering the tip is made round and thin as shown in Fig. 23.3b,c by repetitive heating with a Bunsen burner until the shape of the glass becomes like that shown in Fig. 23.3d. At this stage, the thickness of the glass has to be thin ($<10\,\mu$m) enough to be milled by FIB in several hours. We also check whether the glass layer does not have large pin holes by measuring electrical conductance in an electrolytic solution (5% NaCl). The current through the tip and the solution should be less than 10 nA with an applied voltage of $\pm\mathbf{1\,V}$.

After confirming that the etched tip is fully covered with glass layer, we mill the tip apex with FIB by getting rid of the covering glass layer and sharpening of the tungsten tip (Fig. 23.3e). We basically used the same technique as the case of a metal-tip cantilever described in Sect. 23.2.1. The glass-coated tip is milled from its top by using the doughnut-shape milling pattern, and the size of the doughnut shape is gradually reduced so that the tungsten tip becomes sharp. The ion current and beam size of the focused Ga ion beam are usually set at 0.8 μA and 0.1 μm, respectively.

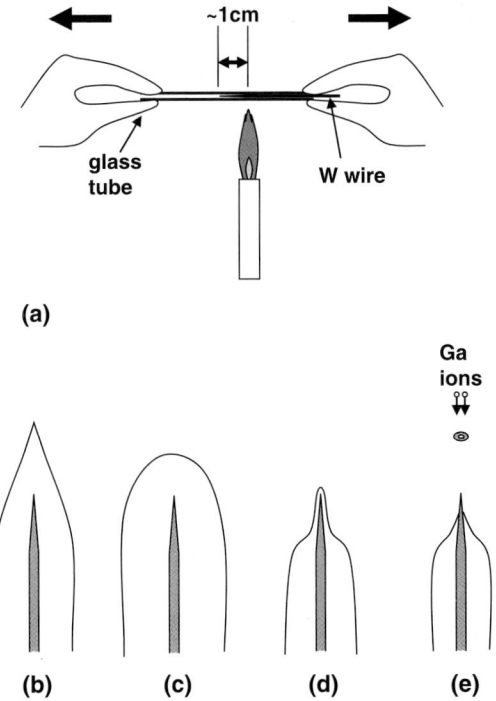

(a)

(b)   **(c)**   **(d)**   **(e)**

Fig. 23.3. A schematic diagram showing how to make the glass-coated tungsten tip for synchrotron-radiation STM. (**a**) A glass tube having a sharp tip inside was heated with a Bunsen burner and an excess part of the glass tube was pulled off; (**b**) the as-pulled tip; (**c, d**) the glass layer was made thinner and thinner by repetitive heating; (**e**) removing the glass layer and sharpening the tungsten tip by FIB milling with a doughnut-shape milling area

23.3 Characterization of the Functional Probes

23.3.1 TEM Observation of the Metal-Tip Cantilevers

We observed the tip apex of the tungsten tip cantilevers, which was made with the method mentioned above, by transmission electron microscope (TEM). The curvature radius of the tungsten tip is found to be as small as 3.5 nm (Fig. 23.4a). Compared with usual metal-coated tips, which have curvature radius more than 50 nm, the tungsten-tip is very sharp, a great advantage for microscopic imaging. From the TEM images, we found that the tungsten tip was covered with a thin layer, which we believe oxide layer since Ga was not detected in an energy dispersive X-ray spectroscopy (EDX) analysis. A diffraction pattern of the tungsten tip tells us that the crystalline structure of the tip apex was not damaged by the FIB milling. The apex of the tungsten tip has a bcc structure pointing in the <110> direction.

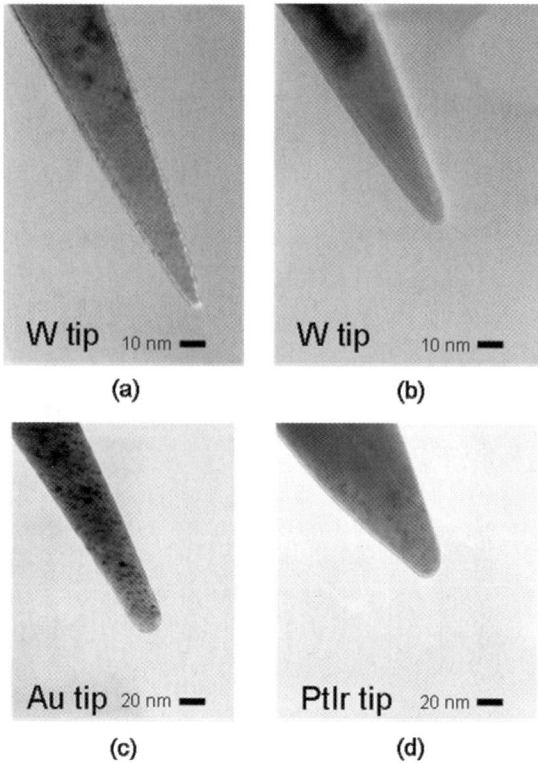

Fig. 23.4. TEM images of a tip apex of various metal-tip cantilevers. (**a**) Tungsten tip, (**b**) further milled tungsten tip, (**c**) Au tip, and (**d**) PtIr tip

In fact, we tried to make tips thinner than the tip shown in Fig. 23.4a by milling with much smaller doughnut-shape patterns. The observed TEM images showed thicker oxide layer covering tungsten tips. Although overall curvature radius of the tip is larger than that shown in Fig. 23.4a, which is determined by the thickness of the oxide layer, the curvature radius of the tungsten tip itself is as small as 0.8 nm as shown in Fig. 23.4b.

To make the best usage of the sharpness of the metal-tip cantilever, we have used it for AFM lithography, in which Au tips are used instead of tungsten. The purpose of our AFM-lithography work is to draw nanometer width metal wires, possibly less than 10 nm, narrower than the previous works [3, 4], in which the Au-coated tips are used. To achieve it, we fabricated Au-tip cantilevers in the same fabrication process as the tungsten-tip cantilever and characterized their tip apex by TEM. The curvature radius was 11.2 nm (Fig. 23.4c). The radius is large if compared with the case of the tungsten tip, but it is still smaller by a factor of 5–10 than Au-coated tips. We thus believe that AFM lithography using our Au-tip cantilevers can draw narrower wires than the previous works. The difference in the curvature radius between tungsten and Au is, we believe, due to softness or large surface energy of Au. For example, deposited Au often forms grains with a size of $\sim$100 nm on substrates. This implies that Au tends to form round shape instead of keeping sharpness after the FIB milling.

As a material of the tip for KFM, PtIr was also chosen because of its inert nature. For the KFM probe, a sharp tip is required for high spatial resolution and a metal tip is required for quantitative measurement. In the case of semiconductor tips, field penetration into the tip may hamper the quantitative measurement. We can make very sharp tungsten-tip cantilevers, but their tip apex was covered with oxide and thus some treatments to remove the layer are required. Being inert and thus not covered with oxide layer, PtIr is an ideal material for the KFM measurements. A PtIr-tip cantilever was fabricated using the same FIB method as the case of tungsten-tip cantilevers. The apex of the tip is estimated to be 13 nm by TEM (Fig. 23.4d). It is similar to that of Au-tip cantilevers probably because of soft nature of PtIr.

23.3.2 SEM/EDX Observation of Glass-Coated Tungsten Tips

The FIB-milled glass-coated tip was characterized by scanning electron microscopy (SEM) with an EDX analysis. Figure 23.5a is an SEM image of a glass-coated tungsten tip. The image shows us that there is a protrusion whose

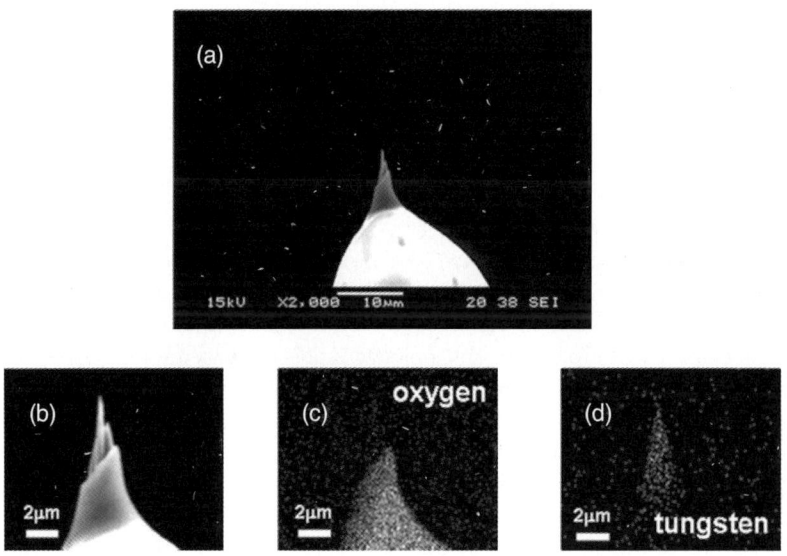

Fig. 23.5. (a) A SEM image of a glass-coated tungsten tip. (b) A zoomed image of a tip apex of the glass-coated tungsten tip. (c, d) Corresponding EDX images made with (O Kα_1, 0.525 keV) and (W Lα_1, 8.398 keV) X-ray peaks. The image in (c) represents a distribution of the glass. These images demonstrate that the tungsten tip is covered with the glass, except an area less than 5 μm from the tip apex

length is around 10 μm from the area coated with glass. This protrusion was made after repetitive heating of the tip apex to make the glass thin. Figure 23.5b is a zoomed SEM image of the tip apex, and Fig. 23.5c,d is corresponding EDX images of O Kα_1 (0.525 keV) and W Lα_1 (8.398 keV) X-ray peaks, respectively. The image due to the O peak represents area of glass coating and the image of W represents area of a bare tungsten tip. Since main elemental peaks of W (Mβ, 1.835 keV) and Si (Kα_1, 1.74 keV) are too close to be separated with our SEM, we chose the O peak for glass and the second strongest W peak for the tip. The two elemental mappings clearly reveal that the tungsten tip is covered with glass, except an area of less than 5 μm from the tip apex. It should be noted here that the X-rays are generated within a region of about 2 μm in depth. The area where both the oxygen and tungsten signals are detected is indeed covered with the glass whose thickness is less than 2 μm.

23.4 Results and Discussions

23.4.1 KFM Observation with a Tungsten-Tip Cantilever

To demonstrate capability of the tungsten-tip cantilever in KFM measurement, we chose Ge/Si(105) surface as a sample, on which slight potential difference among surface dangling bonds are predicted due to charge transfer among their electronic states [5, 6].

Ge/Si(105) surface is a facet plane of dot structures formed on Ge deposited Si(001) surface. The Ge/Si(105) surface is a kind of tilted or stepped (001) surface and thus its structural model was proposed based on a dimer structure by Mo et al., who observed the dot structure for the first time. After their proposal, more than 10 years later, Fujikawa et al. proposed a new structural model called rebonded step (RS) model to explain empty-state STM images taken on the surface and induced tensile strain on the surface, both of which are not explained by Mo's model. They found that the STM images depend strongly on the polarity of the bias voltage, consistent with a result of first principles calculation. From the polarity dependence and the result of first principles calculation, charge transfer among the surface dangling bonds was expected. To confirm the charge transfer on the surface, we performed KFM observation to measure distribution of electrostatic potential on the surface in high precision.

Before the KFM observation of the Ge/Si(105) surface, we checked the performance of noncontact (NC) AFM by observing the atomic structure of the surface [7]. Different from the STM images, which were affected by the electronic states of the surface, NC-AFM images faithfully represented the atomic structure of the Ge/Si(105) surface. Especially, extra dimers added in the new RS model were clearly observed as two protrusions. Those were observed just as one broad protrusion in empty-state STM images.

After checking that the NC-AFM worked well, we performed KFM observation of the Ge/Si(105) surface [7]. For the KFM observation, the tungsten-tip cantilever was used after in-situ electron beam heating at 900°C in the UHV-AFM system. Simultaneously with an NC-AFM image showing the atomic structure of the Ge/Si(105) surface (Fig. 23.6a), an atomically resolved

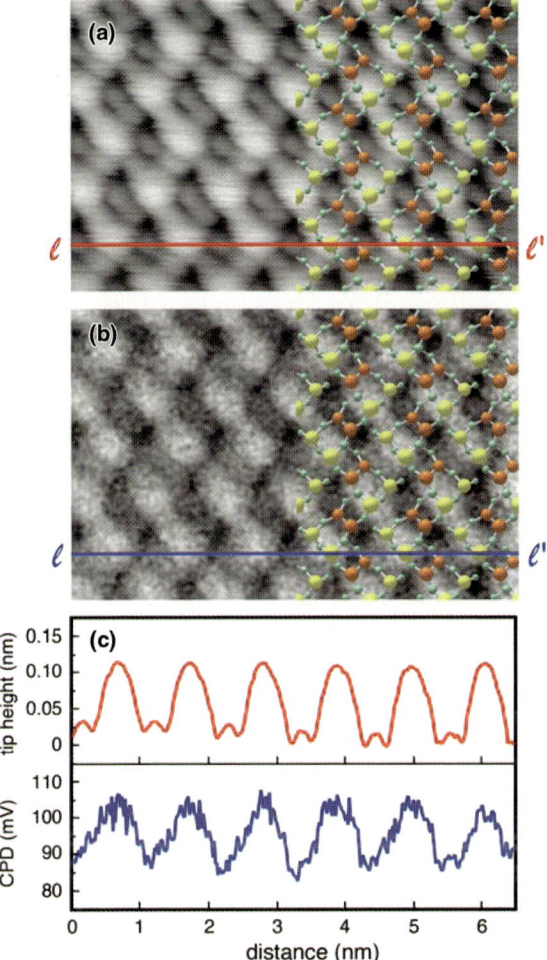

Fig. 23.6. (a) Topographic NC-AFM and (b) electrostatic potential (KFM) images taken simultaneously on the Ge(105)-(1×2) surface superimposed with the RS model. The resonance frequency, the spring constant, and the oscillation amplitude of the cantilever were 273,646 Hz, 44 N m^{-1}, and 0.9 nm, respectively. The frequency shift was set at -125 Hz for the AFM imaging. For the KFM experiment, a modulated voltage with a frequency of 1.5 kHz and amplitude of 250 mV$_{rms}$ was applied on the sample bias voltage. (c) Cross-sectional profiles measured on the topographic (a) and potential (b) images along the red and blue l-l' lines, respectively

KFM image showing a distribution of the electrostatic potential (Fig. 23.6b) was obtained on the surface. From the potential image, we confirmed the charge transfer among surface dangling bonds. Potential difference between the two kinds of atoms is estimated from a cross section $(1 - 1')$ of the KFM image as 15–20 meV (Fig. 23.6c, blue line). This is consistent with the results of first principles calculation. These results clearly demonstrate capability of the tungsten-tip cantilever as a KFM probe by revealing atomically resolved potential distribution quantitatively with high energy resolution less than 3 meV. This high energy resolution was achieved at room temperature although it may sound strange. We observed the electrostatic potential based on the zero balancing method, which detects difference of Fermi levels of the tip and sample and adjust the bias voltage so that the difference becomes zero. Thus, smearing function of the Fermi distribution due to room temperature does not affect energy resolution of the KFM measurement.

In our KFM image, the potential distribution coincides with the topography, in other words, the potential is high on the topographically high atoms. There might be naive question if the topography influences the potential measurement or not. To make sure the point, we have checked our KFM measurement system using Si(111)-(7 × 7) surface on which the potential of the adatoms is known to be lower than the rest atoms due to the charge transfer from adatoms to rest atoms while the adatoms are topographically higher than the rest atoms. We clearly observed an inverted contrast in the potential image compared with the topographical image.

Our KFM study on the Ge/Si(105) surface obtained with the tungsten-tip cantilever is satisfactory to us since it gave us atomic resolution and quantitative potential information. There have been several KFM studies reported so far although they often have poor spatial resolution and no quantitative information. Our goal is to establish a method to obtain atomically resolved KFM images constantly with quantitative potential information. To do that, we plan to use the PtIr-tip cantilever, which is not oxidized and thus may not require pretreatment before KFM observation. We are now working on making sharp PtIr-tip cantilevers for quantitative KFM images with high spatial resolution, and trying KFM study using them.

23.4.2 AFM Lithography with a Au-Tip Cantilever

Making the best usage of the sharpened Au-tip cantilever, which has a curvature radius of 11.2 nm, we have used it for AFM lithography to draw a Au wire with a width less than 10 nm. By applying a voltage pulse between Au-coated tip and a sample surface in tapping-mode AFM, Au atoms are field evaporated from the tip apex to form nanoscale dots on the sample surface. By connecting those nano-dots, a Au wire can be drawn. This method is called AFM lithography. So far, Au-coated tips have been used for AFM lithography, in which the width of the drawn wires typically ranges from 50 to 100 nm, most probably limited by a large curvature radius of the Au coated tips.

Because of rapid scale down in semiconductor devices, drawing metal wires whose width is less than 10 nm is one of the challenging issues in a field of lithography. Currently photolithography is mainly used to draw lines for the devices since it is suited for mass production, but the width of the drawn line is limited by the wave length of light. By changing light into one with shorter wave length, the line width might be reduced although facility for the technique costs too much. If we use e-beam lithography, the problem of wave length can be solved but the line width is still limited around 10 nm because of scattering of the incident electrons. Therefore, AFM lithography is a very promising technique to attain the line width less than 10 nm. Lithography utilizing scanning probe microscopy seems to be not appropriate for mass production since only one probe can be usually utilized for drawing lines. But the invention of "millipede" [8] enables us to use many Au-tip cantilevers in parallel for drawing many lines simultaneously.

Based on a point of view mentioned above, we have developed a new AFM lithographical method using NC-AFM instead of tapping-mode AFM with a Au-tip cantilever. Using NC-AFM we can regulate the tip to sample distance better than tapping-mode AFM. Using Au-tip cantilevers, which are sharper than Au-coated tip, we believe that the 22 nm width limit achieved by conventional AFM lithographical method can be overcome and wires with width less than 10 nm can be drawn.

As a first trial of AFM lithography, we used Au-coated tips, which can be easily made in NC-AFM for making nanoscale dots. By applying a voltage pulse between the tip and a Si substrate covered with native oxide, we successfully formed nanoscale Au dots on the substrate (Fig. 23.7, within red circle). Since we confirmed the dot formation using NC-AFM, we are now performing AFM lithography using Au-tip cantilevers in NC-AFM to draw Au wires with the width less than 10 nm.

The final goal of our AFM-lithography work is to draw metal wires connecting a nanomaterial, such as a single molecule, with micron-scale metal electrodes for electrical conductance measurements of the nanomaterial (Fig. 23.8). Conductance of single molecules has been measured with break-junction method [9]. But in the measurements, it is difficult to see how molecules are there between two metal electrodes. For instance, the number of the molecules between the electrodes is not certain and usually just assumed

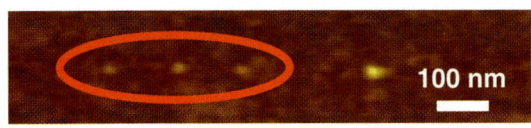

Fig. 23.7. Three Au dots (within a red circle) formed on a Si substrate by AFM lithography with a Au-coated tip in NC-AFM. The oscillation amplitude and frequency shift were set at 8 nm and −10 Hz, respectively. A voltage pulse applied between the Au-coated tip and the Si substrate is −25 V in height for 100 ms

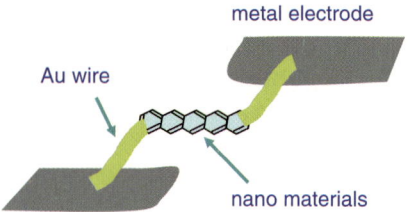

Fig. 23.8. A schematic showing our method for measuring electrical conductance of nanomaterials. After depositing nanomaterials, we will make a connection between the materials and metal electrodes by drawing Au wires using AFM lithography with the Au-tip cantilever in NC-AFM

from a statistical analysis. Furthermore, contact condition between the molecule and metal electrodes is not well characterized, and is probably changing at each measurement although it is regarded as one of the most important issues in conductance measurement of a single molecule.

To the problems, we believe our method to measure the conductance of nanomaterials will give perfect solutions. Since we utilize NC-AFM with a sharp Au tip as a method for lithography, we can observe the molecules on the surface in high spatial resolution. Then, we can draw nanoscale Au wires from a specific site in the target molecule to metal electrodes (Fig. 23.8). In this way we can clarify the number of the molecule. The contact between the molecule and Au wires can also be well defined from an observation by NC-AFM. It might be possible that the Au atoms at the tip apex are gradually consumed after the AFM lithography and the tip apex becomes dull. But a previous work [4] showed that the tip apex becomes even sharper after the lithography. So high resolution AFM imaging is possible, we believe, even after the lithography.

23.4.3 SR-STM Observation with a Glass-Coated Tungsten Tip

One of the challenging issues or ultimate goals of scanning tunneling microscopy is to detect elemental information of the observed atoms. In spite of its capability of high spatial resolution enough to resolve atoms, it cannot tell the element of them. Thus other techniques like Auger electron spectroscopy (AES) or X-ray photoemission spectroscopy (XPS) are often combined with STM to obtain the chemical information but they provide only averaged information over the whole surface.

Considering such situation, we combined STM with synchrotron-radiation light to obtain element specific images of surfaces. In other words, we tried XPS with a functional STM tip as a detector simultaneously with the STM imaging to get the elemental information of observed surface atoms. In the synchrotron-radiation STM (SR-STM), synchrotron radiation light is irradiated just under the STM tip and emitted electrons from the irradiated surface are detected by the STM tip. In fact, electrons are emitted from a large area of the sample surface since the beam size of synchrotron radiation light is in an order of millimeter. Therefore, to shut out those electrons and detect electrons emitted just below the STM tip, the tip must be covered with insulator except its apex. For the purpose, we developed a glass-coated tungsten tip to block the electrons injected to the side walls of the tip and reduce background noise.

Indeed, insulator-coated tips have been used in electrochemical STM with a purpose of blocking ion current and thus reducing background noise, which is quite similar to our cases. Actually we first made polymer-coated tips, which have been used in electrochemical STM. Our observation is, however, performed in ultrahigh vacuum, different from electrochemical STM, and it turned out that the coated polymer on the probe easily contaminated clean surfaces during scanning because of its high vapor pressure. X-ray irradiation may cause decomposition of the coated polymer and contaminate the surface. To avoid such problems, we chose glass as a coating material.

So far, we have used glass tube to cover the tungsten tip and made it thin to be repetitive heating. One of the good point of using glass tube is easy handling, but the problem is that the thickness of the glass at the tip apex cannot be controlled well by repetitive heating. Sometimes large pin holes are formed at the apex of the tips. To avoid the hole formation we are thinking of a new method; depositing insulating material on the electrochemically etched tungsten tip and milling the apex of the tip using FIB, for next-generation coated tips.

The purpose of fabricating the glass-coated tungsten tip is to reduce the background noise caused by electrons hitting the sidewall of the STM tip. To check the performance of the glass-coated tungsten tip, we took X-ray absorption spectrum using the probe, that is, spectra of photo-induced current detected with the tip as a function of photon energy injected on the sample surface. We used the sample having 1 mm-square Ni dots on Au-covered Si substrate fabricated with e-beam lithography. The energy of the X-ray is set around the Ni L edges (852.7 and 870.0 eV). Comparing the spectrum with the one taken with the uncoated tungsten tip, the detected current was fairly reduced down to $\sim 1/40$ by the glass coating [2]. This value is similar to the one obtained with the previous polymer-coated tungsten tips, which is in a range of $1/10$ to $1/100$.

We also checked performance of the glass-coated tungsten tip; whether it works well as a normal STM tip to resolve atoms and it does not contaminate clean surfaces in UHV. Different from the polymer-coated tips, the glass-coated tungsten tip did not contaminate the surface and we successfully observed the atomically resolved Si(111)-(7×7) surface images using the glass-coated tip.

Finally we try to observe element specific images of Ni on the Ni-dot sample under the soft X-ray irradiation. We used $1 \,\mu m^2$ Ni dots fabricated on a Au layer covering a Si substrate as a sample. Under the irradiation of X-ray whose photon energy is just above the Ni absorption edge, we successfully

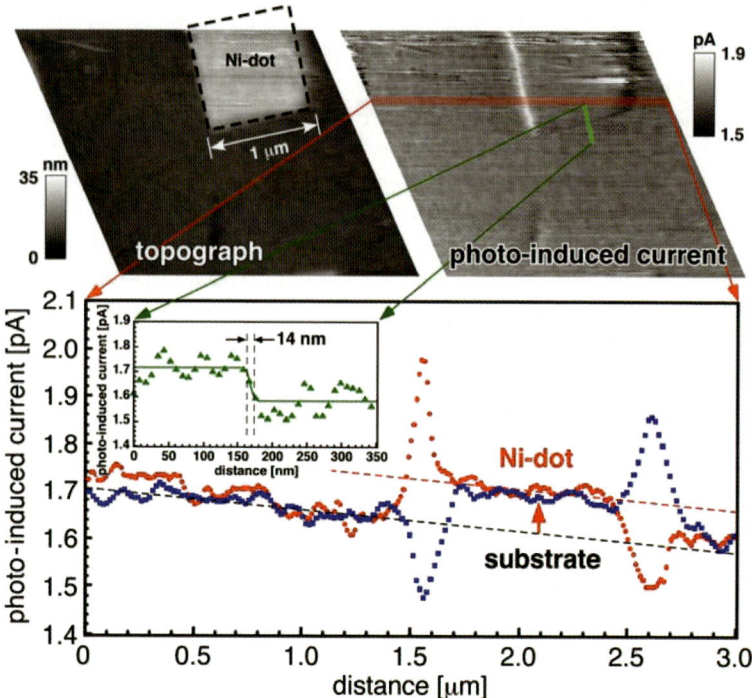

Fig. 23.9. (*Upper*) A STM image of a Ni dot and simultaneously obtained photo-induced current image. The STM image was taken in a constant current mode with the sample bias voltage, tunneling current, and scanning speed of -10 V, 55 pA, and 7.2 s line^{-1}, respectively. The scanning direction is from left to right, and the lateral scanning moved from top to the bottom. These images were taken under the irradiation of light whose photon energy was 855 eV, slightly higher than the Ni L3 edge (852.7 eV). (*Lower*) A cross-sectional plot averaged in a red shaded region of the photo-induced current image is drawn with red circles. Blue squares represent a cross section of the photo-induced current taken in the opposite (from right to left) scanning direction. An inset shows a line profile taken across the bottom edge of the Ni dot. With an assumption of a sharp step convoluted with a Gaussian due to a limited instrumental resolution, lateral resolution is estimated at around 14 nm

obtained a photo-induced current image, in other words, element specific image of Ni simultaneously with a topographic image (Fig. 23.9) [10]. Since we could not estimate spatial resolution from lateral cross-sectional plots due to scanning artifact, we estimated the spatial resolution from a vertical cross section (Fig. 23.9). The estimated spatial resolution is 14 nm, even better than the best achieved by photoemission electron microscope resolution (22 nm).

We believe that the obtained high resolution is due to potential barrier reduction around the tip apex induced by an application of bias voltage and image potential. If it is correct, sharper tips should provide higher spatial resolution. We need to care about the curvature radius of the glass-coated tungsten tip much better than the one of usual STM tips. In the usual STM experiment, electrochemically etched metal tip is often used and their curvature radius is usually some tens of nanometer. With the tips, atomically resolved images can be obtained. But in the case of SR-STM, requirement is much severe and we may need very sharp tips like the case of NC-AFM.

In fact, concerning about the glass-coated tungsten tips utilized in this SR-STM experiment, we did not care about the sharpness of the tip so much and thus its sharpness is almost same as that of normal STM experiments. We are now fabricating much sharper insulator-coated tips possibly with the radius of several nanometer similar to the tungsten-tip cantilevers used in the KFM study. The new tips will probably improve spatial resolution of the element specific imaging. We are fabricating insulator-coated tips by depositing insulating material on electrochemically etched tungsten tips since we can control the thickness of the insulating layer much better than the case of glass tubes.

For improvement of the spatial resolution of element specific images, we need to use advanced SR beam lines to enhance the light intensity below the tip. Using advanced SR beam lines, the light can be focused down to $1\,\mu m$ and the light intensity will be increased dramatically by factors of 100–10,000. The spot size of the present beam line is around $1\,mm$.

Our results are the first step for obtaining the element specific images in atomic resolution. This step is, however, worthwhile since it overcomes the weakness of STM in elemental identification of observed surface atoms.

23.5 Conclusions

We fabricated original functional probes using FIB method for scanning probe microscopy and demonstrated their performances by presenting results that are hard to be obtained with usual SPM probes. Our fabrication method of functional probes is based on FIB and the sharpness of the probes is represented by the TEM image of the tungsten-tip cantilever whose curvature radius is as small as $3.5\,nm$. The sharpness is very critical not only in NC-AFM and KFM but also in SR-STM for reduction of the potential barrier around the tip apex in order to achieve high spatial resolution in the element specific images. We believe that fabrication of the functional probes surely opens up new possibilities of SPM.

Acknowledgements. Some of the authors (K.A., T.E., T.A., Y.H.) acknowledge Professor Kazuhisa Sueoka, Yuji Akiyama, and Kei-ichi Takizawa for fruitful discussions and technical assistance on metal-tip cantilevers. We also thank Professor Kazuo Terashima, Professor Kingo Itaya, Proessor Kenji Sashikata, Professor Junji Inukai, and Dr. Soichiro Yoshimoto for fruitful discussions on glass-coated tungsten tips. All the FIB processing and TEM observation were performed at Electron Microscope section of Materials Design and Characterization Laboratory, ISSP, The University of Tokyo. Technical assistance by Fumiko Sakai, Masaki Ichihara, and Tadao Imai is highly appreciated. This work was financially supported by a Grant-in-Aid for young scientists B (No. 19760019) of Ministry of Education, Science and Technology (MEXT), Japan and Foundation Advanced Technology Institute (ATI).

References

1. K. Akiyama, T. Eguchi, T. An, Y. Fujikawa, Y. Yamada-Takamura, T. Sakurai, Y. Hasegawa, Rev. Sci. Instrum. **76**, 033705 (2005)
2. K. Akiyama, T. Eguchi, T. An, Y. Hasegawa, T. Okuda, A. Harasawa, T. Kinoshita, Rev. Sci. Instrum. **76**, 083711 (2005)
3. M. Calleja, M. Tello, J. Anguita, F. García, R. García, Appl. Phys. Lett. **79**, 2471 (2001)
4. M.E. Pumarol, Y. Miyahara, R. Gagnon, P. Grütter, Nanotechnology **16**, 1083 (2005)
5. Y. Fujikawa, K. Akiyama, T. Nagao, T. Sakurai, M.G. Lagally, T. Hashimoto, Y. Morikawa, K. Terakura, Phys. Rev. Lett. **88**, 176101 (2002)
6. T. Hashimoto, Y. Morikawa, Y. Fujikawa, T. Sakurai, M.G. Lagally, K. Terakura, Surf. Sci. **513**, L445 (2002)
7. T. Eguchi, Y. Fujikawa, K. Akiyama, T. An, M. Ono, T. Hashimoto, Y. Morikawa, K. Terakura, T. Sakurai, M.G. Lagally, Y. Hasegawa, Phys. Rev. Lett. **93**, 266102 (2004)
8. P. Vettiger, M. Despont, U. Drechsler, U. Dürig, W. Häberle, M.I. Lutwyche, H.E. Rothuizen, R. Stutz, R. Widmer, G.K. Binnig, IBM J. Res. Develop. **44**, 323 (2000)
9. B. Xu, N.J. Tao, Science **301**, 1221 (2003)
10. T. Eguchi, T. Okuda, T. Matsushima, A. Kataoka, A. Harasawa, K. Akiyama, T. Kinoshita, Y. Hasegawa, M. Kawamori, Y. Haruyama, S. Matsui, Appl. Phys. Lett. **89**, 243119 (2006)

Printing: Krips bv, Meppel, The Netherlands
Binding: Stürtz, Würzburg, Germany